c

I0611830

May 9--June 11, 1819

GOLD RUSH DIARY

Harriett Hildreth Perkins   Elisha Douglass Perkins

Drawings by Robert James Foose after daguerreotypes owned
by Harriett's granddaughter, Isabel Means Humphrey, of Ashland, Kentucky.

GOLD RUSH DIARY

Being the Journal of
Elisha Douglass Perkins on
the Overland Trail in the
Spring and Summer of 1849

Edited by

THOMAS D. CLARK

UNIVERSITY OF KENTUCKY PRESS, LEXINGTON 1967

To Thomas Bennett Clark, Jr.

Preface

THE STORY of the great rush to the gold fields of California unfolds annually in the discovery of new journals and letters, and even of newspaper accounts, long buried in little-known collections. Elisha Douglass Perkins of Marietta, Ohio, has contributed materially to the enrichment of this moment in the history of the westward movement. Almost every one of his daily notations in his trail diary reveals the author's awareness that he was a part of the important folk movement of that era. Here was a phase of frontier development that involved the largest numbers of people of any single migration in our national history. Gold, it was true, attracted men, women, and children into the vortex of adventure, but the ultimate reward was even greater: struggle on the trail and in California shaped a new society beyond the Sierras.

This particular diary conveys a sense of both the rich expectancy of the gold seeker and the doubts that beset him. When the difficulties of travel became real, the emigrant began to wonder whether he had been foolish to leave home. This was evident in the literature produced by the great folk migration of 1849. Perkins, for example, recorded encounters with men along the trail who had given way to their doubts and were returning home on one pretext or another.

In reading numerous diaries one realizes that even for the more presistent ones who follow the trail to its end, gold fever was easily chilled in California. Some men who had decided on the spur of the moment to go to the gold fields quickly returned home and tried to forget their rashness. Others remained to open farms, build cities, develop commerce, and create a new state. Before this happened, however, there occurred an enormous amount of introspection.

Fate denied Douglass Perkins an opportunity to see the Far West grow into anything more than a composite of arid trails, chemical-laden springs, frustrated gold hunters, rag-towns, gambling halls, houses of prostitution, saloons, flooded gold camps, and islands of ill-health. Except for his journal, Perkins contributed little to the settlement of the West. His journal, however, is an eloquent testimonial of the nature of the great migration. Neighbors in the old Ohio Valley and Appalachian settlements had worked together to get a civilization planted, and the emigrants on the trails did almost the same thing. There is a certain naïvete about the endless trains of emigrants stirring the dust and sitting about their camp fires from St. Joseph and Independence to Sacramento. Perkins documented the fact that the roots of the gold rush lay not in California, but back east in the towns and rural communities where men threw reason to the wind and set out in droves to grab quick fortunes. "Quickness" proved one of the most startling relativities in frontier expansion so far as the gold rush was concerned. Perkins' description of travel across the plains, the mountains, and the desert is so precise and full that a modern reader falls into lockstep with him as he made his way westward, pushed on by the pressures of time and an unyielding determination to succeed.

There is romance in the history of the manuscript journal itself. When Douglass Perkins lay dying with dysentery aboard the wharf boat *Orb* on the Sacramento waterfront, he was attended by John Weaver, of Marietta, and his brother-in-law Dr. George O. Hildreth, one of whom took possession of his personal effects. Among them was the trail-rubbed journal. It had lost its front cover, and the first pages were badly stained and blurred. This document was taken home to his wife Harriett as an eloquent testimonial of her husband's struggles. She read in it of his homesickness and concern for her, of his sore disappointment at not finding gold, of the great despondency that settled over him when her brother left the gold camp and when Samuel E. Cross died. Out of deep sentiment Harriett made a fair copy of the journal in a notebook that resembled the original. Writing in a clear, legible hand, she produced a document that had every appearance of being an original. We at the University of Kentucky believed for several years that it was Douglass Perkins' own manuscript copy, largely, I am sure, because it came to us from such a direct and impeccable source.

The fair copy made by Harriett Perkins was often read by the children of her marriage to John Means of Ashland, Kentucky. They came to think of the author almost as a blood part of their own family background. When Harriett Hildreth Perkins Means died on February 12, 1910, at the age of eighty-one years, her personal belongings were given to her daughter and then to her granddaughter Isabel Means Humphrey. Mrs. Humphrey and her brother John Means deposited the Perkins and Means papers in the University of Kentucky archives for safekeeping. With these papers were included Douglass Perkins' earlier diary, kept on a trip to the Lower South, and the journal of Harriett Perkins Means' trip to California with her two daughters in 1882. Unhappily, none of Perkins' manuscript letters are known to exist; those that Elisha Backus Perkins turned over to Beman Gates to publish in the Marietta *Intelligencer* are available, in printed form.

By the most fortuitous of accidents a letter from Dale L. Morgan, prominent historian of the West, addressed to the Kentucky State Historical Society, was opened by mistake with the Department of History's mail in the University and reached the editor's desk before the error was discovered. Mr. Morgan sought information about manuscript overland journals. When I reported the Perkins journal, he informed me that the copy we held in Harriett Perkins Means' papers was not the original. The original, he said, was owned by the Henry E. Huntington Library in San Marino, California. In the summer of 1964 I had an opportunity to compare the two manuscripts and established that the Huntington Library did hold the original.

In 1925 Mr. Huntington had purchased Perkins' journal from a San Francisco antique dealer, who said in rather extensive correspondence that he was acting only as an agent for a member of the "family." Whether the seller was a member of the Hildreth, the Means, or the Perkins families is not known.

When Harriett Perkins Means made her copy, she deleted a few minor passages, the most important of which was the earthy incident about the frightened dog in the tent at St. Joseph. She sometimes simplified Perkins' sentences, and many times, fortunately, she clarified her husband's involved handwriting. Too, Harriett's version contained material that is now illegible in, or missing from, the original copy because of the damaged first pages and the loss of an internal page.

In annotating Douglass Perkins' journal the editor has attempted to verify and extend the journalist's firsthand observations. These additions are the result of a rather extensive reading of manuscript journals preserved in the Henry E. Huntington Library, the Bancroft Library, the California State

Library, and the University of Kentucky Library. No effort has been made to chart Perkins' progress along the trail; Mr. Morgan has done this well in his comprehensive tables (copyright in 1959 by Fred A. Rosenstock) included in his *Overland Diary of James A. Pritchard from Kentucky to California in 1849.*

I wish to express my appreciation to Dr. John E. Pomfret and the Board of the Henry E. Huntington Library for a summer reader's grant that enabled me to collate the two versions of the Perkins journal. Haydée Noya, Isabel Fry, and Anne Hyder were most helpful to me in my search for corroborative materials. I wish to thank especially Dale L. Morgan for pointing out that we did not have the original journal, and for his subsequent interest and assistance. Dr. George P. Hammond of the Bancroft Library in the University of California was most cooperative in helping me find materials, as was Allan R. Ottley of the California State Library. Dr. Jacqueline P. Bull of the University of Kentucky Division of Rare Books and Archives was generous in allowing me to take our copy of the journal to San Marino for comparison with the original. Robert Caton of the Marietta College Library allowed me free access to an excellent file of the Marietta *Intelligencer* to make microfilms of the gold-rush years. Catherine B. Remley, Curator of the Campus Martius Museum in Marietta, not only made important manuscript material available to me, but was able to direct me to useful sources of information in Marietta. Among them was Edith S. Reiter, a former curator of the Museum, who for years has contributed historical articles to the Columbus *Citizen*. Dr. T. A. Larson of the University of Wyoming, Dr. LeRoy Hafen of Brigham Young University, and Professor David E. Miller of the University of Utah generously assisted me on details of Perkins' route and in preparing the map of his journey. Patricia O'Brien Smylie read the manuscript and gave me many helpful suggestions. Modest grants from the Kentucky Research Foundation enabled me to search for materials in Marietta and to prepare the maps which appear in this book.

Without the generosity of Isabel Means Humphrey of Ashland, Kentucky, I would never have known about Elisha Douglass Perkins and his journal. She not only preserved the papers and made most intelligent disposition of them; she generously allowed me to copy the daguerreotypes of Douglass and Harriett that appear in this volume.

—Thomas D. Clark.

Contents

Introduction

THE MIDDLE of the nineteenth century provided opportunities aplenty for adventuresome youth but sometimes they were hard to discover. Elisha Douglass Perkins of Marietta, Ohio, had not quite found his niche. He had traveled into the Lower South in search of a spot to locate a drugstore, and had even contemplated a visit into the Southwest. Perkins in 1848 was ripe for the excitement which news of discovery of gold in California brought to the Ohio Valley. The year before he had married the daughter of a prominent, relatively well-to-do family, and like his neighbors in the oldest settled Ohio community he envisioned getting rich quick in California and coming home to live in ease the rest of his life.[1]

Twenty-five-year-old Douglass Perkins was descended from six generations of Yankee forbears who had grabbled a living from the rocky soils of middle Connecticut. His father, Elisha Backus Perkins, was a native of Canterbury, Connecticut, and a graduate of Dartmouth College.[2] Like his son Douglass he had had difficulty locating a place in which to settle down. His wanderings had taken him to Baltimore, then to Tallahassee in Florida, and out to the frontier in Alton, Illinois. He had studied law, first with Sylvanus Backus in Pomfret, Connecticut, and then with Calvin Goddard in Norwich. The law, however, did not hold his interest for long. In 1828 he moved his family to Baltimore where he learned the druggist trade. Five years before, his son Elisha Douglass was born in Pomfret on March 23, 1823,[3] and the child's first awareness of life about him was the frequent moves made by his parents. In fact the earliest place of which he had a clear memory was Tallahassee, Florida, which the family called "Tall City."[4]

The Perkinses arrived in Tallahassee in 1830 and lived there for the next six years. Douglass made his first friends in this place, a thing he had very much in mind when eighteen years later he visited the Florida city with the hope of establishing a drugstore.[5] In 1836 Elisha Backus Perkins again uprooted his family and moved them off to raw backwoods Alton, Illinois, but Douglass was sent East to study in the Farmington (Connecticut) Academy under the tutelage of Mr. Simon Hart. He remained in this school for three years and was a good student, but the harsh Connecticut climate was injurious to his health, and he returned to Illinois.[6]

Douglass was allowed to remain in Illinois and recover his health. Quickly he regained vitality and strength in the woods of that area. He formed a friendship with an old border hunter and spent much of his time camping in a backcountry

[1] Elisha Douglass Perkins, Diary of a Journey to the South, October 13, 1848, to January 13, 1849, Seaton Papers, University of Kentucky Library, Lexington.
[2] George T. Chapman, *Sketches of the Alumni of Dartmouth College* (Boston, 1867), 168.
[3] Clipping from Marietta *Intelligencer,* January, 1852, in Eliza Hildreth Perkins Means scrapbook, Seaton Papers, Miscellaneous clippings 1840-1900, University of Kentucky Library, Lexington.
[4] Perkins, Diary of a Journey to the South, Dec. 14, 1848.
[5] Chapman, *Alumni of Dartmouth,* 168, and undated newspaper clipping in Means scrapbook, January, 1852.
[6] Undated newspaper clipping, Means scrapbook.

cabin and on the trail. Perhaps he followed the instructions of his backwoods tutor with even more attentiveness than he had those of Master Hart, for he became a crack shot with a rifle, and a good trail scout. He learned to hunt deer, wolves, and other game with great skill.[7] Later when Perkins was preparing to leave St. Joseph, Missouri, for the California gold diggings, he broke a wild range pony to stand while he shot his pistol from its back. In the Illinois woods the delicate boy had learned to sleep on the ground, to eat coarse food, and something of the art of self-preservation.

With his health restored, Douglass moved to Marietta on the Ohio where his father had bought both a "fine" country estate, a mile from town, and a drugstore in the town. Marietta in 1839 was already one of the oldest settled places on the Ohio River.[8] Nestled in the elbow of the confluence of the Ohio and Muskingum rivers, it was splendidly located on the main line of American commerce and travel. Steamboats daily nudged their prows against the steeply inclined rock-lined city water front and exchanged cargoes of merchandise for the rich harvest of Ohio country grains and meats. Occasionally the ravagers of the great virginal woods still fetched barrels of potash to the river to be sold in the New Orleans trade.[9]

Everywhere one turned in Marietta in 1840, one met a member of a pioneer family which had come out to Ohio in the famous migration of 1788 to plant a civilization in the opening Northwest Territory. Among these were the Putnams, Tuppers, Fearings, Dawes, and Cutlers.[10] An equally prominent family, but of much later arrival, were the Hildreths. These were the economic, professional, and social leaders of the town. They had built fine houses on carefully laid out streets, erected a towering Congregational church, like the one back in Canterbury where E. Backus Perkins had studied law, with its keen steeple thrusting a stern Yankee finger straight up to God. They had laid out a public park which they called Campus Martius, organized an academy and a college, and made themselves at home among the fertile ranges of the valley.[11]

In Marietta Elisha Douglass Perkins attempted to take up his disrupted educational career and entered the academy in preparation for college. Recollections of the rugged Illinois woods, however, made it difficult for the sixteen-year-old lad to settle down again to his studies, but even so he did acquire considerable skill in self-expression. He may have had too restless a nature and too much intellectual curiosity and ability to drone his way through a formal academy curriculum. Perkins

7 *Ibid.*

8 This famous city, the first settlement made beyond the Ohio after the adoption of the Northwest Ordinance, was founded on April 7, 1788. Thomas J. Summers, *History of Marietta, Ohio* (Marietta, 1903). The Perkins drugstore, located on the south side of Front Street, was the second one in Marietta. It was later sold to Cevier and Stimson, then to William Glines, in 1864 to Hollister and Allen, and in 1868 to R. L. Curtis. *Ibid.*, 248.

9 Summers, *History of Marietta.*

10 *Ibid.*

11 *Ibid.*

learned to play several musical instruments and to sing with some talent. He organized his young friends into singing clubs and thus enlivened otherwise dull evenings in Marietta. As a clerk in his father's drugstore he sought relief from the tedious labors of compounding drugs by inventing an alcohol-fired steam engine which turned mixers and pounders.[12]

In his subsequent California journal, young Perkins gave the impression of being deeply religious, with the particular outlook of the nineteenth-century puritan. Perhaps this was the result of his academy training in Connecticut under the spiritual direction of Simon Hart. He had subsequently joined the Congregational Church in Marietta and was an active member. Douglass Perkins also joined the Masonic Lodge and the Odd Fellows. Later, on his way west, he noted his excitement over seeing members of these fraternal orders along the Overland Trail.

Among the members of the Congregational membership in Marietta was the much-respected Dr. Samuel Prescott Hildreth with his wife and six children. Among the children were Harriett Eliza and George Osgood Hildreth, who were to be intimately associated with Douglass Perkins in the future. Dr. Hildreth had come west from Haverhill, Massachusetts, in 1806. On his horseback journey out to the Ohio country he kept a highly perceptive travel journal which related more than the bare facts of traveling in the early years.[13] Like his son-in-law a half century later on his way to the gold fields, the Yankee doctor's curiosity ran beyond the boundaries of the road itself, and included also the personalities whom he saw along the way. In keeping this early diary, Dr. Hildreth set an excellent literary example for the future traveling members of his family.[14]

Once settled in the new state of Ohio, Dr. Hildreth adopted the valley as his own. He first practiced medicine in the small river village of Belpre, but later moved upstream to larger Marietta.[15] Day and night he called on patients along the Ohio and Muskingum, and occasionally he rowed across the Ohio to minister to the sick in western Virginia. He combined the practice of medicine with digging into Indian mounds, gathering natural and historical artifacts, collecting local historical information, and writing. He interested himself in the broader subject of Ohio history and in time wrote a description of the state's beginnings. Quickly Dr. Hildreth became a central figure in the Marietta countryside. He bought himself a stately brick mansion in the heart of the town near the courthouse where he lived until his death in 1863. It was in this old

[12] Undated newspaper clipping. Means scrapbook.

[13] Samuel P. Hildreth, *Genealogical and Biographical Sketches of the Hildreth Family* (Marietta, 1840), 60-195. Dr. Hildreth's collections were later installed in Marietta College.

[14] *Ibid.*, 60-293.

[15] *Ibid.*, 184-85.

mansion that his daughter Harriett Eliza and her brothers and sisters were born; Harriett on September 4, 1826.

Elisha Douglass Perkins married Harriett Eliza Hildreth on May 11, 1847.[16] He had not yet established himself in a business, and the income of the Perkins drugstore was too meager to support two families. Now that Douglass had a wife, he had to think seriously of her support. The tall dreamy lad, with his fine yankee goatee, was better suited either to wandering in the western woods or to living the leisurely life of the sleepy river town than to settling down in a confining routine job. Besides, Marietta, then as now, was a rather unpromising place for an unestablished youth to succeed in a new business. The advertising columns of Beman Gates' Marietta *Intelligencer* reflected this fact every week.[17]

But there seemed better opportunities elsewhere. Steamers putting into Marietta brought news of excitement in the old Spanish territories of Florida and Texas. There were even stories that things were "looking up" in Arkansas. On Monday, October 30, 1848, as Perkins recorded in his southern travel diary, he, "turned his back on good old Marietta & took passage on the Schuylkill for Cin."[18] He was off to New Orleans and Florida in search of a place to establish a drugstore. From November until mid-January he searched in vain, and was then forced to return disappointed to his Harriett, a thing that pained him mightily. There is internal evidence in Perkins' writings that his lack of success was also disappointing to the Hildreths. Thus far their son-in-law had not proved himself capable of making a living for Harriett, and a family—for she was now in an advanced state of pregnancy.[19]

Early in 1848 the Marietta *Intelligencer* carried stories not only of the discovery of gold on the American River in California, but also of the political excitement over the organization of that western territory. Perhaps Douglass Perkins saw somewhere Captain Mason's report on his visit to the Sutter's Mill site.[20] Edwin Bryant, the Louisville editor, published his book *What I Saw in California* early in that year, and John C. Frémont's description of his western explorations was nationally a popular seller. On December 8, 1849, the Louisville *Daily Courier* ran the lengthy instructions of Edwin Bryant, and this paper was read up and down the Ohio. Countless other newspapers carried the article also.

There was much excitement around Marietta, Harmar, and Belpre. Men and boys talked of the possibilities of gathering fortunes in the California gold fields as glibly as they spoke of

16 Marriage records of Washington County, Ohio, 1841-1864, Campus Martius Museum, Marietta. E. Douglass Perkins to Harriett E. Hildreth, May 11, 1847.

17 See files for 1847-1853.

18 Perkins, Diary of a Journey to the South.

19 *Ibid.*

20 This appeared in newspapers all over the country, and Perkins either saw it on his trip South or soon after he returned to Marietta. Internal evidence in his journal indicates that he read widely on the subject of California and the discovery of gold.

harvesting wheat or corn from the nearby bottom lands. In Zanesville, Cleveland, Cincinnati, and elsewhere in Ohio by late February, gold-hunting companies were being brought together to make the overland journey to California as soon as grass would sustain teams on the plains. In Harmar a joint-stock company was organized with Yankee thoroughness.[21] Every eventuality of plains travel and gold mining was anticipated and provided for—all except one, human sensitivity and temperament.[22]

The Harmar Company was divided into two divisions. The first was the Home Board, which was to remain in Harmar and provide funds and supplies. The other division was composed of the adventurers who promised to undertake the hazardous overland journey to California and to mine gold. Both members of the Home Board and the trail adventurers assured themselves of the moral fiber of the company by taking pledges that they would refrain from drinking, gambling, and carousing. Members of the company made equal investments to finance the costs, and they were to share equally in the riches.[23]

There was much excitement over the organization of the Harmar Company, and the news which floated up and down the Ohio aboard the steamboats became more enticing with each passing week. In neighboring Marietta, young Douglass Perkins seems to have concluded that if Florida, New Orleans, and the expanding Southwest offered no opportunities for a young druggist, he could garner a quick fortune from the gold pockets along the Cosumnes, Mokelumne, and American rivers in California! He read Bryant's and Frémont's books, two or three of the popular guidebooks to overland travel, including Edwin Bryant's special instructions to gold rushers.[24] With far less publicity than was given his Harmar neighbors, Perkins helped assemble what he called the "Marietta Gold Hunters." Its members consisted of E. Douglass Perkins, Samuel E. Cross, a New Yorker recently arrived, Joseph L. Stephens, Z. J. Chesebro, J. Q. A. Cunningham, and John L. Huntington, a visitor from New Orleans.[25]

On Tuesday night May 8, 1848, the Marietta Gold Hunters rolled two loaded wagons aboard the *DeWitt Clinton* bound for St. Louis where they would transfer to the Missouri River steamer *Highland Mary II* for the journey upstream to St. Joseph.[26] In the latter place they planned to purchase mules and horses for their overland trip, which they thought could be swiftly completed. Perkins carried in his baggage an oblong leather-bound open-ended notebook, about five by eight inches,

21 Marietta *Intelligencer*, April 26, 1849.
22 *Ibid.*, October 31, 1850.
23 *Ibid.*, April 26, 1849.
24 It is evident, however, that Perkins did not see the 1849 edition of Bryant's *What I Saw in California* which contained fuller information on the finding of gold in California.
25 Marietta *Intelligencer*, May 10, 1849.
26 St. Joseph (Missouri) *Gazette*, June 1, 1849, quoted in Marietta *Intelligencer*, June 21, 1849.

containing approximately 200 pages in which he would record his travel experiences. He was now already an experienced traveler and journalist who knew what to describe.

Writing in a slightly cramped but somewhat sophisticated hand, Douglass Perkins was to give a reader of a century later the laborious task of separating his words so as to transcribe and collate his journal into readable typescript. Fortunately he chose to make his daily entries with a lead pencil and the script has remained fresh and clearly delineated. The young journalist had genuine literary ability. He was a close observer not only of the people whom he saw along the way but of the nature of the country, the Indians, and animal life. Fundamentally a romanticist, Perkins described the experiences of the rugged life on the Overland Trail from the viewpoint of an imaginative man. Even though he lacked understanding of the implications of distance, rugged terrain, and the wear and tear of plodding overland travel, he was never frivolous about his undertaking. He sensed the enormous amount of movement about him, the excitement and anticipations of hundreds of other gold hunters, the basic goodness and cussedness of raw humanity, and, most important of all, the perversity of Mexican mules.[27] Most fortunately the industrious young Yankee allowed scarcely a day to pass without making an entry in his journal. Often the reader feels that he is sitting beside the trail itself waiting for Perkins to finish an entry.

There is a youthful grace and enthusiasm about Perkins' journal which at times all but obscures the grinding reality of the trail. Every strange bird, plant, flower, and animal awakened in him an eager question about its nature.[28] He showed interest in the rapidly vanishing Carolina parakeets which he saw on the Kansas plains. He anticipated seeing the plains Indians, antelopes, wolves, and buffaloes. He had a rather good prior knowledge of the landmarks which would appear on the trail from having read Frémont's and Bryant's descriptions. With Yankee frugality he lay on his buffalo robe and tried to calculate the costs of moving a frontier military unit across the plains to establish a new Rocky Mountain outpost in terms of the educational facilities the money would purchase. With boyish friendliness he visited Fort Leavenworth, the Kansas Indian mission, and Fort Laramie. He was always on the lookout for new friends, especially fellow Masons and Odd Fellows.[29]

On June 1, 1849, the St. Joseph *Gazette* carried a story on the Marietta Gold Hunters who had crossed the Missouri and

27 St. Louis *Republican*, April 6, 1849, quoted in Louisville *Daily Journal*, April 16, 1849.

28 See diary for May 29, 1849. The fact that the birds Perkins described did exist is amply verified in the report of Lieutenant J. W. Abert on the *Examination of New Mexico in the Years 1846-1847*. This report is contained in W. H. Emory, *Notes of a Military Reconnaisance*, pp. 419-562. See specifically p. 544.

29 Diary entry, July 2, 1849.

were eager to be off along the Platte Valley part of the trail. Beman Gates added the comforting note in the Marietta *Intelligencer* that all seemed to be well with both local companies. The Harmar Company was already on the trail, and the Gold Hunters would soon be ready to start.[30] St. Joseph, Independence, and Weston were hustling places. Thousands of eager gold seekers poured in daily from everywhere in the eastern United States.[31] Drovers from down the Santa Fe Trail brought up from Texas, New Mexico, and Missouri great droves of mules and horses for sale, and still they were unable to satisfy the demand.[32]

A scarcity of mules delayed the Gold Hunters several days in getting started for California. Too, the Ohio lads were largely unprepared for the intense excitement of the outfitting towns on the one hand, and for the scare caused by an outbreak of cholera.[33] Certainly they were unprepared for the heavy rains which they encountered in what is now Kansas, or for the rough heavy roads on which they eventually began their journey. In order to condition themselves even slightly for their journey, the Marietta boys began living in their tent immediately after arriving in St. Joseph. The rains, however, made this a trying ordeal. Nevertheless they accepted their early discomfort with youthful good nature and anticipated better things, once away from the river.

At last Gold Hunters had their complement of mules and ponies and started. They learned in quick order that the terrain of the western plains was an enormous problem with which they would have to reckon. Their wagons, along with those of hundreds of other travelers, were hopelessly overloaded, their unseasoned teams were unbroken and contrary, and the rains had left roads and stream crossings deep mires. Before the end of the first week of travel it began to be apparent that the great passage over the Sierra Nevada was more than three months and half a continent away.

Despite anxieties about the ultimate success of their overland journey, the Marietta Gold Hunters were thrilled at their approach to the range of the buffalo and antelope. They kept an eye out for Courthouse and Jailhouse rocks, for Scotts Bluff, and ultimately for Fort Laramie.[34] Before they reached the latter place they had learned a hard lesson. Cholera was a bitter threat to human life along the trail as the number of graves proved. In places water was more precious even than grass. In this country travelers in a hurry to get to California could not drag along with them many luxuries which they

[30] June 21, 1849.
[31] St. Louis *Republican*, April 5, 1849, quoted in Louisville *Daily Journal*, April 14, 1849.
[32] *Ibid.*, April 6, 1849, quoted in Louisville *Daily Journal*, April 16, 1849.
[33] Marietta *Intelligencer*, June 21, 1849.
[34] Diary entries, June 27, 29, July 1, 1849.

thought necessities before they started. The road from St. Joseph to Fort Laramie was a sore trial to human relationships. Many a company that started so amicably from one of the outfitting towns in Missouri found its members could not bear one another by the time they reached the north fork of the Platte, and started the hard climb to South Pass.[35]

It is unclear who exercised leadership in the Marietta company. Whether it was Perkins or not is vague. At times he wrote in his journal as if he were the leader, and at other times he seemed to indicate that Samuel Cross or J. Q. A. Cunningham was leader. At any rate the party decided to abandon its wagons at Laramie and travel in pairs from that point on. There is every indication that they made a miscalculation, if not in distance, then in conditions of travel from Laramie and in the amount of time which would be consumed in passing through the mountains and down the Humboldt River beyond. In breaking up their company the boys undertook to construct carts on which they could load a meager amount of supplies. This brought them near to disaster.[36] Quickly Perkins came to realize that his own destiny was closely linked to that of his mules. Their loss could mean loss of his own life.

There was no lack of human company on the trail. Also, there was plenty of advice from men of judgment who knew something of the treacheries which a traveling party might expect to face, even though they themselves had never exactly faced them. On July 4, 1849, the St. Joseph *Gazette* reported that there were on the plains bound for California approximately 5,000 wagons, which meant there were about 16,772 persons, and 33,544 mules and oxen; in all the *Gazette* expressed the opinion that approximately 27,000 people were on their way to California.

A month later the St. Louis *Republican* reported that *"Five thousand and ninety-two wagons,* at sundown last night, had moved past this place towards the golden regions of California, and about one thousand more, I think, are still behind."[37] Behind them emigrants left a trail of abandoned foodstuffs, equipment, clothing, and even household furniture as tangible evidence of enormous ignorance of the nature of the great overland adventure.[38] Samuel E. Cross and Zebulon Chesebro followed Edwin Bryant's advice and example and discarded as much of their goods as they could do without and moved on at a good clip. They made the crossing in eighty-five days, almost record time. They apparently suffered only a limited hardship in doing so, and were no worse off than thousands of

35 In earlier years Edwin Bryant noted this fact, and he organized a small pack party to hasten on to California ahead of snowfall, Bryant, *What I Saw in California*, 68-69, 96-97.

36 Diary entry, July 2, 1849.

37 The Marietta *Intelligencer* quoted the St. Joseph *Gazette's* estimate on July 5, 1849, and the *Republican's* statement on July 25, 1849.

38 Charles Glass Gray, An Overland Passage from Independence, Missouri, to San Francisco, California, June 11, 1849.

others who arrived in California without another meal in their packs. Perkins was not so diligent, and he and Johnny Huntington sometimes dawdled along the way. They stopped to graze their mules. Sometimes a seductive spot tempted them to linger longer than they intended to do. They came up with pleasant company and straggled along with the ox teams, losing precious time. On the Greenwood Cutoff they undertook to follow a second shortcut and found themselves involved in far more traveling than they had expected.

If Perkins had traveled as rapidly as Chesebro and Cross no doubt his journal would have suffered. There seemed never to have been a day on the whole route when the author was too hurried to make an entry. Sometimes he lacked enough information to make a significant entry, but he was never too pressed at least to enter a line.

That part of the overland journey from the Sweetwater River to the "Sinks" on the lower Humboldt just before the Salmon Trout passage into the eastern shoulder of the Sierra Nevada was indeed exacting. All the thrill of traveling along the great trail had worn away. A tremendous amount of shuffling of parties had occurred. Graves lined the trail as monuments to the harshness of the country and the deadliness of disease. Carcasses of oxen and mules all along further revealed the cruelty of nature. Wolves, coyotes, and bears by night threatened peace of mind if not body. Somewhat more subtlely the Shoshones, Snakes, and Digger Indians added their own particular brand of drama to life in the hard country. Failure to guard teams for even a single night often resulted in tragedy. Indians drove animals away to the mountains and killed or hamstrung them so they could not be used. A company left without its teams was at the bitter mercies of a barren country and were doomed to almost certain destruction.[39]

Independence Rock, the Devil's Gate, the great frowning walls of the snow-capped Wind River Mountains, and the deep crossings of the icy mountain creeks and rivers startled the flatland Ohio travelers. If they had wandered onto another planet, the world about them at that point could not have seemed more strange. Birds, plants, animals, and even fish resembled those back along the Ohio and Muskingum, but they were not quite like them. The raven, the jay, the buzzard, and even the flowers were different. When Perkins came on the great clump of conifers along the Sublette Cutoff he experienced the same delight which John C. Frémont had described on his visit several years earlier.[40]

[39] William E. Chamberlain, Diary, July 26, 1849.

[40] John C. Frémont, *Report of the Exploring Expedition to the Rocky Mountains in the Year 1842*, 52-79.

It was the Humboldt Valley, however, which finally tested Perkins' mettle. The lack of fresh water, the coarse roughage, the stinging alkali dust, and the terrific heat gave a fatalistic timelessness and endlessness to the great trail. Beyond this was the chancy crossing of the great mountain divide ahead of winter. The cool flow of the Salmon Trout with its widening marshes and fish weirs tempted the travelers, homesick for the broad bottoms of the Ohio, to wander dangerously afield. By the sheerest good fortune they did not encounter hostile Indians who could have wiped them out in a moment. Despite this carelessness the pine-studded crossing from the arid valley of Nevada to the oak-dotted slopes of eastern California proved to be one of the most exciting experiences of the whole journey. Before the cabin ruins of the Donner-Reed winter camp site the introspective Perkins was reminded how fleeting human fortunes could be.[41] This gloomy thought was quickly diverted, however, by the high drama of teamsters, who like those of some dark saga beat and cursed their overloaded animals, stage by stage, up the rocky face of the high pass. It reminded the reporter of the stirring scenes which a transient panoramist had rolled off before audiences back in Marietta.[42]

Penniless and virtually without food and other supplies, Perkins and Johnny Huntington began the long descent down the Yuba and Bear rivers toward the rising city of Sacramento. Like Edwin Bryant's party of three years before the Ohioans made their way to Johnson's Ranch, then to Marysville, Fremont, Vernon, and Sacramento. Sorrowfully they learned two bitter facts: food prices were exorbitant, and gold was not nearly so plentiful in California and in September as they dreamed it would be 2,000 miles back on the trail at St. Joseph in May.

The history of Douglass Perkins' life in California was, as far as it can be pieced together, one of abysmal failure. He found little gold, suffered from ill-health, and finally left the gold fields to find employment as a steamboat captain. Unhappily the shock of failure to find gold, the death of two of his trail companions, and the desertion of the gold camp by his brother-in-law, Dr. George O. Hildreth, and the raw biting climate of the Sierra Nevada foothills frustrated the journalist. He summed up his life through February 1850 and discontinued his writings.[43]

There seems to be no way of determining what Perkins did with a part of his time in California in 1850, 1851, and 1852. Sometime during these years he became captain of the steamer *Marysville* which plied the Sacramento River trade.[44] In

41 Diary entry, September 15, 1849.
42 *Ibid.*
43 *Ibid.*, February 28, 1850.
44 Sacramento *Daily Union*, December 20, 1852.

December 1852 he became ill with dysentery and when the *Marysville* landed against the wharfboat *Orb,* an old New England bark which had brought a load of gold seekers around the Horn, Perkins was removed from the steamer.[45] He was attended by John Weaver of Marietta and Dr. George O. Hildreth. There is a contradiction as to the circumstances of Perkins' death. The Sacramento *Daily Union* on December 20, 1852, said Perkins died aboard the *Orb,* and that he was attended by Dr. Johnson Price.[46] In a letter to Dr. Samuel P. Hildreth, George Hildreth said Douglass had been ill for five weeks, the last three of which he was confined to his bed. "His last days were free of pain.— Great kindness was extended to him by his acquaintances during his sickness."[47]

Perkins' body was buried in Lot 1302 in the northeast corner of the newly opened Sacramento City Cemetery. In fact he was one of the first persons buried here. His grave is at the corner of Riverside Boulevard and Broadway, just across the street from the old Edmond's Field baseball diamond. There under an arch of spreading elms the adventurer who had traveled so far to court dame fortune was laid to rest virtually penniless and far removed from his Harriett and Marietta.[48] He had found no gold, and worst of all he had not been able to support his faithful young wife in the manner of the Hildreths.

On October 25, 1854, Harriett Eliza Perkins married John Means, the successful iron smelter of Catlettsburg, Kentucky. She bore him six children, three girls and three boys.[49] In 1882 she set out by steamboat for Cincinnati, accompanied by two of her daughters, to begin a railway journey across the west to California. As the train roared westward it passed through the rugged stretches of the Rocky Mountain passes, and across the arid country similar to that which Douglass Perkins had crossed in 1849. Harriett had not lost her interest in him. In Sacramento she and her daughters went in search of his grave on a May afternoon. Unhappily they went to the wrong cemetery. Harriett wrote in her journal that "We reached Sacramento at 10 a.m. and came to the Capitol Hotel. In the afternoon we rode out to the New Helvetia Cemetery—we found graves of those who were buried there as early as 1852 & went all over the Cemetery & tried to find Douglass' grave but we could not find it. It is a lovely spot of trees & flowers and singing birds & I was glad to find it so,—(though I have since learned that he is probably buried in a cemetery still older than New Helvetia)."[50] Soon after Perkins' death Harriett had bought a small marble slab on which was carved the brief information of his

[45] *Ibid.,* December 27, 1852.
[46] Marietta *Intelligencer,* February 5, 1853.
[47] Dr. George O. Hildreth to Dr. Samuel P. Hildreth. *Ibid.,* February 5, 1853.
[48] Vital Records, California State Library, 1852-1853.
[49] Hildreth, *Sketches of the Hildreth Family,* 314-15.
[50] The New Helvetia Cemetery was organized in 1857 and was in active use until 1917. In 1955 the remains were removed and a junior high school was built on the site. This information supplied by Allan R. Ottley of the California State Library, November 12, 1964.

home, birth, and death. In California Harriett saw the broad stretches of brown arid fields dotted with their stands of scrub liveoaks and other undergrowth. She saw the American and Sacramento rivers, and viewed the scenes about the handsome state capitol where Douglass had seen "Ragtown" take form on the muddy flats. She had transcribed descriptions of those places from his stained and worn travel journal. Life was so easy and pleasant for her and her daughters in 1882 that it seemed almost a desecration of her first husband's memory. They had traveled to California with such ease and at such great speed. Where Douglass' western journal had measured a day's travel in terms of twelve to thirty miles, Harriett spoke of crossing the continent in less than five days, and arriving back in Ashland, Kentucky, on May 20 aboard a local Ohio packet.

Douglass Perkins failed to find gold at the end of an arduous journey; he even lost his life in California, but the production and preservation of his journal has prevented his name from being lost to the history of the Gold Rush. In many respects it was fortunate that Perkins and his party were late in starting westward from Marietta. By the time they reached St. Joseph the great body of emigrants was on the trail. Behind they left a scarcity of mules and supplies, ahead along the Platte and in the mountain passes they left even more important evidence of the fact that they had rushed away ill-prepared for the journey to California.

Not only did their animals crop the grass disastrously short along the trail and their guns drive game far back on the plains, their graves created psychological barriers. Cholera, accidents, and other diseases took their toll. New-made graves became monuments to folly and the frailty of human beings in the face of so vast a natural challenge as crossing the continent to get rich. Travelers who lost their nerve and returned told discouraging tales of their trail experiences. These were parts of the saga.

Perkins was only one of many diarists. Fortunately there were others within a few miles of the Marietta party and a day or two ahead or behind it. Among these was Charles Glass Gray, Elijah Farnham, J. Goldsborough Bruff, John Banks, J. Elza Armstrong, and David DeWolfe. All of these kept diaries. They viewed the same landmarks that Perkins saw, met most of the same people, saw the various Indians, found game in the same spots, and had the same difficulties with mountain grades and stream crossings. Occasionally one of them confided to his journal the same fears and doubts as beset Douglass Perkins. This was especially true of the New Yorker, Charles Glass Gray. It may be within reason, however, to say that Perkins was more intimately observant than most of his fellow travelers. There is both a compactness and subtleness about his journal which gives it unusual value. All of the diarists mentioned above recorded many simple statements of their conditions and that of the country through which they traveled. Perkins, however, went further. His journal is complete from beginning to end. He missed observing no part of the journey, and never failed to make an entry in his diary. He made interesting comparisons of conditions in the West with those in the Ohio Valley; at times he speculated on the country, the Indians, plant life, and animals. He often revealed a fuller knowledge than others of what he saw, and in making entries in his diary he revealed a deep emotional involvement with everything around him.

In a penetrating recapitulation Perkins looked back on his trail experiences and suggested where emigrants had erred on their long journey west. This perhaps is a unique feature of gold rush diaries. This summary revealed the fact that crossing the continent had become an end within itself. Perkins came to believe there was an easier and safer way to make the crossing than that used by the gold rushers. He also hinted that he believed that emigrants in 1849 had opened a way west which in time would feed a sizable population onto the Pacific slope. Even if emigrants found little gold this latter fact would be sufficient compensation.

As emotionally involved as Perkins became in the Gold Rush there is no hint on his part that it involved a mystique. Traveling westward to California in 1849 was sweaty hard work. It challenged the courage of men and women, and of draft animals. Death lurked in many forms on every stretch of the trail, and death meant failure and defeat. Although hundreds of feet had stirred the dust on the trail, and hungry mouths had snatched at every blade of grass, there was always optimism that an individual or party would escape punishment.

Materials in 1849 describing the route and the rigors of western travel were meager, and more often than not greenhorn travelers failed to read what was available. The journalists of 1849 no doubt were somewhat motivated by this fact. They wrote as though they were offering information and advice to a still larger army of travelers to come. Perkins often gave evidence that he sensed that he was involved in one of the great adventures in American history. He wrote at a time when conditions provoked men to exaggerate their exploits; this Perkins did not do. There are too many corroborative materials to doubt the validity of his observations. His diary is a rather comprehensive chapter in the great saga of pioneering which was touched off by the discovery of gold in California.

The argonauts of Marietta found little gold and much frustration. Only one of the Gold Hunters, Joseph L. Stephens, returned home. They and the Harmar Company however, left behind an unusually full story of a community's preoccupation with the gold rush. Beman Gates published their letters and the reports of his special reporter in the *Intelligencer*. Perhaps few towns in the eastern part of the United States were kept so well informed about the fortunes of their wandering sons as was Marietta, Ohio.

As more Gold Rush material comes to light it becomes clearer that this great surge of the westward movement was of broader interests in the East than perhaps in the West. The purchase of supplies and equipment gave local economy a lift, but the loss of thousands of men disrupted families and labor forces. No doubt the great body of emigrants straggled back to their homes, but many of them died either on the trails or in the gold fields. They did, however, gather an enormous amount of experience which was sobering and prepared them for assuming responsibilities in the immediate years of national crisis after 1860.

May 9–June 14, 1849

1 E. Douglass Perkins seems to have been the unofficial leader of the Marietta Gold Hunters Company. Samuel E. Cross had recently come to Marietta from New York, and John L. Huntington was from New Orleans. Of the six only one, Joseph L. Stephens, ever returned to Marietta. Cross died soon after arriving in the gold fields; Zebulon Chesebro and Perkins also died in California. Huntington apparently either remained in California or went back to New Orleans. J. Q. A. Cunningham remained in California the rest of his life. During the Civil War Stephens was a volunteer member of the Marietta Silver Grays, a home-protection organization. Marietta *Intelligencer*, May 10, 21, 1849, May 2, 16, 1850, February 5, 1853. Lot 1302, Sacramento City Cemetery, and inscription on Perkins' tombstone; Sacramento *Daily Union*, December 20, 1852; Howe, *Argonauts of '49*, 204-21. Vital Records, 1852, California State Library, Sacramento.

2 The year before Perkins had kept a very detailed travel diary of his trip to New Orleans and Tallahassee, Florida.

3 Although the Marietta Gold Hunters possibly were able to avoid contact with individuals who had cholera, the *Highland Mary II* had aboard persons ill with the disease.

4 The *Highland Mary II*, a rather ornate steamer with a lassie dressed in a Scottish plaid for a figurehead, was said to have had a lively band and an excellent bar and culinary corps. Ferris, "Steamboat Art, Decoration, and Elegance," *Bulletin of the Missouri Historical Society*, XVIII (October–July 1962), 143. This boat was commanded by Captain John Atchison, who died of the cholera in the summer of 1850; after his death his boat disappeared from the river. *Western Journal*, VII, 398. The distance by river from St. Louis to St. Joseph was estimated to be 450 miles. Cabin fare was $6.00 and deck fare $4.00. Ware, *Emigrants' Guide*, 2-3.

5 St. Charles, the capital of Missouri from 1820 to 1826, was a well-known frontier post. The town was for a long time the "jumping off" place for the vast Far Western frontier, Williams (ed.), *A History of Northwest Missouri*, I, 218.

6 The first capital building of Missouri, designed to have been the governor's mansion, was destroyed by fire on the night of November 17, 1837. In 1840 the new capital,

HAVING DECIDED upon undertaking a journey across the Plains by way of the South Pass of the Rocky Mtns. to the far famed valley of the Sacramento, I left my pleasant home & dear friends in Marietta, Ohio—on steamboat DeWitt Clinton—Wed. May 9, 1849, in company with five good fellows. S. E. Cross, Z. L. Chesebro, Jos. L. Stephens, J. L. Huntington & J. Q. A. Cunningham.[1] Our trip down the Ohio & up to St. Louis was without incident worthy of record. The scenery of these rivers has been over & over again described every one has heard of their rapidly growing cities & towns.[2] In St. Louis we met with an abundance of that scourge of the world the cholera, but were ourselves preserved from any personal acquaintance with it.[3]

We left St. L. after a stay of two days on the *Highland Mary* for St. Joseph.[4] The Missouri River is much less known & less written about than its great rivals above named. I myself had no idea of what I was to see in my five days journey on its waters. I find that it has more diversified scenery than either of the others and you see in one days sail the level bottoms of the lower Mississippi, the rolling country & hills of the Ohio the abrupt bluffs & jutting precipices & wild rocks of the upper Miss. Some of the projecting rocky bluffs are most sublime & it requires no great stretch of the imagination to see Castles with their battlements & towers in their fantastic outlines.

St. Charles the first town is a little collection of houses on a side hill, nothing about it to notice.[5] Jefferson City the Capital of the great Missouri State, is built on a succession of small bluffs & has a very irregular appearance.

The State House standing on a bold elevation about 100 feet above the river is built of the Missouri marble a cream colored stone which polishes well but seems to crumble on exposure to the air, & I noticed many of the stones in the building were broken into perhaps a dozen pieces. The State House has a semi circular portico with white marble pillars on either front & with its dome which more resembles that of the St. Charles Hotel in New Orleans than any I ever saw presents altogether a fine appearance.[6]

At Jefferson City we saw a most melancholy monument of the ravages of the cholera in the steamer *Monroe*. This vessel left St. Louis last week with about 100 cabin passengers & 50 deck, all well, at the mouth of the Missouri the Cabin passengers began to be attacked & before reaching Jefferson City 150 miles —70 of them died; at the city the boat was deserted & all fled for their lives. Of the 100 passengers only *three* are now living

& well! This is most appalling & unaccountable. The strangest thing about it is that none of the Deck passengers were attacked. The citizens account for the mortality by the fact that the cabin of the *Monroe* was painted just before leaving & smelt so strongly of the lead. Whether this is so or not officers of the Boats should avoid paint as a probable cause at least.[7]

All the officers of the M——— died but the mate. I shuddered in spite of me as we passed close by her & could look at her only as a great floating coffin. The graves of her passengers ranged along the shore for about half a mile. May I be spared the contemplation of another such an object. One company from Indiana were on board, of 27 men with their wagons, baggage &c only two of them were spared to tell the sad tale to relatives & friends, whom they left only two weeks before in good health & spirits. How many broken-hearts & widowed wives & fatherless children are made, God only knows. But when we reflect that these unfortunate wretches were all from one district or township, must not nearly every house be a house of mourning.

About 100 miles below Independence we had the—to most of us—first view of a prairie on fire.[8] Wapintaw [Wahcondah?] Prairie borders the River for several miles and from the hurricane deck we could see the almost interminable plain stretching far away to where a faint blue outline indicated a range of hills. Some of the old grass was still remaining & was burning forming a line of fire about 1 mile long. The grass not being very long or rank did not make so great a blaze as the fires of the great prairies of the frontier, but still twas a fine sight & served very well as an introduction.

The far famed Independence we did not see as it stands back 4 miles from the River. Its landing place is a collection of warehouses & livery stables under an overhanging bluff. The town itself I was told by some of our passengers who went out, is of about 1500 inhabt. & looks like most country villages, Except that most all the houses are Hotels & mule stables.[9]

Weston some 50 miles above Independence is a neat looking town built on a rocky hillside.[10] This is another of the starting points for the journey across the plains. Just above here we met a train of some 20 wagons with oxen going down to Fort Leavenworth to go with the troops. Rain was literally pouring down but on they splashed seeming perfectly reconciled. We passengers got out on the guard of the Boat to cheer them on, but when we imagined ourselves in their places, I think we involuntarily stepped in out of the damp! We expect "to see

constructed at a cost of $300,000, was occupied. This building, which was destroyed by fire in 1911, was the one that Perkins saw and somewhat resembled a Greek-revival temple. Shoemaker, *Missouri and Missourians*, I, 437.

[7] In this connection Perkins reveals the primitive state of medicine in 1849. Writing earlier, Joseph Sedgely of the Sagamore and California Mining and Trading Company of Lynn, Massachusetts, said that on May 5, 1849, "The steamer *Mary* [*Highland Mary*] stopped within pistol shot of our camp, and buried eighteen men who had died of cholera." *Overland to California*, 12. For a discussion of the disease, see Chambers, *The Conquest of Cholera*. Chambers suggests that the disease came into this country from France. The steamship *Swanton*, which sailed from there in October, 1948, had discharged infected passengers at New Orleans. From there the disease spread up the Mississippi and Missouri rivers.

[8] Perkins little knew how prevalent grass fires would be on the plains, or how destructive of grass and game. Charles Glass Gray seemed to have been no more conscious of this fact. On May 15, 1849, he wrote, apparently undisturbed, "We at length arrived safely on the opposite side & about 2 miles from the creek encamped, & within 100 yrds of a prairie, *which was on fire* & burning with a tremendous noise, but we had no fear of it, as we were on the opposite side of the road." *Overland Passage*, 17, hereinafter cited as Gray Diary.

[9] The Independence *Western Expositor*, March 19, 1849, said that the town had twenty wagon and blacksmith shops actively engaged in manufacturing wagons, besides large droves of mules, herds of cattle, and provisions—everything the traveler needed. Independence then had 1,600 inhabitants, thirty drygoods stores, six or seven of them doing $100,000 to $300,000 worth of business annually. There were two large hotels, and other necessary accommodations. *Missouri Historical Review*, XXVIII, 252-53.

[10] Weston, Missouri, founded in 1836 by Joseph Moore, ex-soldier from Fort Leavenworth, had 400 inhabitants in 1840. Barry, "Kansas before 1854; a Revised Annals, Part Ten, 1838–1839," *Kansas Historical Quarterly*, XXIX, 143-89.

11 St. Joseph was first settled in 1831 as a trading post by Joseph Robidoux. After the Platte Purchase in 1837 the ferry landing and trading post at "Blacksnake" Hills began to grow. In 1849 it was a booming county seat where gold-rush emigrants fitted out. Perkins saw here a village of part-log, part-frame houses reaching up the hill from the river. On the hilltop stood the courthouse, with its high dome and spire. Later St. Joseph was to become famous as the eastern terminus of the Pony Express. Williams (ed.), *History of Northwest Missouri*, I, 335-56.

12 Emigrants felt that it was a good idea to camp out a few days at St. Joseph while they were getting organized so that by the time they were ready to travel, they would have conditioned themselves somewhat for the long period of camping out across the plains. Edwin Bryant and his 1849 party camped out on the Neville Ross farm before they departed Independence in April. Louisville *Daily Journal*, April 16, 1849.

R. C. Shaw described the gloomy days his party spent in a pretrail encampment in which several members came down with the cholera. "It is remarkable that, notwithstanding the depressing circumstances under which we were laboring and the gloomy prospects of the future, not one of our party was disposed to abandon the enterprise and return home." *Across the Plains in 'Forty-Nine*, 26.

13 "*Californians*: The St. Joseph *Gazette* of June 1st contains the following notice: 'The following gentlemen crossed the river yesterday, with twenty mules and horses, and are now on their way to California:—Dr. R. B. Riggs, Wm. DeCamp, E. E. Willis, Chas. Stansbrough and Robert Henderson, driver, of New Jersey; and Samuel E. Cross, Douglass Perkins, J. Q. A. Cunningham, Zebulon Chesebro, John Huntington, and J. Stephens from Marietta, Ohio.' We understand that letters have been received from some of this company stating that some men who had returned from Fort Laramie to St. Joseph, met the Harmar Company some distance out, all well and in good spirits. Grass was abundant, and the prospects for a safe journey encouraging." Marietta *Intelligencer*, June 21, 1849.

Perkins and his companions had met the New Jersey company aboard the steamboat

the Elephant but we had rather see the wooley horse first," or in other words we would rather commence our experience of camp life with mild rains & have the *storms* deferred for a time.

Fort Leavenworth just below Weston is no fort at all, but more properly a barracks, there not being a defensible building on the grounds. The situation is delightful being on a slope receding from the River & on the top of the hill. The officer dwellings as usual were handsome & the grounds around them were tastefully arranged. Here we saw some of the troops & wagons destined for the Oregon expedition. They start soon.

We arrived in St. Joseph on Monday May 21, & from this I shall commence keeping a regular daily journal of our haps & mishaps.[11] We had a slight touch of the beautiful on landing our goods. A heavy thunder shower came up while our breadstuffs, sugar &c were lying unprotected on the wharf. There was no time to get them under cover so we out with our tent & spread it over the lot & with our India rubber goods on stood & took it, & succeeded in keeping everything dry, but at one time when the storm was driving ahead & fast coming down on us & nothing taken care of, we thought all our little stock must be sacrificed & we considerably set back in consequence. Having been up town and looked round we concluded we would live as comfortably & quite as well to camp out at once. So while part took the tent out & set it about a mile from town, the rest had the goods loaded & hauled out & here we are commencing our camp Experience.[12]

We have ascertained that there are but few mules to be had & those of poor quality. Last week there was a large lot for sale low but a day or two before our arrival the order arrived from Ft. Leavenworth for them & so we are likely to stay here till more come in & that may be two weeks.

Our friends at home would have been much amused could they have looked in on our cooking operations & supper this evening. Quincy (Cunningham) acted as chief cook, Cross 2nd. cook, & between them we had a comfortable meal of fried Ham & coffee & with Zebs jam & biscuits we made out finely. Our table was a board laid across two of our camp chests & as we squatted & knelt round it many were the jokes passed on the novelty & ludicrousness of our situation.

Our two friends of Doct Riggs company with whom we have made an arrangement to travel, are with us & we shall all wait here probably till we hear from the Doct.[13] Meantime our goods are piled up with the tent of our friends spread over them to protect them from the weather. We shall sleep tonight five of us in our tent and two in our wagon, or rather we shall try to sleep. Its an old saying that a strange pillow never sleeps well & it will be a little strange if so strange a pillow as ours sleeps at all.

Tuesday May 22. Had a rare time last night, we are all now more or less uncomfortable & finding we could not sleep, amused ourselves by laughing at each others troubles. Our Buffaloe robes in the rain yesterday were considerably wet, our addition tumbled off the dray as they were being hauled up & of course got muddied also. We having no other beds there was no way but to pick out the dry corners & curl ourselves on them & then we had it, first one would bounce up cursing the Buffaloes, he'd got on a wet spot & his clothes were soaked thro, & he couldn't sleep. Then another had about 50 sticks & stones under him & his bed must be tumbled on & everything cleaned out. One of us laid across a couple of chests & the one being some higher than the other but the sharp edges cut considerably into his arrangements & he was exclaiming against camp life half the night. Having determined to make the best of Everything. Every fresh outbreak was saluted with bursts of laughter & many jokes at the Expense of the unhappy individual. "Reuben" "our big dog" not liking the appearance of sundry persons passing on the road & sundry pigs seeking their living among our leavings. Kept up a most faithful watch all night barking like all possessed. So that between everything we slept about five or six hours. Got up this morning stiff & chilly & with a sore throat. This I shall expect for some time to come tho' until I get hardened to this kind of life. There may be fun in camping when you get used to it but we have not yet discovered any.[14]

We have made considerable progress today towards getting off. Cross & Chesebro bought a mule & pony & we learned down town of some good mules just in which we went to see. They were trained Mexicans but were Entirely too high being 100 dollars. This tho is the first asking price & in a day or two can be bought for two thirds, in fact Stephens who has just come up, says three of them were offered him at 75$. Then again we have seen two Santa Fe mules for which 120$ is asked but which we can & probably shall buy at 100.[15] So that we may be off in three or four days yet.

Tomorrow morning it is decided that a deputation be sent down to buy the large mules if possible & then get the St. Fe.

coming up from St. Louis and perhaps had entered into some kind of agreement with them to travel with or near one another. The New Jersey men were led by Dr. R. B. Riggs.

[14] Despite the fact that Americans were great outdoors people, remarkably few of the Forty-niners seem to have been conditioned to outdoor life. Perkins himself had lived in the open when his family resided in Alton, Illinois, and he was the companion of an old hunter. Nevertheless, the experiences of the boys from Marietta were little different from those of the vast majority of their fellow campers. David Rhorer Leeper described the predicament of his party in the rain. "I remember that on one such occasion in Missouri, when after trudging all day long through the mud, night overtook us in the middle of a wide prairie. We had no alternative but to chain our oxen to the wheels of our wagons, make our couches beneath the wagon covers as best we could, having no fire and no food for man or beast. As I lay on the rude pallet reflecting on the situation, the winds meantime keeping up an ominous refrain without, my thoughts naturally turned toward home, its blazing chimney fire, its generous cupboard, and its other creature comforts." *Argonauts of '49,* 8.

[15] The reporter of the St. Louis *Republican* wrote from Independence on April 5, 1849: "In my last, you received the rates at which stock was being sold, from which there has been no change—say for American mules from $55 to $65, as in quality, very inferior Spanish mules have been sold at $35.00 to $75.00. Others suitable for trip, will range from $40 to $75. There are mules in the market, both American and Spanish, considered to be superior quality to those quoted, which are held by their owners at much higher rates—say $80 to $100. The supply is good, and at present exceeds the demand. The supply of oxen is fair, and the demand good, at from $50 to $65 the yoke as is the quality and condition. The most extensive wagon establishments in the western country are in operation at this point, capable of supplying almost any demands; while the facilities for such an outfit as is necessary for the trip cannot be surpassed in any town of similar size." Louisville *Daily Journal,* April 14, 1849.

& so have enough to be off. We shall probably hear from Doct R. tonight & will be ready to start as soon as he can make his arrangements. We left him at Weston to look for mules with two of his men while the other two came up & are now with us. They are from New Jersey and are fine fellows. We shall probably get along together finely & with our expedition than if there were 100 instead of 10 of us.

We find we have twice as much of everything as we can carry & shall sell nearly half of our provisions & send most of our clothing home or round by the cape to meet us at San F. Hardly a company has left here yet but that had too heavy a load & every train before travelling 50 miles throw away quantities of provisions & articles not absolutely necessary. Wagons go out from this place & come back loaded with Ham, flour, pilot bread, Beans, sheet iron stoves extra axletrees & wheels, medicines Tools of all kinds & personal clothing of all descriptions. All which are picked up within 30 or 40 miles having been thrown away by Emigrants who could not haul their loads. Then again there are some 40 or 50 wagons broken down & deserted within the same distances. Some of them splendid & expensive ones but being entirely overloaded could not stand the terrible roads.[16] The greatest amount considered by those who know as safe to start with is about 1800 (pounds) to six mules in not very heavy wagons. We have only about 2000 for the same team! We are curtailing our stock considerably & shall reduce it to the standard & travel fast.

The guide books for Emigrants are all humbugs as anyone who has seen the Elephant will testify.[17] One I remember the pages of which I perused with great interest when preparing for our outfit stated that a six mule team might start with 5000 lbs beside the wagon! as the roads were good! & the load constantly decreasing.[18] If I had the property which has been sacrificed within 50 miles of here by those who followed the directions in the "Guides," I never would go on to Cal.

Very few Emigrants are left. The majority left three weeks ago.[19] We are well off in being two or more weeks behind the crowd as the second growth of grass is much better than the first, so say the frontiersmen. Weather—nights cold, day warm. Thundershowers, wind today from South.

Wed. May 23. Slept rather better last night tho' not very warm. Woke up this morning with bad cold, & head ache from it which disappeared under the influence of the sun which

[16] The first two-hundred miles of the journey over the California Trail saw heartbreak and frustration. In almost a common refrain diarists told of their early breakdowns. Wheels, tongues, and hounds gave way. One notation reflects the experience of almost every wagon train. Joseph Sedgely said, just after his train crossed the Blue River, "After, traveling about two miles the axle of the foremost wagon broke, and the train was detained for some time." Ofter trains were held up a day or more. *Overland to California*, 14.

[17] Perkins had seen Frémont's *Report*, Edwin Bryant's *What I Saw in California*, and Hastings' *Emigrants' Guide*. He may have seen Joseph E. Ware's *Emigrants' Guide to California*. He might even have seen Bryant's article of December 9, 1848 (Louisville *Daily Courier*) specifically advising against loading a light wagon, pulled by three or four yokes of oxen or six mules, with more than 2,500 pounds. Bryant was specific as to the food and equipment that should be carried on the various stages of the routes to California, and of the territory itself, but he was never specific as to the details of preparing for the journey. It is hardly possible that Perkins saw the second English edition of Bryant's *What I Saw in California*, which contained a chapter, not of Bryant's writing, describing in most extravagant terms the possibilities of getting rich in California. Included in this volume was a badly mangled version of the instructions Bryant had given in the article in the *Daily Courier*. Only general instructions were given as to weight, but rather specific caution was urged in the selection and care of utensils, equipment, food, vehicles, and teams in Hastings, *New Description of Oregon and California*, 143-48.

[18] Perhaps some obscure guidebooks gave this advice, but those based on their authors' experience recommended 2,500 pounds, a more realistic figure.

[19] In July, 1849, the St. Joseph *Gazette* contained "an estimate of the number of persons now on the plains bound for California." The paper said that "1,508 wagons crossed the river at this place, 685 at Duncan's Ferry [four miles above St. Joseph] and at Bontown [*sic*], Savannah, and the ferries as far up as the Bluffs, 2,000 more—making 4,139 wagons." The Marietta *Intelligencer*, July 5,

came out hot. In morning cloudy with some rain 10 A.M. clear & wind from north. Owing to our overloading ourselves & being obliged to send home so much. I shall be compelled to give up my Thermometer & Barometer which I hoped to carry & shall only keep the direction of the winds & remark on the weather.

We heard today of the atrocious conduct of some Pennsylvania Company in insulting the Indian graves & attempting to violate their women. The Indians in self defense fired & their fire was returned. Some seven of the Indians were killed & 3 of the Pennsylvanians. The latter were arrested by some U S. troops & will be taken to Fort Leavenworth for trial for murder. The general opinion seems to be that some of them will stretch hemp.[20] It is to be hoped they may, for by their reckless conduct they lay the thousands now crossing the plains liable to a combined attack of all the Indian tribes.

Our mule buyers today bought 4 of the large Mexican Mules with pack saddles & lariats complete for 395.00$ a very high price but taking into consideration the size of the animals & their being completely broken into harness or packs, they are probably cheaper than any others for us. We have now five mules & a pony. We want another mule & pony & then will be ready for a start. No news yet from Dr R. but expect him up tomorrow & we hope with mules.

Thursday May 24. Last night hard rain, & cool wind from West. I slept rather comfortably. We as usual spread our India Rubber goods on the floor of our tent with our bedding on them. About 10 oclock the Squall came up with violent wind from W & a heavy fall of rain & hail. The hail exceeded anything I ever saw, being as large on an average as pigeon Eggs. We were apprehensive that our tent might be cut thro by it, but being of prime quality it stood thro the storm finely not leaking a drop. We neglected digging a trench round the upper side of our tent, & the water running off it like a great umbrella came down through underneath like a mill race, I heard, or I thought the sound of running water *under* me, & putting my hand outside of my India rubber spread, found water enough to float a cork! This discovery started us all rather suddenly, & raising our pillows we ascertained to our satisfaction that tho the water was streaming all over the ground our India Rubbers were on top & our beds still dry. We however sat up an hour or more watching till the storm was over, & being satisfied that we had fortunately escaped a wetting, lay down

1849, said the *Gazette* believed a fair average would be about four men and eight mules or oxen to each wagon. "There are 16,772 persons on the Plains," said the *Gazette,* "besides 33,544 mules and oxen. A number of emigrants, anticipating some difficulty in getting through with wagons, went with pack mules, which would probably increase the emigration to at least 17,000, and the number of cattle and mules to at least 34,000. From the best information we can get, about ten thousand persons have left Independence, which will increase the number of persons to 27,000." These statistics explain the amount of supplies that Perkins and his party saw along the way.

20 No other traveler known to me noted this fact, but this story could easily have been so. In the great horde there were emigrants low-principled enough to violate women and despoil graves. The number of travelers who noted having seen Indian women indicates at least a passing interest in them.

& had a good nights sleep. We live to learn, & today we shall take care to provide against a similar occurrence by digging a good trench round us.

This morning clear & cool and wind fresh. We occupied ourselves today in drying sundry damp articles of clothing & bedding & in selecting such articles from our chests as we could do without preparatory to sending them back. We also sold 130 lbs of our pilot bread, 140 lb meal, 2 Bush. Beans, 2 Bush. dried apples, 20 lb soap, & some little items all which goes towards reducing our load.[21] We shall get along without any trouble with our big mules & reduced burden.

We were surprised this morning by Doct Riggs & his men coming into our tent. He arrived in the stage last evening & did not succeed in getting any mules at Weston, & we may be detained waiting for him. He will have to depend as have we, on chance purchases. We bought today another Mule well broke, & a beautiful Indian pony, to ride at 55, & 50 dollars. With one more mule we shall be fitted out. Chesebro goes tonight some four miles up in the country to buy some second hand harness, which can be had cheap.

Friday May 25. Last night another perfect deluge of rain all night. Notwithstanding our trench the water found its way in quanities under our tent wetting various articles & discomposing our sleeping arrangements as usual. Having determined to take things as we found them, our nights disaster furnished us with more food for mirth some of which will last us our journey thru. Cross was the most persecuted. His bed was pretty well wet, & of course not very comfortable. Doct R. has a little dog with him, which not liking the appearance of the weather found his way into our tent, & unknown to any one ensconsed himself at C.'s head. During the storm Sam felt a stream running down his face & neck very unlike the rain water to which he had become accustomed, it being *quite warm.* This naturally excited his curiosity somewhat & he was not long in tracing it to its source, between his rage, & our amusement! with dilated eyes muscles distended he braced himself for a terrible kick at the unfortunate offender. 'Tis delivered, but missing his mark some two inches Sams leg is nearly jerked off by its monstrous turn, & he capsised backward, having made a very good attempt to throw his heels as high as his head, & then the fury of the second attempt. When the poor doggy "catches it" with a vengeance getting a succession of pedal applications from Sam, with various blows & hoist

[21] Lansford Hastings advised emigrants to carry with them a good gun, five pounds of powder at least, twenty pounds of lead, and, possibly a brace of pistols. Pistols would be fine for shooting buffalo. The emigrant should carry 200 pounds of flour or meal, 150 pounds of bacon, 10 pounds of coffee, 20 pounds of sugar, 10 pounds of salt, and other provisions as preferred. He cautioned that it would be fatal to skimp on meat because of a belief that enough buffalo meat could be had on the way. Hastings, *Emigrants' Guide,* 143.

Catherine Margaret Haun gave a good description of the limited idea most emigrants had of how much weight they could haul across the continent. "Two wagons were filled with merchandise which we hoped to sell at fabulous prices when we should arrive in the land of gold. The theory of this was good but practice—well we never got the goods across the first mountain. The flour ground at our own grist mill and bacon of home curing filled the large, four-ox wagon while another was loaded with barrels of alcohol. The third wagon contained our household effects and provisions. The former consisted of cooking utensils, two boards nailed together, which was to serve as our dining table, some bedding and a small tent. We had a very generous supply of provisions. All meats were either dried or salted, and vegetables, and fruits were dried, as canned goods were not common sixty years ago. For luxuries we carried each a gallon of wild plum and crabapple preserves and blackberry jam. Our groceries were wrapped in India rubber covers and we did not lose any of them, in fact we still had some when we reached Sacramento." A Woman's Trip Across the Plains in 1849, 3, hereinafter cited as Haun Diary.

from the rest, was finally ejected. The thing was altogether so rich that we were all repaid for all our discomfort from the wrath.[22] Stephen thinking he heard some rattling among our boxes outside bounced up & his first step toward the door was into a hole full of cold water ankle deep & the suddeness of the immersion made him catch his breath & chatter his teeth as if in a shower bath. Notwithstanding our troubles & exposure we all keep our health & spirits, barring some slight colds & sore throats.

This morning on getting up found our cook stove full of water which delayed our cooking operations somewhat, but about 8 we sat down to our Ham, Crackers & coffee with good appetites. Wind from West, night & morning quite cool. Drizzling rain till 10 A M. When the clouds blew over & we have had a beautiful day.

Chesebro came back this morning with 4 mule harness, good & strong. Today we have been packing our wagons preparatory to a start. Our crackers have been put into sacks to save the weight of Bbls, & they will probably be ground to fine powder before we have progressed far. This tho will be quite a saving to our teeth for they are as hard as the "nether mill stone." Our Hams & Beef are stored in their proper places. Condiments of various kinds got out and placed where they can be easily handled, tool box rigged & wheels tarred. We hope to get off Monday or Tuesday.

Chesebro bought two more harness down town which makes our set complete. Four of our animals were harnessed up this morning & taken down to haul up Doct Rigg's wagon, which they did in beautiful style. Our mules & ponies are picketed round us, & Doct Riggs tent pitched by ours. With our two wagons & their white covers. We present quite a camplike appearance.

Today we saw a St Joseph paper containing a card from H. Chapin who went out in command of the Harmar Company, stating that two of his men had died of cholera & two more had been very sick but were cured by some physician whom he lauds. We called at the hotel where they stayed & ascertained the names of the two who died, Gage Drown & Abijah Hewlett, we knew them both & the tidings of course cast a gloom over our mess. We have been particularly fortunate thus far in avoiding this terrible disease. We did not see a case of it on our journey round & here it has almost or quite disappeared. The Harmar Company must have been here during the worst of it.

22 This incident was deleted by Harriett Eliza Perkins when she carefully transcribed her husband's journal. See above pp. vi-vii.

Reports come in from the plains of rather a disagreeable nature of the ravages of the cholera among Emigrants even so far out as Grand Island in the Platte some 400 miles, tis said there are more than 500 graves between here & that place.[23] A company which left here some two weeks since sent back yesterday for medical aid, having lost two & some sick. Still I think with proper care it may be avoided & do not apprehend any difficulty in checking its first stages.

John Huntington's funds which he expected at Independence not reaching him, we supposed he would be obliged to turn back. Today however we concluded to carry him through for 110 dollars which he could muster & he not to own anything but go only as a passenger—of course helping when needed & standing watch in his turn.[24]

Saturday May 26. Early part of night clear, & new moon visible. Abt 12 M clouded up & this morning there was every appearance of rain. Wind from West quite cool & chilly. We have hardly seen a warm dry day since we arrived here. The weather has been more like March than June, & of course in our Exposed situation is very uncomfortable. We hope for warm dry weather when we get on the Plains if not before & shall no doubt have it to our satisfaction. Day hazy. P M. little sprinkle of rain, Evening cloudy.

Nothing particular occurred or was done. Finished packing Wagon & made tool box, Doct Riggs & party started out into the county to buy some Mules this A M. If they are successful we shall leave Monday.

I amused myself by breaking one of our Indian ponies to stand shooting off his back. He pitched round considerable at first & in the most Expressive terms his vocabulary admitted of objected to all such proceedings. However I finally succeeded in firing a six barrelled revolver at a mark from his back & hitting it. Pretty good evidence that he had "given in."

"One of our Mules stampeded" today. Something startled him & pulling up his Picket away he went like a comet, the lariat & Picket trailing astern. Three or four men were soon mounted & after him & the fugitive was brot back all O.K. in half an hour or more. Apparently considerably mortified at the failure of his projected Expedition. In view of this fact we have procured a lot of shackles, which fasten around the mules legs by a very similar arrangement of rings to the famous puzzle over which many a person has spent hours of deep study, & which will bother an Indian's brain to unlock so long as to prevent all attempts at driving off.

[23] Many diarists commented on the graves that dotted the way to California, and there was much discussion of cholera and of accidents along the trail. See, for example, Gray Diary, 9-11, and Buffum Diary, 49. David Jackson Staples left one of the most vivid descriptions of what these graves meant to the emigrants when he wrote of the last illness of George Winslow on the early morning of June 8, 1849, out on the Kansas plains. "I watched with George and there was a perceptible change going on all night. I laid down this morning to sleep was waked at 8 o'clock and we thought he was dying. He did not know us and we gathered round him in the tent to witness the last struggle of breath leaving his body and a most solemn sight it was at fifteen minutes past nine he ceased to breathe and God only knows the feeling in our camp and particularly his friends away from home & kindred with but 3 exceptions as melancholy an event as was ever witnessed by man but all and everything was done to save him but it was no avail and we must content ourselves with the thought that God's ways are best yet we cannot see as he sees our ways are not his ways and our thoughts not his thoughts the company feel it hard.

"Uncle Jesse [Winslow] and I searched this forenoon for a stone that could be engraved we found a large sandstone and we engraved his name and where from and age on it in good deep letters. His grave was dug in a conspicuous place on the main road 30 miles west of the Big Blue River on the right side of the road. . . ." Taggard (ed.), "The Journal of David Jackson Staples," *California Historical Society Quarterly*, XXII (June, 1943), 119-50, hereinafter cited as Staples, "Journal."

[24] It was not unusual for companies to take along a passenger. Sometimes these passengers were travelers who had been left stranded by rascality or by tragedy. Perkins later referred to footpads he saw along the way. Huntington turned out to be the most loyal member of the company.

Our camp life has settled down into a regular routine of Ham, crackers, Molasses, coffee, & sometimes eggs for our meals, taking care of Mules between times & fixing our private stock for the long journey before us. We are getting some tired of the sameness & must be off soon to keep clear of Ennui.

Sunday May 27. Morning clear & pleasant. Wind from West. Day hot & clear. Our clothes not being comatable [*sic*] on acct of being packed for shipment we could not well go to church, so spent the day in reading & writing. Our two parties are agreed to rest on the sabbath & spend the day in the best way our circumstances will admit of. An arrangement the Policy of which is evident to those who know anything of frontier travelling, & very gratifying to those who have been taught to reverence the day at home, whether from moral principle or not. We think we shall get to California quite as soon in the increased vigor & health of our animals, & all experience has proven that both man & beast require rest one day in seven.[25]

Doct. Riggs returned this morning having bought six mules to be here tomorrow. Two more will make out his lot which he expects to get tomorrow.

The town of St Joseph which is now quite famous as a starting point for Emigrants & will continue to be, has entirely grown up in 5 years, there not being any vestige of it at that time. Now it numbers some 2000 inhabitants, has its blocks of brick stores & Main St presents more of a business appearance than any of its length I ever saw. I should not be surprised if in the course of time it was second in population & importance only to St. Louis.[26] Its favorable situation & fine back country warrant this supposition. It is already ahead of Independence in its own peculiar advantages, being a better starting point. The latter town by the way as I was lately informed about & from the description must be very beautiful & beautifuly situated. Its rows of trees & neat dwelling houses were compared by one who had seen both [to] Marietta Ohio, one of the most beautiful places under the sun.

Monday May 28. Day pleasant & hot completed our Preperation for starting & determined to leave tomorrow. Doct Riggs Mules, being some of them quite wild we assisted him in breaking them & before night had the whole six in the wagon working together beautifully. They were not as difficult to break as I had supposed. All things being ready & both teams loaded—we are determined to start tomorrow.

25 Ware's guidebook cautioned, *"Never Travel on the Sabbath;* we will guarantee that if you lay by on the Sabbath, and rest your teams, that you will get to California 20 days sooner than those who travel seven days a week." *Emigrants' Guide,* 10.

26 The hustle and bustle of the gold-rush companies made St. Joseph appear much larger to Perkins than it really was. To the overland travelers it was the last anchorage of civilization. In 1960 St. Joseph had a population of 78,370, and St. Louis' population was 740,434.

Tuesday May 29. Struck our tent this morning and packed it with the saddles &c, on our pack mules, harnessed up our team & by 12 oclock were ready to move. Doct Riggs not being ready we waited for him till 8 when we finally bid adieu to St. Joe & took up our line of march for the far distant gold region. Day hot & pleasant, Wind West.

Our commencement was not very flattering. We had hardly gone 1 mile when we came to a little hill, but very steep almost perpendicular, & there gave it two fair trials our mules could get only half way up when they would give in & commence backing down. There was no way but to unload so at it we went & after getting one wagon up the hill packed our goods up on our backs & loaded again.[27] Twas about the hardest days work I ever did & gave us a very fair insight into our future. Out of this scrape we went on without trouble to the ferry, four miles where we arrived about sundown & finding we could not get acrossed camped.

The country through which we passed today is Missouri bottom very rich in all kinds of finest vegetation. Occasionally a red squirrel would bounce away to some gigantic tree & I was much surprised by a flock of genuine Carolina paroquets fluttering & chattering through the woods.[28] I supposed they were never present so far north by several degrees. We camped tonight in a deserted log cabin. Our mules are picketed in the woods around & there being no grass we cut down small trees for them to browse upon.

Wednesday May 30. This morning clear, wind west, day pleasant, slight rain in P M. We threw out some 300 lbs from our load which we thought we could dispose with, among which was our vinegar, Stephens cot bedstead, Brass Kettle & molasses, over which last I shed some tears of regret. We shall have to do without most of the little comforts & luxuries on which we had been depending & that we may travel more expeditiously, do it less comfortably.

The goods we threw out we sold to the ferry man for 5$, & were glad to get even this, as many before us have had to lose entirely hundreds of pounds of valuable articles.[29]

We crossed the ferry about 10 oclock, took over the mules first & ran the wagons in the flats by hand. Once over the swift & muddy stream we harnessed up thinking our troubles for the present were over, when lo about 1/4 of a mile from the ferry up rose before us the bluff, steep to almost perpendicularity, & the road running directly over it! & about 200 feet higher!

[27] Emigrants started for California with little knowledge of the roads and trails beyond what they gleaned from guidebooks. Gray's party had experiences that almost paralleled those of the Marietta Gold Hunters. He wrote, "May 1, 1849, started in earnest at 10 o'clock, beautiful day, but very cold at sunrise, road good with exception of a few steep hills & some sloughs, two of our 8 wagons broke their tongues." Gray Diary, 3. Another diarist, John A. Markle, wrote: "We started about 1 o'clock traveled over 17 miles of hilly rough roads, and stuck fast three times, the third time we unloaded part of our baggage, got our wagon out of the road and encamped all night." Markle Diary, April 19, 1849.

[28] In approximately the same area Lieutenant J. W. Abert wrote in 1847, "Paroquettes were sweeping in large circuits among the topmost branches of the ancient denizens of the forest and their screams shrill and grating echoed through the lofty arches of bows [sic], now shorn of their summer glory." Emory, *Notes of a Military Reconnoissance,* 544.

[29] "One of our wagons gave indication of breaking down this afternoon, so we stopp'd at 4 o'clock & threw out & burned up many articles valuable to us, but impossible for us to carry about—large wooden boxes to pack provisions in, iron chains, many instruments of iron, all had to be abandoned to lighten our loads about which we are apprehensive." Gray Diary, May 21, 1849.

Here we were almost "bluffed," but finally harnessed both teams to Doct R's wagon & started for the ascent. The mules pulled nobly & by blocking often & letting them rest, they finally took us up safely. They were then driven down and made fast to our wagon & took it up in the same style, and on top the view was most beautiful.

Far off could be seen St. Joseph with its elevated courthouse. Underneath us was the River, for so steep was the bluff that it appeared to me to overhang the Missouri, & I could hardly understand how the mules could get themselves up alone much less with a loaded wagon behind them, & far in front spread out before us in all its beauty & grandeur lay the broad Prairie. Soon out of the timber and on our way across the plains, a part of our trip we had been looking forward to for some time on the commencement of our lonely journey.

Now we were out of civilization & the influences of civilized society entirely, & cut loose from the rest of the world to take care of ourselves for a while. I confess to a feeling of lonliness as I thought on the prospects before us, & all we were separating ourselves from behind. Henceforth we shall have no society, no sympathy in our troubles, & none of the comforts to which we have been accustomed, but must work across these vast wild wastes alone & go in our own strength & his who takes care of us all.[30] So be it. Gold must be had & I for one am willing to brave most anything in its acquisition.

After surmounting the "big hill" at the ferry we travelled ab 5 miles & are encamped tonight by a spring with some smaller trees around it, furnishing us with wood & water pitched our camp in a hollow picketed out our mules & are comfortably fixed for the night. There we shall wait for Doct Riggs whom we left sick in town. He will probably join us tomorrow.

Thursday May 31. Morning cloudy. Thunder shower & heavy blow last evening, day pleasant & warm, spent most of it in writing letters home as this is the last point when we can send them to St. Joseph.

Wrote 4 myself & after dinner Stephens & myself prepared to go to town. Took two of our ponies & started. In crossing the prairie to cut across a bend in the road. My pony got into a deep mire, by jumping off in double quick time, & throwing the bridle over her head, I succeeded in time, pulling her out with a good deal of floundering tho & considerable sprinkling of mud both over herself & me. We went some six miles on

[30] Perkins was like many another forty-niner in imparting to his diary at this point his feelings at cutting himself off from the society and world he knew. Joseph Price wrote his wife Elizabeth on April 28, 1850, "I hope the Lord will bless us to meet once more on the Shores of time on this side of the boundary line of time that we may once more be happy. We are well-pleased with our place of Rendesvose [sic] and . . . if I could heare from home sweet home . . . it would be a great deal to one that has been used to heare from home every day." Marshall (ed.), "The Road to California," *Mississippi Valley Historical Review*, XI, 244.

Madison Berryman Moorman fell into a somber mood.—"I thought of the changes and vicissitudes through which I had gone in so short a time and considered the stupendous undertaking before me and the uncertainties —hazards, hardships and the like connected therewith.—The mind of man—the ever busy and immortal part! How strange and inexplicable! With inconceivable rapidity it surveyed all my past life and with electric velocity recoinoitred [sic] the almost interminable road to California and told of the multiple thousands of restless mortals wending their way to the Land of Gold. . . ." Paden (ed.), *The Journal of Madison Berryman Moorman*, 14.

Charles G. Gray, however, was more lighthearted. "At about 9 o'clock this morning a very sorry array of faces presented themselves —very long faces indeed—a barber certainlly would have charged extra to have shaved them, as for my humble self I suppose I look like the others. So I fired my gun at a mark, whistled some opera tunes as I was setting in the wagon, gazing at vacancy, thought of HOME determined to return when I reach'd Fort Laramie, endeavored to make out how it was *I* ever was such a fool as to come *this route*, & transferred it all to my friends at home, & sat looking at the 'empty, vast & wandering air' till quite stupid—more so than ordinary & then by way of satisfaction took another whistle." Gray Diary, May 31, 1849.

PENNSYL

OHIO

MUSKINGUM RIVER

MARIETTA
MAY 9 / 1849

WEST
VIRGINIA

DIANA

O I S

CINCINNATI

N

I N T O

OHIO RIVER

LOUISVILLE

KENTUCKY

T T

C I N

I L

CAIRO

R m

TENNESSEE

May 9–June 14, 1849

our way, taking the lower Ferry road & had about half crossed the river bottom when we met the Doct & one of his men coming out. They told us they had been trying all day to get over the ferry, but the River was so high that they had nearly given it up & that twould be no use for us to go farther as we could not get over so we turned back with them & all arrived in camp about sundown.

We shall have to do without one mule which we have ordered. Without our extra trace chains & trust to chance to get our letters put in the office. We have determined to make an early start tomorrow morning, so shall go to bed early. Was taken with diarrhea this afternoon rather severely when I met the Doct. Got from him an opium pill & some brandy which I hope may check it. Feel rather slim tho now.

Friday June 1. Morning clear & pleasant. Woke up feeling very dizzy & weak & on attempting to rise was very sick at my stomach. Sent for Doct Riggs who said I must take some calomel & morphine & left some powders. Had a bed made up in the wagon & all being ready was helped into it & we started at 8 oclock. As we left it clouded up & a drizzling rain fell which continued at intervals thro the day.[31] The jolting of the heavy wagon did me no good & my sickness increased, but I said nothing & took my powders every two hours. Calomel gr v, morphia 1/4 gr. In afternoon owing to some bad & dangerous places in the road was obliged to get out & crawl along myself for a short distance.

Our road today thro a beautiful country sometimes heavy rolling prairie & at others through woods & bottom lands travelled till 5 P.M. when we made the Indian Misnahamcha, a Presbyterian charge I believe, where we camped on the bank of a small creek. Distance today 20 miles. Here we first saw wild Indians, a small portion of the Sacs being in the vicinity, who came galloping round us, & with their fine horses & red blankets made quite an imposing & military appearance. They dismounted & threw themselves flat on the ground to watch our proceedings, leaving their horses to graze at their pleasure. In general they were a fine looking set of men & rather exceeded any previously formed opinion of the tribe of Black Hawk the Prophet.[32] The mission at this place have some 30 or 40 Indian youth whom they are educating & some of our boys went up & took supper with them. They represented them as being very orderly & well behaved & at the supper table everything was like clock work.[33]

[31] The spring of this year saw bad weather in eastern Kansas. Perhaps none of the emigrant companies was prepared to endure so much rain. Added to this was the danger of becoming seriously ill in trying to make hurried adjustments to hard camp life. The recurring storms were very much on the minds of the travelers throughout the season for making starts to the West. On May 30 J. Elza Armstrong noted, "Raining, raining very hard." John Edwin Banks confirmed his statement. "It rained very hard in the night. Commenced at two and rained until daylight. Grass poor." Scamehorn (ed.), *Buckeye Rovers,* 13. The reporter of the St. Louis *Republican* in Independence corroborated this. "For several days past," he said, "the weather has been very disagreeable. Several of the emigrants, from the exposure they had been forced to undergo, have been made sick, and considerable fear is manifested by many in consequence of reports of death from cholera in this vicinity." Louisville *Daily Journal,* April 14, 1849.

[32] Joseph Sedgely reported that on May 29, 1849, his party was being accompanied by Indians. *Overland to California,* 20. The Indians Perkins saw could have been Pawnees. Stansbury reported seeing the remains of an abandoned Pawnee village near this spot. Pawnees, however, might not have been so friendly; these Indians had been recently driven away from Fort Kearney by troops. Stansbury, *Explorations,* 29.

[33] Gray mentioned a similar mission on the Independence route. He recorded in his journal on May 13, 1849, "Started at 8 o'clock & came on a fine country—more rolling than usual, nearly all day in sight of wood & brush, something unusual for us, road quite dusty, passed a settlement of the Pottawattomie Indians, some 20 or 30 buildings. I gallop'd up to one of the houses where there were quite a number of people, thinking it a tavern of some kind (not thinking that it was Sunday, for we don't know what day of the week it is half the time) & asked as loud as I could, 'Have you any whiskey?' When a *black white* man, who spoke English with the French accent, informed me that it was a church! So I told him I hoped I had not disturbed them by my question, when he replied, 'Oh! no, they all like to hear of whiskey.' There were quite a number of Indians in the house, so I rode off thinking

My sickness had so reduced me & weakened me that I was glad as soon as the tent was made up to turn in & so saw nothing of the mission myself. Evening clear & pleasant, Wind West.

Met today two or three returning Emigrants on horseback & they undertook to deposit our letters at St. Jos. for which we were very grateful. Said they turned back because they wanted to, a reason which there is no arguing against.[34]

Saturday June 2. Woke up feeling some better but very weak. Some dizzy & had to ride in the wagon again. Morning pleasant & clear, started at six oclock & travelled till 12. When we lay by two hours to rest & feed our mules.

Just before noon we crossed the junction of the Fort Leavenworth trail with the one we were in & here we passed one of the government trains going out to Bear River to build a fort.[35] The troops were some days ahead. There were twenty-five wagons with five or six yoke of cattle to each & as they stretched out across the Prairie for nearly a mile. They looked like some huge serpent, "dragging his slow length along."

At 2 P M. we harnessed up & went on till 5, when we found our camp on the bank of quite a creek, but no timber of course, some stunted wild plums & bushes just enough for fire wood was all. Distance today 30 miles, through a much more level prairie than yesterday, & almost entirely destitute of wood.

We travelled for hours with nothing but the horizon to bound our view. The feeling was more like being at sea than anything I ever knew before on land & with my weakness & dizziness, it is easy to imagine myself again on the green waters & laid up with that most distressing of disagreeable complaints, sea sickness.

Wild flowers are scattered over the prairie beautifully, one much resembling the Sweet William of our gardens, being most abundant.[36] We also saw many Prairie Hens & an abundance of plover, but had no time to hunt so killed none of them.[37]

Rode most of the day in the wagon, but towards Evening got out & rode Stephens pony, feeling much better, am quite relieved of my complaint, but think I had a narrow escape from cholera. Most of our men & Doct R's have been troubled more or less with the first symptoms of this disease, but are now well. Cunningham was quite sick yesterday but is better this morning.

Met today more returning Emigrants who have been out about 200 miles, but were compelled to turn back on account

how we at home were doing this day, whilst the poor Indians were worshipping here in the wilderness." Gray Diary, May 13, 1849.

[34] Some emigrants had begun to return home before Perkins and his group had left Marietta. The local reporter for the St. Louis *Republican* observed, "Some of the boys have seen the elephant, and intend returning; others are already enroute for home. Some have attempted to whip the 'tiger,' but, instead, have been fleeced of their all, and unable to obtain an outfit." Louisville *Daily Journal*, April 14, 1849.

[35] Orders to establish a post on the Bear River were issued in April, 1849. The expedition was described in a letter from Lieutenant D. P. Woodbury to General Joseph G. Totten, Chief Engineer in Washington. Williams (ed.), *History of Northwest Missouri*, I, 258.

On the same day that Perkins saw the train, Major Osborne Cross wrote: "This day was passed in making out papers and arranging the train. The command moved four miles above to change their encampment, get better grazing, and be nearer the water. An order was issued by Colonel Loring separating the command into three divisions of two companies each. They were to march at an interval of five miles between the first and third divisions, and encamp in the same order until otherwise changed." This command was en route from Fort Leavenworth to Fort Boise. See map in Settle (ed.), *The March of the Mounted Riflemen*, 62-63.

This train was a common sight to emigrants along the way. On May 20, Gray wrote: "Since our accident 300 teams have passed us, amongst them a U.S. Government train & company of dragoons." Gray Diary, May 20, 1849.

Actually the Bear River expedition established Cantonment Loring on the Snake River about three miles above the trading post of Fort Hall. The cantonment was established in July, 1849, and abandoned in 1850. Prucha, *Military Posts of the United States*, 86.

[36] In his entry of June 17, 1842, John C. Frémont described the flowers on the Kansas plains, giving botanical names for most of them. Frémont, *Report*, 14.

[37] "In the afternoon went out a short time with my gun shot 11 large gray plover, in a flock, killing 5 with one barell," Gray Diary, J. 6.

of their cattle giving out. This may be the reason, but I rather think most of these excuses are made up to justify the conduct of those who become discouraged or frightened at the prospect before them.[38] We shall according to previous arrangement remain in camp tomorrow it being Sunday.

Sunday June 3. Morning clear, day pleasant & warm. Strong breeze from the south, we all, mules & men, spent the day quietly getting over the fatigue of our trip. Most of us took a wash in the creek, which will be for our health if we can do every week.

No incident to record, except that the government train which we passed yesterday passed us this morning, in our Sunday quarters & this afternoon another train, the remainder of the supplies for the Bear River works, went by. We shall see them tomorrow again.

Monday June 4. Morning clear, very heavy dew almost equal to rain. These will however be frequent hereafter. Day pleasant, wind South, our men up at 4 oclock, had breakfast & broke up our camp & started at 5.

At 8 came up with the government train which passed us yesterday, & had my first view of a regular "corral," as they were still in camp. There were in the whole 57 heavy wagons drawn by 5 to 7 yoke of cattle each making the number of oxen about 680 or 90, a very large herd. The wagons were ranged side by side in an ellipse with an opening at one end like a gate across which one of the lighter wagons could be drawn at pleasure. Thus forming a fine pound in which the animals were driven at night. The tongues of the wagons projected outward so that teams could all be yoked to their places & all being ready to start one after another forming a train as we afterward saw them nearly or quite two miles long. Being to our eyes a novel and splendid sight. The whole were under the charge of an officer with whom I had some conversation. He told me he expected to be 3 months at least getting to his destination. Not much more than half as far as we have to go. The average days travel is 12 miles.

I tried to make some rough estimate of the cost of this immense caravan with its freight, but could only come to the conclusion that Uncle Sams preparation for war of defense must cost him a great amount of treasure. As a protection to the nation all this may be necessary, but as a general thing how useless a waste of property all warlike preparations are? There

38 Newspapers carried stories of the return of gold rushers from both the trails and the sea. One group came back from Panama owing to a lack of west coast shipping. The editors advised, "Let no man who wishes to reach the gold diggins in the next six months set out." Louisville *Daily Journal*, April 21, 1849.

was more money probably in this one expedition than are yearly spent on common schools! Perhaps this may be an overestimate but any one seeing the huge train with its cargo would be very apt at first to think very much as did I.

We made a stoppage of two hours at noon, to refresh ourselves & mules & camped this evening on the banks of a beautiful stream rippling over a rocky bed, more like some of those seen in New England, than anywhere else. The water was very fine & we all drank with a relish far surpassing that of the pampered Epicure over his choicest mixture, just before reaching our camping ground we met three men with a mule train returning disgusted with the hardships & difficulties they had encountered.[39] Our party came to the conclusion that twould be safer for us to have another mule so bought one of their mules for 75.00 rather high considering the size & appearance of the animal, but probably our funds had better be invested in movables than any other way.

We found today on a rock by the wayside a card with written on it "Cook & Co of Cin. passed here yesterday 12 oclock." This is one of the companies that came up the River with us carrying their oxen with them.[40] They started soon after landing, so we have gained fast upon them. We shall probably overtake them tomorrow & shall meet them with pleasure as they were very pleasant fellows. Cattle cannot travel anything like as fast as mules. Many persons to the contrary notwithstanding. The supposition that they will crawl up to their more speedy rivals every night & so keep up with them is a notion or a story of those having them to dispose of, for they cannot travel as long or as many hours per day as mules. All the teams we have as yet met, except yesterday, returning have been oxen, & the general excuse for turning round has been "Cattle gave out." We heard so much about the relative merits of these two means of locomotion across these plains, that I have taken some pains to observe & inform myself & am satisfied that my above expressed opinion is Entirely correct.

The travel since we left the Missouri River has so far been tedious enough, good roads to be sure but day after day the same monotonous unbroken view of flat prairie, no game, & nothing to make variety or cause Excitement. The grandeur which strikes one on his first view of these vast expanses soon wears off & settles down in a tedious sameness. An old guide with the U.S. train told me that he always disliked the trip from St. Joe. to the Platte River more than all the rest, & encouraged us that in a few days we should come to a different

[39] This was the real point of decision on the whole journey to California. Before emigrants reached Fort Laramie, they could turn back with every expectation of reaching home safely. Beyond that point distance began to count against them.

[40] The California Club of Cincinnati, which left that city on March 10, 1849, aboard the *Bay State*, included an R. W. Cook and possibly is the company to which Perkins referred. Cincinnati *Gazette*, March 13, 1849. There were, however, several companies from Cincinnati on the road; Pritchard mentioned a joint-stock association with fifty members, driving eight mules to a wagon. Morgan (ed.), *The Overland Diary of James A. Pritchard*, 69-70, hereinafter cited as Pritchard, *Diary*. See also below, Perkins' entry for June 12.

country with an abundance of antelope & smaller game & probably plenty of Buffaloe, which of course cheers us much. Distance today 25 miles.

Tuesday, June 5. Morning clear, day cool & pleasant. Stiff south breeze. Started from camp at 6 oclock last night stood our first watches as we are now in the Indian country. The Nahamehaw [Nemaha], where we camped last night is a kind of dividing line between the non stealing & stealing tribes & tonight we shall probably camp on the Vermilion after crossing which we shall be in the Pawnee country & they are notorious thieves as well as desperate fellows.[41]

Made a tedious days travel today & camped tonight on a little creek. There had been so many camps near the stream that the grass was short & we had to camp nearly 1/4 mile & carry our wood & water that distance. Distance today 30 miles.

Wed. June 6. Lay in camp this morning till 10 oclock, to rest our animals. While waiting a couple of men rode into camp who proved to be Fort Laramie traders! One of them, a Frenchman had traveled the road twice a year for sixteen years & was as pretty a specimen of a bronzed & sun burnt frontiersman as I ever saw. He gave us considerable information as to our route &c, & said we had come over the dreariest part of our journey. He had some 4 or 5 wagons loaded with furs which we met in course of todays travel.

At noon we came to the Vermilion, a very pretty stream of considerable size & rapid current.[42] In fording we had to take a straight course about to the centre & then follow a bar down stream till opposite the landing. Here we filled our water casks & took wood for our suppers cooking as we would not be able to make another stream that day.

The Vermilion is about 50 yards wide & water clear & sweet. We just escaped wetting our breadstuffs as the water touched the bottom of one wagon body. Here I had to do some wading to free the traces of the mules which had become somewhat snarled by their repugnance to Entering the stream & though my boots were some of the shortest, they were not long enough by considerable & I took on water Enough in them when I waded ashore to last our company for cooking some time. As they declined my proposition to transport their drinkables for them, I lightened them some 5 lbs a piece by raising my foot & straining out the liquid through the knees of my pants.

Our trip today has been over a very level prairie except when

[41] The Pawnees occasionally caused trouble along this section of the trail. "This morning we heard that some trains had 68 head of oxen driven off by the Pawnees, who have been seen prowling around the roads, so the Gen'l orders the guard to be doubled tonight." Gray Diary, May 20, 1849.

[42] The big Vermillion River was one of the first landmarks reached by trains departing from St. Joseph. James A. Pritchard, who took the Independence road and reached the river farther south, said, "The Vermillion is the largest that we have crossed since we left the Kansas—it watters are clear bold & beautiful with stomy bead [stony bed]. Its banks are steep. The bottoms are about one mile in width & finely timbered in places for some distance out." Pritchard, *Diary,* 58.

in the Vermilion bottom. Thus far no game except prairie snipe of which there are & have been thousands & a few prairie hens. Doct Riggs killed one of the latter & we have killed many of the others. We have seen no Indians & very little occurs to relieve the monotony of the journey. Day pleasant & hot. Wind East.

Thursday June 7. Last night we camped in the level prairie & were just getting supper when a terrific thunder storm came up, compelling us to catch up our plates & kettles & scud into the Tent in a hurry. The rain poured down in a torrent & the thunder & lightining were grand & awful. The flashes of lightning were so brilliant that everything would be as light as sunshine, & followed by total darkness blinding us completely, & the crashes of thunder shook the very Earth beneath our feet. We kept perfectly dry under our snug tent, & enjoyed our supper as we best could under such circumstances. While I enjoyed the grandeur of the scene much more than anything else this is our first thunder storm on the Prairie, often have I read of them, but the thing cant be described.⁴³

This morning pleasant & cool Wind North. We stood our regular watches last night & in my watch 4 of the mules broke loose & were straying off, but by following some of them a mile or more in the dark I succeeded in capturing them again. If Indians had made their appearance just at this juncture we should have made a sorry ending of it.

While getting breakfast Early this morning two large deer passed within 75 yards of our camp but before any of our multitude of guns could be got out they were beyond reach. They did not seem much afraid of us & were cantering along without apparently noticing us.

Started this morning at 1/2 past 5, found the roads quite slippery from last nights rain & being more unsure than yesterday, our mules had a hard days work. There being no water nearer them than Turkey Creek some 28 miles we pushed for it, & arrived here at 1/2 past 7.

After the hardest time we have yet seen, I would have given anything for a drink during the day, but no water was to be had for love or money, so on we jogged, mouths parched & throats cracking. Once we came to a puddle where the rain water had been standing till green on top & so muddy that if there had been hogs about I should have set it down at once as one of their wallowing places, yet this stuff which would have been rejected very suddenly by my stomach at home I

⁴³ Many travelers from the wooded areas of the eastern part of the United States were seeing for the first time great expanses of land and sky, experiencing the full drama of the roll of the clouds, and of lightning and thunder. "Shortly after supper, the lightning became very vivid, but no thunder was heard, and the black & gathering clouds, rolling up in immense masses 'gave dreadful note of preparation.' We accordingly las'd all down tight & gather'd our things together for a storm, & all our anticipations were exceeded ten fold, by the severest storms we have yet encountered. All the thunder, lightning & rain of half a dozen great storms seem'd united in this one. The earth seem'd to shake, the heavens were one continual glare, & the rain came down like a young deluge. Such sheets of fire, such bursts of horrid thunder, as Shakespeare says in Lear, 'it never was before my lot to witness.'" Gray Diary, May 29, 1849.

drank with considerable relish, by shutting my eyes & holding my breath. This is what is called seeing one of the Elephants tracks.[44]

The creek on which we are camped this evening is rapid, banks steep & crossing bad, not much water in its bed. Here we overtook an ox train of Fort Laramie traders, which camped on the opposite side of the stream. During our noon rest today the U. S. Mail from Fort Kearney in charge of 4 dragoons came by. I had considerable conversation with the orderly, whom I found a very pleasant & intelligent fellow & like myself from Ohio. He goes the route twice every month, staying only two days at each end. He told us we should see plenty of antelope & deer before two days more travel. He killed an antelope yesterday on the Little Blue. He told me also that we need be under little apprehension from the Indians as the Pawnees & Sioux were at war & the latter had driven the former entirely beyond the Vermilion so that we had passed through them Entirely & the Sioux seldom disturbed any one traveling through their country.[45] He & his men were camped Every night without setting watches merely picketing their horses & turning in, we however do not feel as if this would be quite safe for us. If we lose our animals we could not get a fresh supply from Uncle Sam by merely representing our case to that individual. Day has been hot & oppressive owing to steam rising from damp ground, distance 28 miles.

Friday June 8. Crossed Turkey Creek this morning at 8 oclock, as our mules were so hard pushed yesterday we concluded not to start early as usual. Morning clear & warm, strong breeze from south.

I was beginning to feel very sensibly the effects of loss of sleep. We usually getting through camp duty so as to retire about 9 or 10, & being obliged to turn out every morning to get breakfast & take care of the animals at from 2 to 3, besides Every other night each standing watch. So that to me who had been accustomed to 8 hours of regular sleep, 4 hours did not suffice. Huntington & myself after the wagons were started took a couple of horses & went up the creek some distance in the woods determined to have a couple of hours nap so selecting a spot we laid our pistols & bowie knives by our sides & went to sleep. Our slumbers were broken in upon however by a mule train from New York which was a few hours behind ours coming down to cross. There were 9 wagons & some 40 men all dressed in uniforms & their Capt. with the insignia

44 More than one person drank from the stagnant holes. "Yet we had nothing to drink during the evening when we were thirsty— nothing to wash in the next morning, or cook with till 12 oclock the next day, when I luckily fill'd 2 canteens with dirty water to drink. This may appear a very small matter but it was a great annoyance & I merely mention it to show how great a deprivation the loss of a gallon of water is on the plains." Gray Diary, May 16, 1849.

45 Lansford Hastings informed emigrants: "The only hostile Indians, that are to be seen between the States and Fort Laramie are the Pawnees, who are a powerful and warlike tribe, and who are, generally very troublesome to the emigrants; yet they are, generally south of this route, at the season, at which emigrants pass through that portion of the country." Hastings, *Emigrants' Guide*, 135. "Today we lay by in the morning 60 Sioux Indians came down the river and from what we could understand they were in pursuit of the Pawnee Indians, as the two tribes were at war with each other. They were very friendly and asked for something to eat. We gave them some biscuit which they ate, then they turned back and went up the river and crossed, a short time after an old Indian came up the river and showed us a scalp off an Indian which we supposed was the Wounded Indian [a Pawnee that the party had just seen on the trail]." Markle Diary, May 20, 1849.

of office on his shoulders. He had been in Mexico & had his men in the regular order & discipline of a company of U.S. troops.[46] They expect to go through in from 80 to 90 days, if they do we shall certainly do the same for they cant catch us. I hope we may but confess I have my misgivings at times about our success in crossing this year at all, & the reflection that we may from want of food for our animals or cold weather in the Sierra Nevada be compelled to turn part way back again gives me no very pleasant sensations.

If the thousands who have gone should find themselves in the mountains without provender or caught in the snow storm what would become of them & us? We could not find sustenance sufficient for all at any of the stations on the road or at the Mormon settlements & our stock of provisions could not last us back to the U.S. & I dont see how great loss of life could be avoided. I cannot keep clear of the blues sometimes when I look at the possible result of this Expedition & think that I may have seen my last of those dear ones at home & enjoy my journey much less than if I was relieved of all uncertainty as to its termination. But I must away with such gloomy anticipations & trust to Providence to take care of me & restore me alive & well to the home I have left.[47]

Our route today has been over a rather rough prairie with innumerable little gullies ten or twelve feet deep, through which our wagon had to plunge & which rocked it terribly. Our crackers were all jammed into "spoon victuals" already by the shaking they have undergone thus far, & our clothing, guns, &c &c, are all more or less damaged by being jolted up with boxes & bags. There is no help for all this, over such roads as we went, & we make the best of our things and as they are.

This afternoon we overtook another ox train of Emigrants & among them we found two or three of our travelling companions of the *Highland Mary* who left St. Joseph a few days before us. We of course were naturally rejoiced to meet Each other. A man on this California route soon learns to love anyone that reminds him of home, or civilization. We took a cut off this afternoon which saved us some five miles, & which brought us to the Little Blue at six oclock & here we are camped. The ox teams soon followed us & we all camped together.

The grass here is very fine & our poor animals appear to enjoy it much after their days work. While picketing out our mules this evening a couple of antelope were seen & Stephens

46 This may have been the train of which Charles G. Gray was a member. Gray referred frequently to "the General," a General Darcy of Newark, New Jersey, who ruled his highly organized train with an iron hand. Gray Diary, 1849. It could also have been Captain McNulty's uniformed New York company, which was conducted like a military company. It was seen at various places along the trail. Read and Gaines (eds.), *Gold Rush, the Journals, . . . of J. Goldsborough Bruff*, I, 26, 118, II, 1180, hereinafter cited as Bruff, *Journals*.

47 Many a traveler who went along the California Trail in 1849 seems to have questioned his own judgment in beginning such a difficult journey on what amounted to little more than a whim. Even before leaving home Catherine Margaret Haun had some melancholy thoughts: "When I awoke the next morning a strange feeling of fear at thought of our venturesome undertaking crept over me. I was almost dazed with dread." Haun Diary, 6.

A much earlier traveler, a Methodist minister named Joseph Williams, had also noted grave misgivings about his journey across the Kansas plains en route to Oregon. He wrote, "Pursuing my journey that day, I tried to give myself up to the Lord. I could scarcely follow the wagon tracks, the ground was so hard in the prairie. I had almost concluded, at last, to turn back, and got down on my knees, and asked the Lord whether I should do so or not. These words came to my mind: 'The Lord shall be with thee, and no hand shall harm thee.'" Williams, *Narrative of a Tour*, 7.

started out after them but without success. These are the first we have seen. They are a pretty little animal some larger than a goat & not unlike one in shape, with a black stripe from the neck down their back & a pair of short horns standing straight up. They are much more fleet than a deer & it is said to be entirely useless to chase them with even the best of dogs & horses.

After supper we went over to our Emigrants camp found among them some ladies young & old & were quite cheered by the sound of a female voice.[48] The party have for a guide an old frontiersman named Taplin who was with Frémont in his first & 2nd Expedition & in his last ill-fated attempt to cross to California in Mid-winter.[49] He gave me some very interesting particulars of this expedition & their sufferings. He was one of those left behind while Frémont went to seek his relief party & for ten days lived on a few buds & leaves of the wild rose which they dug out of the snow. He describes Frémont as a slender slight built young man of about 31, & thinks his equal does not breathe. He is going on now to join him intending to live with him the rest of his days. The guide was a man of about 40 bronzed & hardy looking, but neatly dressed in frontier costume & intelligent in his conversation. He had a long & very handsome rifle which showed evident signs of much & hard usage, a revolver with Bowie Knife & hatchet stuck in his belt & Mexican spurs some three inches in length, completed his attire. His horse was one of the hardy mustangs & had a Mexican saddle, high fore & aft with large wooden stirrups nearly large enough to cover the entire foot. He was altogether quite an object of interest to me as I have almost as great respect for Frémont as had he. Day has been hot, wind southwest, distance 20 miles.

Saturday June 9. Started this morning at 5, intending to make a long stretch before lying by for Sunday. Morning clear, day warm, cool breeze from north. Our course for 52 miles is along the little Blue with some variations one of which was made this morning. We left our camp ascended a rise & for six hours travelled over the most perfectly flat country I ever conceived of. Nothing was in sight for this time, but the green of the Prairie & the blue of the sky, & reminded me forcibly of a calm at sea. We struck the bottom land of the Blue about noon & stopped two hours to rest.

While here the leader of the ox wagons said since we left them in the morning they had killed a fine antelope after this

[48] "Our caravan had a good many women and children and although we were probably longer on the journey owing to their presence —they exerted a good influence, as the men did not take such risks with Indians and thereby avoided conflict; were more alert about the care of the teams and seldom had accidents; more attention was paid to cleanliness and sanitation, and lastly but not of less importance, the meals were more regular and better cooked thus preventing much sickness and there was less waste of food." Haun Diary, 9.

[49] Charles Taplin was employed, apparently in St. Louis, to accompany the second Frémont expedition in 1843. Frémont, *Report*, [105]; Frémont did not mention Taplin in the 1842 expedition (see *Report*, [9]). It is evident that Perkins here refers to Frémont disastrous fourth expedition of 1848.

we traveled on to the place where we have found our camp for Sunday, on a bank of the stream in a level meadow & shaded by some fine trees growing near the water, the Little Blue called I think by Frémont, Republican Fork is about 50 feet wide, current rapid with about three feet tolerably good water.[50] Its course thro the Prairie is marked by a belt of trees & vegetation which abounds in antelope & other game.

This afternoon Doct. R. & myself were riding in advance of the wagons & to save distance, started across some irregular ravines in the Prairie. Out of one of these we started a couple of Prairie wolves which went scampering off over the next ridge. While galloping up through ravines to head them I saw grazing at me on a ridge an antelope which as I came within 50 yards started off & I after. I had a very fleet mare & I put her to her best speed & for a short distance held my own but was finally completely distanced. The race was quite pretty & exciting & the first thing like sport I have yet met with. Camped this evening at six oclock, distance today 28 miles.

This morning a large wolf trotted past the camp but out of our shot, he appeared to be on the trail of something. Stephens & Chesebro started out to find some game but have just returned unsuccessful. They saw a large deer & antelope also a wolf but could get no shots at anything. We have been very unsuccessful so far in procuring game & all have a hankering after fresh meat which an antelope would satisfy very satisfactorily.

Sunday June 10. Morning cloudy & raw chilly wind from north. Day clear & clouds flying. Wind north & air pleasant. Spent the day as usual in cleaning up reading & writing. Found the water of the Little Blue rather cold bathing. Carried our mules over the stream & picketed them on the other side in better grass. This if there were Indians about would be very unsafe but with some watching will do very well now, thanks to the Sioux for driving off the thieving Pawnees. We kept our rifles in readiness for any sudden Emergency & my belt with knife & pistol is called into requisition Every other night in standing guard, the only time thus far when I have considered it necessary to wear it.

Monday June 11. Was waked this morning by gusts of wind & distant thunder, which increased in violence rapidly. Breakfast was about ready & we were going into it when large drops

[50] Perkins was correct in saying that the Frémont expedition had been on the Blue River. Frémont reported, "We were moving forward at seven in the morning, and in about five miles reached a fork of the Blue, where the road leaves that river, and crosses over to the Platte. No water was to be found on the dividing ridge, and the casks were filled, and the animals here allowed a short repose." *Report,* 16. The Blue and the Republican were sometimes confused. The name Republican was first drawn on a map by the Spaniard Antoine Soulard. A confusing sentence in Robert W. Baughman, *Kansas in Maps,* 16-17, states, "The resurgence of trade and exploration in Spanish Louisiana is reflected in a striking map (opposite page) drawn by Antoine Soulard at St. Louis in 1795. The Kansas portion introduces place names never before seen on a map, with a showing of the Blue fork of the Kaw as the Agua Azul (Blue Water), and the Nation Republique in the very act of giving name to the Republican river." The map shown, however, bears the date "1785."

began to fall & our chest which serves for a table was hurriedly moved into the Tent & we commenced operations. Suddenly there broke upon us a regular hurricane accompanied by very heavy rain & increased flashes of lightening & loud thunder. Our tent stood for a short time shaking & grieving & did its best to keep up against the fury of the gale but at last a treacherous rope gave way & then one after another, crack, crack away they all went & over upon us came the canvas capsizing men & meat & coffee crackers &c. all in a heap, & the flood of water drenching us thro thoroughly. We scrambled out in the most expeditious manner & while the rest of the boys scud for the wagon I got some new ropes & belayed on & soon had our house on its legs again & with it I staid during the rest of the storm, my back braced against the center pole to support it as much as possible & picking out some clean bits from the wreck of our breakfast with which I managed to satisfy my appetite.

The storm finally cleared off about 7 & at 8 we were off. Morning till 10, cloudy. Wind & storm from N.W. Day hazy & warm. P m. breeze from W. Roads today very bad, slippery & mud in some places deep.

Afternoon Chesebro went out after antelope & coming up with one got off to slow leaving our fleet & spirited Buffaloe mare by herself & taking advantage of her freedom off she started straight for home in the States. When Zeb came out ahead of us in the road & told his misfortune both teams were stopped & the three best horses of both parties started in pursuit but they had a hard race of six miles before they succeeded in coming up with & capturing the fugitive.[51] The antelope probably owed his life to this freak of the mare & we also lost our fresh meat, than which nothing could now be more acceptable.

Camped at six on Little Blue. Distance today 18 miles.

Tuesday June 12. Start 5 A.M. Morning cloudy & chilly. Wind N.E. day pleasant. Having just as we started from camp a view of two wolves scampering over a hill I started in pursuit & was followed by one of Doct R's company. We ran them close some miles till we lost them in a ravine.

I noticed today more by far of flowers of various kinds, than we had yet met. There has been all the way numbers of wild roses, a species of marigold, a flower much resembling a hyacinth & morning glories, Sweet Williams &c. Today we met with vast numbers of all of these & many more, beautiful & gay in their appearance & in a sandy ravine with what surprised me

[51] Among the most vexing problems the Argonauts faced was that of keeping their livestock safe. Mules and oxen were obstreperous from the start. They were difficult to adjust to the traveling and grazing routines, and at the slightest opportunity made off toward Independence or their place of origin. D. R. Leeper reported in his journal that a train from Columbia, [*sic*] Ohio, lost every animal it had through the "inexplicable fright known as a stampede." *Argonauts of '49,* 16.

much the prickley pear covering the ground with its curious fruit & also the Southern Bear grass in full bloom.[52] Of the latter we gathered two large spikes with 25 or 30 cup shaped flowers of a faint but most luxurious fragrance. I did not suppose either of the last named plants were found so far north by several degrees & could almost imagine myself back to bright beautiful Florida on seeing them. I thought how my Father would enjoy being in this valley for a little while & what a load of roots he would carry home.

This afternoon we met near where the trail leaves the Little Blue, Maj. Belcher, U.S.A. with escort.[53] He was going home having carried to Fort Kearney a government train of 200 wagons. He said he met between here & the Fort some 500 Chian [Cheyenne] Indians, who were very insolent & fired at his men with their arrows & advised us to be on our guard. In consequence of this news we have been very busy today getting out our rusty pistols & guns & cleaning & preparing them for action. We shall probably see the rascals tomorrow unless they have gone on. We shall if we come up with them endeavor to push right through without stopping or having anything to do with them. Should they try to stop us we must fight our way along & 11 men well armed can drive a host of these cowardly scoundrels.

This afternoon we overtook a company of very pleasant fellows from Cin. who came up on the same boat with us & designated themselves as Cook & Co.[54] We were much pleased at overtaking them as they started some days before us & were much pleased also to meet them & have a chat with our old friends of the Highland Mary. They travel slow with cattle & we had not much time to stay with them. They think they will overtake us again in the Mts. I wish them all good luck but in this I hope & think they will be much mistaken.

At 4 oclock this afternoon we left the pretty Little Blue whose course we have followed for 50 miles & ascend the hills to the high prairie again. For 22 miles now we have neither wood or water then we strike the Platte. Camped this evening on the bare Prairie, having not wood & water with us. Distance today 30 miles. Roads good, level.

Wednesday June 13. Another violent thunderstorm with wind early this morning. The rain was like a torrent & wind furious from the East. This time though our Tent stood the storm & our only damage was some wet corners to our buffaloe robes where too near the edge of the Tent. Morning quite chilly

[52] The bear grass mentioned by Perkins was a fibrous yucca type plant. On June 27, David Jackson Staples saw the cactus or yucca described by Perkins. "We noticed some new kinds of cactus, the most abundant kind is of a pale yellow blossom and they grow in large beds as many as a hundred buds and blossoms from one root." Staples, "Journal," 130.

[53] Perkins misunderstood the officer's name. This was James Belger of New York, who had been promoted from the rank of captain to brevet major in recognition of especially meritorious performance of duty in the Mexican War. Heitman, *Historical Register of the United States Army.* The best account of Belger's expedition is contained in Settle (ed.), *The March of the Mounted Riflemen.* Major Belger was relieved by Major Osborne Cross at Fort Kearney.

[54] It is highly probable that by this date R. W. Cook of the California Club of Cincinnati had organized a smaller company.

many black flying clouds. Wind round to the S.W. Day chilly. Wind in P.M. round again to N.W. Got off at 8 oclock, roads slippery & muddy.

At 11, we came in sight of the range of sand hills which borders the valley of the Platte & at 12 we passed through them & came out in full view of the broad stream & its fringe of timber. Travelled some six miles up the valley. The road keeping about the middle of the Prairie or valley which is here from 4 to 6 miles wide. Camped this evening in the valley. Distance today 16 miles.

Thursday June 14. Morn clear, wind W. Start 6. Were most eaten up by mosquitoes last night. Never in my life did I have the misfortune to fall into the hands of such a pursuing merciless set & they came in such perfect clouds & their sting was so sharp & "itchie" that sleep was out of the question. I got up & tried to pass away an hour by walking around the camp. I found the poor mules in even worse situation than we having no means of defense & being perfectly covered with the torments. Some of them were almost frantic rolling & kicking, & will be not much refreshed for tomorrows journey. Our pony a kind gentle fellow when I came near him put his head under my cloak & was quite loth to have me go beyond the reach of his lariat.[55]

This morning antelope are running round us on all directions & such running! Without apparently any effort they glide over the Prairie at a bird's wing pace that scorns the effort of the fleetest horse to come up with them. We have not yet succeeded in killing though two of our party have snapped caps at them.[56] Several wolves also were seen sneaking away from our vicinity as daylight advanced. These rascals have surrounded us with their unearthly howling for several nights.

Our course this morning lay up the valley of the Platte about midway between the stream & range of sand hills. The valley is like any Prairie perfectly level & free from all vegetation but grass & averages 5 miles wide. The range of Sand hills referred to are quite a curiosity. Running parallel with the River keeping the same distance from it they form a dividing line between the high Prairies & the lower bottoms or Prairies of the Platte. The appearance too is singular, presenting irregular collection of mounds, peaks & ridges ranging from 30 to 100 feet high in some places abrupt fronts of white sand where there had been slides & cavings in & between a scanty wiry grass finds nourishment on their sides & summits.

[55] John C. Frémont had found mosquitoes in abundance in this area. In the long grass of the bottoms of the South Fork of the Platte, Frémont wrote, "We roused myriads of mosquitoes and flies, from which our horses suffered severely." *Report*, 22. Encamped on a small creek emptying into the Platte, Edwin Bryant wrote that the mosquitoes were especially troublesome. He said, "They collect about our animals and ourselves in immense swarms, and bite with the most ravenous eagerness. The slightest puncture of their proboscis, inflames the skin and produces a most painful sensation." *What I Saw in California*, 82.

[56] Although it was difficult to get close enough to an antelope for a fair shot, occasionally an emigrant caught a young one. "Saw a little girl by the name of Virginia catch a young Antelope, and her father could hardly get her to leave it." Markle Diary, June 4, 1849.

They resemble more than anything & can compare them to the highlands in New Jersey below Staten Island. The narrow valley winding through these ridges are very fertile from the concentrated washings of the hillsides probably & they would make the most luxurious grazing spots for our mules were they near enough our trail to be readily available.

We came in sight of Fort Kearney at 9 A.M.[57] & as we came nearer & nearer it seemed as though some vast army was encamped before it so great a multitude was there of wagons, Tents &c of California Emigrants who had stopped here for various repairs, restocking &c &c. There must have been two hundred wagons with their mules & oxen scattered over the plain & nearly all in motion some just starting forward on their way, some just camping, others driving their animals to water, &c. The scene was a very animated & exciting one.

We came to a halt opposite the Fort & turned out our animals to graze while some went over to see what could be done in the way of a Little Blacksmithing. From the Fort came the sound of the merry drum & fife & the hum of many voices & I felt my spirits quite elated at the appearance of an inhabited & civilized community again. I'm tired already of this life of solitude & long for new faces & new scenes.

Close by our camp was Doct Jackson[58] & his party who came up on the same boat with us from St. Louis & who left some days before us with pack mules. We were much rejoiced to see his pleasant company & had quite a pleasant hour in their tent, we shall endeavor to travel in company hence.

In course of the forenoon I took a horse & rode over to the Fort from which we were encamped about half a mile, to deposit some letters to the loved ones at home, & to take a view of things. Found the establishment though quite extensive to be in a primitive state as to buildings &c. With one exception the houses were all made of blocks of turf taken from the neighboring prairie & put together with split laths running lengthwise through them. They make I was told a very comfortable & durable house & some of the officers quarters looked quite neat. The Sutlers Store I had a curiosity to peep into. It was built in the same style as the others & had a ground floor. Around were ranged his shelves with a pretty good stock of all sorts of notions, cigars, sardines, & some few extras for officers use, & in front his counter of rough plank. I enquired his prices for some articles out of curiosity. Sugar only 25″ per lb. Cigars 10″ ea. muslin 25″ per yd. Common Calico 50″, &c. Well might he say as he did to me that if he had had on hand

[57] In July, 1836, Congress passed a law providing for the establishment of a fort and the opening of a military road in the West. In 1838 Colonel Stephen Watts Kearny and Nathan Boone were commissioned to select a site for this post somewhere near the Missouri River, and the first Fort Kearney was built near what is now Nebraska City. This post is now referred to as "Old Fort Kearney." In May, 1848, Lieutenant A. J. Smith, with thirty dragoons, was sent to begin building a new fort above Grand Island on the Platte. When Perkins saw this new fort it was still being constructed. At the time Captain C. F. Ruff was in command. Willman, "The History of Fort Kearney," *Publications of the Nebraska State Historical Society,* XXI (1930), 213-326, and Prucha, *Guide to Military Posts,* 82.

[58] Other trail diaries and journals also frequently use the title "doctor" in a way that seems rather casual, one hundred years later. Edwin Bryant was called "Dr. Bryant" simply because he knew something about medicine. There were many legitimate medical doctors traveling on this road, but the chances are that many so-called "doctors" were mountebanks looking out for the golden chance in California.

all the articles Californians enquired for he could have made more money than by going to the Gold region itself. As it was I could see in the short time I was in, that few merchants in the states do better than those appointed Sutlers.

At the Fort I had some conversation with Maj. Chilton the commandant, a very pleasant man.[59] I found him out shooting at a mark with one of Colts revolvers & pretty good shooting he made too. The Fort stands on the south bank of the Platte & not on the Island as we had been informed in St. Joseph & in the present California Emigration forms quite a stopping place for travellers where repairing can be done &c, & also is quite a protection against Indian hordes who but for their fear of U. S. Dragoons would ere this have committed their depredations upon the numbers of small parties now crossing the Prairies.

On my way back to the wagons went round through some other camps. Found one from Lynn Mass.[60] The men composing which had been more than a month from St. Joe. being detained by breakdowns, sickness, &c. They were throwing away a variety of articles. They could get along without such as lead crucibles, gold washers, extra axles, &c, & also a lot of clothing which they could not carry. We were so fortunate as to think of the latter item before leaving & by shipping round I shall save all.

All being ready we left Fort Kearney & moved on at 2 P.M. Day clear, wind west, warm, road level & good. We begin to feel the scarcity of water at times of which we read before leaving home. Today from leaving the Fort till we camp at six we could not get a drop of anything to wet our parched throats, & trudging along in the hot sun & dust is rather conducive to thirst than otherwise. We have seen whole days on the prairies without any water except mud puddles by the road side or green slimy water standing at the bottom of some ravines.[61] This however, we drink with great relish, our thirst not admitting of nice distinctions. What would my wife say to see me with a cup of mud slime & water to my lips holding my breath shutting my eyes while "putting it down!" Yet so far from injured by such drinks, we actually grow fat upon it! or upon something else, for one thing is evident we are improving in bodily condition. The exercise in the open air, plain food & good appetites no doubt will make men of us before the trip is over.

Camp tonight at 6 oclock. Distance today 18 miles.

59 Major Robert Chilton commanded at least two companies of the Sixth Infantry at Fort Kearney in 1849. Willman, "History of Fort Kearney," 230.

60 This was the Sagamore and California Mining and Trading Company, which left Lynn, Massachusetts, on March 29, 1849. This company traveled over the plains in close proximity to Perkins and his party. Sedgely, *Overland to California*, 3-66.

61 On the stretch of road described by Perkins, Edwin Bryant had suffered for lack of water. Travelers could read in his book, "Nature in this region, is parsimonious in the distribution of such counties, and consequently when met with, their value is priceless to those who have suffered through a long day's march under a burning sun, and whose throats are parched with dust and heat. . . ." *What I Saw in California*, 93.

June 15–July 1, 1849

FRIDAY, JUNE 15. Morning clear, cool, wind East. Start at 5 oclock. Met some returning Emigrants who confirmed the reports we have previously heard that a large body of Chians were on our route, & also told us that some 4 or 5 men had been killed by them, stragglers from camps whom they caught alone.[1] We met the Postrider from Ft. Laramie today also & he repeated the same story. All the effect upon us will be to prevent much leaving the camp alone & will keep the wagons in close company through the days travel. It will not be long the Dragoons will make the Indians pay dearly for these & other depredations.

Camped at 11 oclock & as no water was to be had we dug a well about 4 feet & soon had a supply of pure cool water. We pass every day more or less of these wells & always find good water in them though it has something of an alkaline soapy taste. The ground thrown out is white sand & on drying in the sun a crust of some salt is deposited of what character I cannot ascertain. The water when poured on a smooth place & left to dry in the sun also leaves a thin deposit of the same appearance. The crust cannot well be lime as the water is quite soft.

This afternoon we caught up with & passed a train of 28 ox wagons from Missouri & Tennessee. They started some two weeks before us & will hardly at their present rate of travel get through this year. Deliver me from following after cattle on a journey.

Day has been cool & pleasant, Wind round to S.E. Camp at half past six. Distance 30 miles.

Saturday June 16. Morn clear & cool, air raw & damp. Strong S.E. wind all night & this morning start at half past 5. Our course is still up the valley of the Platte.[2] Sometimes on the River bank & at others 4 or 5 miles from it. This River is in its way quite interesting. It runs through the low level Prairies in a broad shallow channel with banks not more than three feet high. At times fringed with a border of small trees & then nothing at all for miles & miles on its banks but Prairie grass. It is without exception the straightest stream I ever met with, its course being somewhat south of east & varying nowhere two points of the compass. We have traveled so far on its banks in a direct course by our compasses & in many places the water of the River can be seen on the Horizon like looking out to sea. Its waters are muddy though not so much so as the Missouri & it is as large to appearance as the upper Ohio, being in places over a mile in width.

[1] Captain David DeWolfe wrote his wife from the Platte River, "We are now among the Pawnees [perhaps Cheyennes] & they are a thieving set of Devils. There was not one that left his train to go out & hunt had not got out of sight when the Indians were watching him shot him dead on the spot & then robbed him & got on their ponies & run like the Devils & another Company they stole their cattle in the night & drove them off & the Californians followed them the next morning & overtook them in their Wigwams having killed three of the oxen & was cooking the meat." Diary, June 17, 1849.

[2] A rather precise notion of this route can be had from the T. H. Jefferson Map, "Emigrant Road from Independence Mo. to San Francisco, California," 1849. A reproduction of this map is folded in the pocket of Morgan (ed.), *Overland in 1846.*

Another curious & in places beautiful feature of this stream & owing probably to its shallowness, is the great number of little islands scattered along its channel. In one place I counted nearly 30 & no one of them larger than the foundation of a good-sized house & from that down to little points along the surface all covered with a rich growth of grass & bushes, vines & weeds. The appearance in these places was very charming & only wants a heavy forest of shade trees to make it really enchanting.

Its a pity that this River is not deep enough to admit navigation as it runs through a country that one day will be the garden of the world so rich & fertile. The only things made by man that float on its waters are the Indians canoe & the Makinaw boats of the Am. Fur Company of which latter we have seen a number.[3] They are barges some 40 or 50 feet in length roughly & slightly put together loaded with furs which are piled 'up in the middle 10 feet high & covered with tarpaulins leaving a space at each end for oarsmen & cooking operations. In this trim they are manned by some of the hardy Employees of the Company & sent to St. Louis where the Cargoes are disposed of & the boats knocked to pieces for wood.

We overtook today a solitary traveler on foot leading a mule heavily packed. He left Independence with a company of six & went as far as the Big Blue, where he says "the wind blew too strong for them one night," & they all turned round next morning for home. He was not to be so easily deterred however & so taking his share of provisions & one mule off he started to travel alone to California & had come thus far with no company but his animal & his own indomitable perseverance to support him.[4] We advised him to keep along with us, which he was grateful to be allowed to do. Such a spirit deserved success certainly & will pretty certainly lead him to it.

This afternoon has been the most Exciting & interesting of any since we commenced our journey across these plains. Some of our men were riding a little in advance when all at once we were startled with "huzza, huzza! Buffaloe!" & sure enough there was a dozen of these animals grazing quietly on the Prairie some miles off. Our best horses were instantly put into requisition & several of both parties started off in pursuit but returned unsuccessful owing to their thinking they discovered Indians which would have rendered a lengthy chase dangerous.[5] The Indians turned out to be a train of Emigrants who were crossing a hill in the distance & whom we passed before night.

In the course of the afternoon we saw two more droves & I

[3] On June 28, 1842, John C. Frémont's party met the American Fur Company trappers making their annual pilgrimage down the Platte. "They proved to be a small party of fourteen, under the charge of a man named John Lee, and, with their baggage and provisions strapped to their backs, were making their way on foot to the frontier. A brief account of their fortunes will give some idea of navigation in the Nebraska. Sixty days since, they had left the mouth of Laramie's fork, some three hundred miles above, in barges laden with the furs of the American Fur Company. They started with the annual flood, and, drawing but nine inches of water, hoped to make a speedy and prosperous voyage to St. Louis; but, after a lapse of forty days, found themselves only one hundred and thirty miles from their point of departure. They came down rapidly as far as Scott's bluffs, where their difficulty began. Sometimes they came upon places where the water was spread over a great extent, and here they toiled from morning until night, endeavoring to drag their boats the sands, making only two or three miles in as many days." Frémont, *Report*, 17.

[4] All along the way parties disintegrated. Men grew anxious, believing they could move faster in smaller units or alone. "We see the cards & notices which we find on every tree & also on the *bleached skulls & shoulder blades of the buffaloes* that our jackass friends who deserted their party are only a few miles ahead of us, which quite animates us, as we have a little pride to beat them to California." Gray Diary, June 10, 1849.

[5] All of the guides carried a warning about the dangers of hunting buffalo. Lansford Hastings wrote, "In hunting the buffalo, the greatest precaution should be observed, as the hunters are not, unfrequently, attacked and robbed, of both their meat and horses; hence, it is advisable, that they should, always go out in sufficient numbers to insure their protection. . . ." Hastings, *Emigrants' Guide*, 150.

mounted our fleet animal of which I have spoken before & with Doct Riggs gave chase to one. As we galloped up to within 200 yards before they discovered us, I had an opportunity of taking a first & good view of these strange creatures, & when they at length raised their heads & looking for a few moments at us started off with their heavy rolling gallop I discovered now from the impressions received from descriptions & pictures of the animals which roam over these plains. There were 14 in the drove led by an enormous bull & as they "lumbered" off for the sand hills, the long hair on their shoulders & forelegs & the enormous "goatee" which in some instances swept the ground, blew out in the wind & gave them an appearance savage & strange.

As soon as the Buffaloe began to move we put spurs to our horses & after them with a regular yell of excitement. I left all behind me with my mare & was soon close along side of a large shaggy fellow & dropping my reins I levelled my double barrel, but just as I was about to pull the trigger the Buffaloe turned & made a half rush at me which so startled the animal I rode that she gave a sudden shie & over her side went I, not fortunately, thrown entirely though, one leg was over the saddle & by this I scrambled into my seat just as the "Buff" came across in front of me. I was soon after him again & running on his other side fired one barrel into his huge bulk as near "center" as the circumstances would admit of. On he went with increased speed & on went I after. I could see the blood running from his side & he began to fall behind his companions, I followed him some three miles back through the sand hills & until he had entirely lost sight of the rest of the drove I was all alone & could not tell how much farther he might go. I had snapped the cap on my loaded barrell & it was of no service, & remembering the stories of Indians & their depredations in this vicinity I concluded 'twould hardly be safe to go farther so reluctantly bidding farewell to my anticipated prey I turned my horses head for the River & threading thru the valleys came out some little above where I had entered. I felt very sorry to give up my chase as I am confident he was mortally wounded & regretted having fired at all. I regretted more than this though that I could not show my "Buff" of my own killing. My chance may come yet though.

On reaching the bottom I found Doct Riggs who had followed up & killed his Buffaloe. He fired some ten or a dozen balls into him before he dropped. I was surprised at his enormous size when I stood by his side & had a fair opportunity of examining him closely. His head was a great mass of shaggy hair with nothing to be seen but little eyes & horns & the body I should think would stand 6 inches higher than a large ox. For so peaceable & harmless an animal they have a very ferocious appearance.

Stephens stayed with the fallen carcass while I went over to the teams to procure bags, hatchet &c to cut up & carry the meat, & we soon had some hundred weight of his tenderloin steaks, heart, liver &c, safely to camp, & then the luxury of eating it! Such delicious tender juicy meat I never before put under the operations of my masticating organs. All that has been written in the praise of "Buffaloe hump" falls short of the reality. Oh if I could only send this great tender piece of tenderloin to my friends at home! would not they luxuriate over it? & this enormous heart which lays before me, would I could eat a piece of it boiled under the superintendence of my Mother! We all eat largely, in fact making our entire supper of our prize & without feeling any ill effects.

Bryant says that a man can eat 15 lbs of Buffaloe meat at night for supper & get up at midnight & eat 15 lbs more without hurting him. I almost believe him for it seemed as though I would not get enough. We have at last had some little variety to our hitherto monotonous journey, & our spirits are elated a hundred per cent in consequence, & as Buffaloe promise to be plenty I hope we shall for some time have our fill of sport & good eating.[6]

Day has been pleasant with strong breeze from S.E. Camp at 5. Distance 28 miles.

Sunday June 17. Camped all day on Platte near a clear brook on the one bank of which is a spring of clear cold water.[7] Colder than the majority of wells. This is the greatest treat we have yet enjoyed always excepting our Buffaloe meat of course, & we have been drinking as if to make it last us through. Morning pleasant. Day do, with strong breeze from S.E., as usual we have spent the day in cleaning up, reading writing &c, a mode which at home would be highly reprehensible but here is the best we can do. Our Sundays we prize far more than do four fifths of those at home though perhaps from different motives. We do look forward to a Sundays rest with great pleasure.

Last evening a single wagon camped with us which had been traveling with some ox teams but found them too slow & wished to keep with us in future to which arrangement we agreed. The owner of the wagon is from New Orleans, a very pleasant

[6] Buffalo hunting was a highly exciting sport, but by midsummer, 1849, the gold-rush trains had taken a heavy toll of the animals and had driven the remnant back from the trail. Branch, *The Hunting of the Buffalo*, 116-17. Stansbury noted the retreat of the buffalo from the emigrant trail. "Today the hunters killed their first buffalo; but, in order to obtain it, had to diverge some four or five miles from the road, and to pass back of the bluffs, the instinct or experience of these sagacious animals having rendered them shy of approaching the line of travel." Stansbury, *Explorations*, 34-35.

[7] Bryant's party stopped at this spring in 1846, and some of the men overdrank themselves. Bryant, *What I Saw in California*, 92.

fellow Adsit by name, & his companion's name is John Coolidge Hildreth, I must find out his history & see whether any relationship can be raked up.

We have been luxuriating on Buffaloe Steak, tender loin, liver, heart, &c, & every delicious mouthful I swallow I think of those at home, who would so much enjoy a taste of what we have in such abundance. We have cut up a large quantity of the lean meat in thin strips & strung them on a rope to try & make jerked Buffaloe whether they will dry good & sound without any salt is yet to be proved, but I think the Indians put up vast quantities in this manner. We have been enjoying the luxury today also of most delicious water having found a spring than which few wells can be colder, & we are by this time capable of enjoying & appreciating anything of this kind. Day pleasant.

Monday June 18. Morn clear cool. Wind strong S.E. Start 1/2 past 5. We have seen several rattlesnakes today & Cross stepped on quite a good sized one which he killed, had seven rattles.[8] The Prairie is said to abound with them.

Our course has been today quite a chapter of accidents. We went on finely till abt 10 oclock when our leading mule becoming frightened at something by the side of the road shied suddenly to the right & turned so short that we had no time to stop him till, crack! crack! and away went the tongue of our wagon! broken nearly off but fortunately in long splinters. Here was a fix!

Doct R's wagons came up soon after but we concluded to send them on & take our time to repair the damage. We cut up raw hide into strips & first nailing a piece of hickory on the broken side, we wrapped it well with the rawhide which as it becomes dry will shrink & tighten & over this by means of a contrivance I learned at sea wound a new rope over the rawhide as tight as the hemp would bear straining & to all appearance the tongue is about as strong as ever. We were ready & off again in about 2 hours after our accident occurred & camped tonight with our party. This was accident No. 1 & we have no desire for a repetition of anything of this kind.

Some two miles further on as we drove into a steep gutter one extra axletree which slung under the wagon broke loose at the front end & ploughed into the ground while the other end steered by the rope which held it up stove right through the bottom of our wagon body into a bag of flour! & we had to dig a hole deep enough to let it down into before we could take the axle out. Here was No. 2.[9]

[8] At about this point on the trail many other diarists commented on the presence of rattlesnakes. Joseph Sedgely mentioned killing "a large rattlesnake, from which I took seven rattles." *Overland to California,* June 24, 1849. William Banks noted on May 30, 1849: "We are now above Grand Island. Plenty of snakes, green spotted and rattle. Six of the latter kind killed in less than a half acre." Scamehorn (ed.), *Buckeye Rovers,* 13.

[9] Accidents to wagons were a source of recurring annoyance and delay to the emigrants. "One of Mr. Jenkins' wagons broke an axle about nine o'clock; we waited six hours, then went on." Scamehorn (ed.), *Buckeye Rovers,* 10.

Then a little farther on Stephens driving lost off the lash of his whip, which being the only one in our possession may inconvenience us considerably at the first steep hill. Misfortunes tis said never come single and it certainly holds true with us today. However, here we are all damages repaired & feeling better than ever that we got off so easily.

Have seen several large herds of Buffaloe today & Doct R & some of our men rode in among them, getting quite close to a number but refusing to shoot them on the ground that we have plenty of meat for the present. I do not think my forbearance would have gone so far had I been one of the party.

Day pleasant. Roads good. Distance 25 miles. Camped on Platte at 1/2 past 5.

Tuesday June 19. Morning cloudy, breeze south. Started 1/4 before six. The wind has at last changed its direction & character. Four days it has blown from S.E. violently so much so as to be very disagreeable. The general direction of the winds thus far has been S & W. This morning saw more Buffaloe, but our Capt. thinks we have not time to spare to hunt them, as we must get ahead. In this we agree with him pretty well, but still its very hard to have these huge monsters come close up to you & snuff at you shaking their enormous goatees without giving them chase. We shall probably though have opportunities enough yet & will be content to wait a little.

Have met several returning wagons today & their drivers give us great accounts of the difficulties & danger we shall meet with. We of course make all allowance for their disappointment & necessity for excuses for their conduct, but we expect to meet hardships & shall go on till we come to them.

Saw this afternoon three graves. Before leaving St. Joe. we heard doleful tales of the amount of deaths on the road & the graves we should see &c. But so far the number who have laid their bones on these plains has not been at all a fair proportion of deaths compared with some parts of the South & Southwest. Taking into consideration the numbers on their way to California & quite as large a proportion would have died had they stayed at home.[10] The air of these plains seems remarkably healthy & it is almost impossible to take cold. I have gone through exposure of all sorts by night & by day, sleeping in wet clothes &c, which at home would have laid me up for a long course of sickness, without feeling any ill effects.

We have been annoyed excessively for several days & nights by swarms of buttes or what we used to call June Bugs—a harmless insect, but crawling & tumbling into everything. We cant

[10] Perkins may have been right in his conjecture, if he was thinking of the raw edge of the frontier in the South and Southwest, or in terms of a fever epidemic. He had reason to change his mind before he reached California. Another diarist was less sanguine. "This morn we were visited by a man who had left St. Jo on May 29. He informs us that at least 1500 have died on the road & that thousands will not be able to reach their destination this season. Families on the road are numerous & perhaps many of them left behind without a protector. How thankful we should be for our success. No one sick, no loss, no breakdown, our cattle in good trim." William Steuben Diary, June 26, 1849, hereinafter cited as Steuben Diary.

fry meat or boil coffee without spooning out half a dozen or more, & last night I picked out of my hair in the course of the night 8 of them. They pulled my hair so hard as to wake me quite suddenly & would cling tight enough to bring away their little hands full!

We reached the Platte ford today at 2 oclock, P.M., & immediately commenced preparations for crossing.[11] We found here 4 trains crossing & waiting their turn. The Platte here is about 3/4 mile wide & varies from 1 to 3 feet deep with part sand & part gravel bottom.

Mr. Adsits team was to go over first & we contributed each a pair of mules & harnessed them to him & with some from each party on horseback as drivers in they went & over they went finely without any stoppage. Our team was to go next & as soon as the mules were brought back we fastened to & started in not without some anxiety lest our heavy hard running wagon should stick, & our apprehensions were not without foundation. Half way over all went well enough, but running into a deep hole the mules began to lag & finally stopped short, & no efforts of ours could start them. Some of the drivers were dispatched to shore for more mules & returned with two pairs which made 14 altogether, a pretty long team & difficult to manage. Just at this time a heavy black cloud which had been hanging over us for some time muttering & flashing began to pour down a perfect deluge of rain, India Rubber goods were in great demand & hurriedly put on, but as we were all on foot wading nearly waist deep some of the time we could expect only to keep our shoulders and bodies dry & this was all we did do. With our additional struggles we walked ashore without much trouble & landed a little after sundown.

There was some little consultation whether it was best to go back for Doct Riggs wagons tonight or not which was ended I believe by my jumping on one of our mules & heading on into the river & the rest soon followed. When we got over twas quite dark except when flashes of lightening would illumine everything for an instant leaving us in total darkness after. However we tramped on & plunged into the Platte governing our course by a lantern hung up at our camp. We went only about 20 yards from there tho when the mules stopped again I could not move tho we tried them at "gee" & "haw" & with all our whips in full play we were all very much exhausted, wet, thro & chilly, hungry & so forth, & so determined to abandon the job for that night and leave the wagon sticking in the sand. No sooner decided upon than we all broke for our camp

[11] Crossing the Platte anywhere was an ordeal for emigrants because of the width of the stream. On July 2, 1842, Frémont's party crossed at the forks, and he wrote, "I had left the usual road before the mid-day halt, and in the afternoon, having sent several men in advance to reconnoitre, marched directly for the mouth of the South fork. On our arrival, the horsemen were sent in and scattered about the river to search the best fording places, and the carts followed immediately. The stream is here divided by an island into two channels. The southern is four hundred and fifty feet wide, having eighteen or twenty inches of water in the deepest places. With the exception of a few dry bars, the bed of the river is generally quicksands, in which the carts began to sink rapidly so soon as the mules halted, so that it was necessary to keep them constantly in motion." Frémont, *Report,* 21. On the same date in 1849 Howard Stansbury wrote, "After traveling up the river for fourteen miles, it was determined to make the crossing of the South Fork by fording. In preparation for this movement, one of the wagons, as an experimental pioneer, was partially unloaded, by removing all articles liable to injury from water, and then driven into the stream; but it stuck fast, and the ordinary team of six mules being found insufficient to haul it through the water, four more were quickly attached, and the crossing was made with perfect safety, and without wetting anything." *Explorations,* 39. At the time Perkins crossed the river on June 19, 1849, the level of the stream was at a seasonal low comparable to that described by Frémont and Stansbury.

leading our mules and arrived safely in the midst of another storm of rain & hail which all admitted exceeded anything of the kind in their experience.

I do not exaggerate at all when I say that I picked up several hail stones as large as a common pullets egg, one of these large ones struck one of our men on the head & for a few seconds he was almost unconscious.[12] We had quite a treat of iced Brandy & water & after our mules were picketed out & our clothes changed & after a heavy supper of rice, beans, & crackers, & tea, I believe I never felt so well in my life. If this is what they call "seeing the 'Elephant'" I dont see anything very bad or ugly in his appearance. Tomorrow we will get Doct Riggs wagon over, dry our clothes & goods which were wet & be all just as well as ever.

Day pleasant till 5 P.M. when heavy rain commenced with hail & still raining, Wind West. Distance today 16 miles.

Wednesday June 20. Slept rather late this morning for which the fatigues of yesterday are a sufficient excuse. We worked last evening getting our mules taken care of & changing our wet clothes till 10 oclock when I did get between the Buffaloes, didn't I sleep.

I never enjoyed a night's rest so much in my life before.

Turned out this morning at 7 & breakfast over began to make preparations to get the Doct over & the first & most uncomfortable thing was to take off our nice warm dry clothes & put on our cast off wet ones of yesterday. Was not there some shivering & chattering of teeth just about that time? The rain fell all night heavily, & an immense quantity of water must have fallen. The Platte is a little swelled & we were in something of a hurry to get the wagons over before it should rise into the body. While getting up our mules a drove of Buffaloe came up close to our camp, to cross the river. The shaggy rascals snuffed at us a few minutes & then plunged into the water not before though some half dozen bullets had been sent after them apparently without effect.

About 8 we commenced crossing. Doct Riggs had decided that twould be best to unload as much as possible & pack over on mules, & each taking an animal we forded over to the wagon, which we found sunk in quicksand over the wheel hubs on our side, loading abt 50 lbs each time we succeeded in carrying over some 900 lbs safely, tho twas a tedious & tiresome business. After this lightening we rigged purchases under the wheels & lifted the whole concern out of the sand, when lo!

12 Emigrants were astonished and frightened by the terrific storms that came up on the plains. In 1841, the Methodist missionary Joseph Williams, described one of these storms: "The next morning we continued up this river, along smooth banks, without any timber. That afternoon we had a very severe hail storm, accompanied by thunder; one Indian was knocked down with a hail stone, about as large as a goose egg. We soon discovered a water spout, which came down into the river. When it struck the river it made a great foam, and then passed off in a dreadful tornado." Williams, *Narrative of a Tour*, 10.

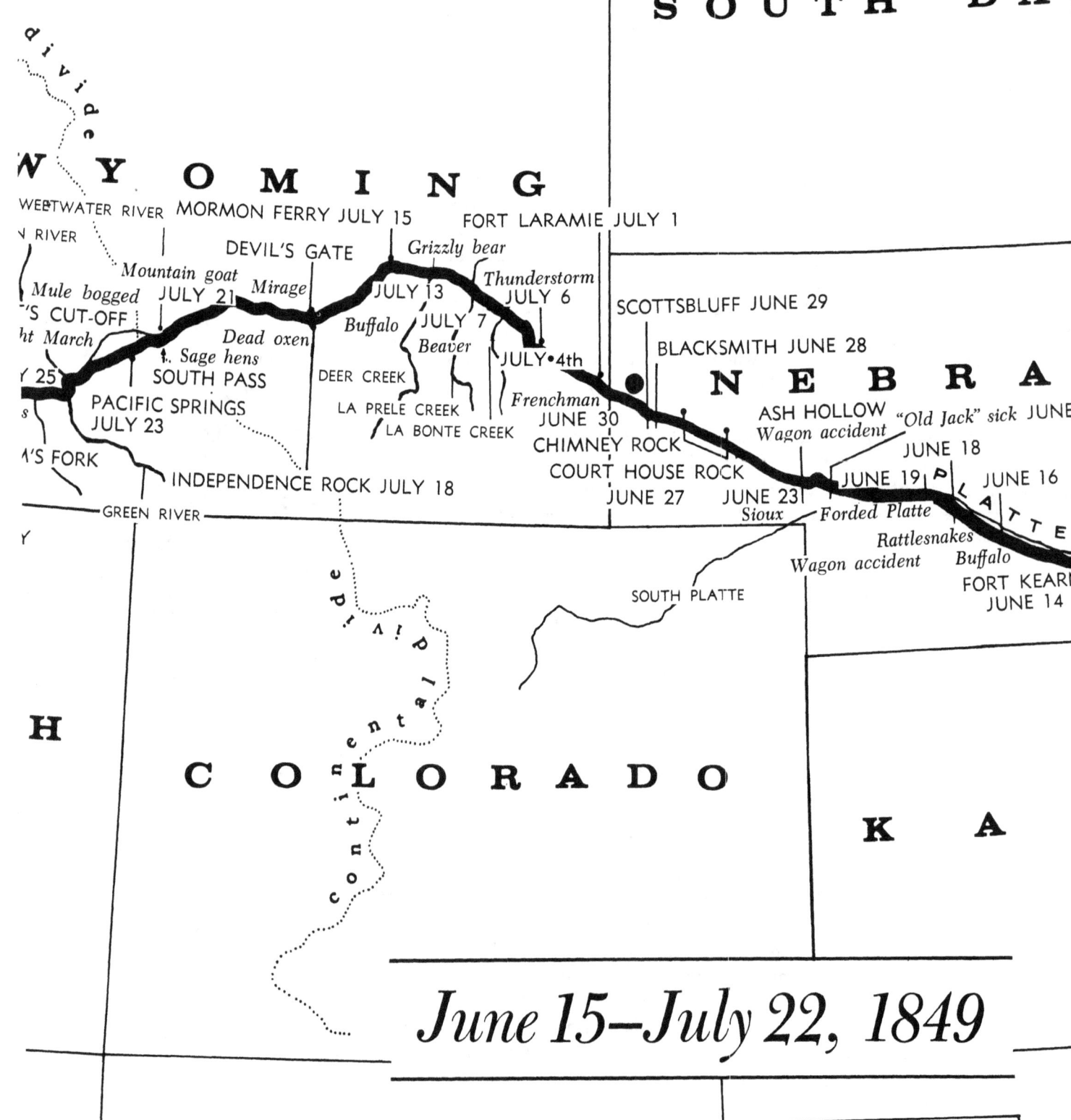

SOUTH DA[KOTA]

divide

WYOMING

WE[ST]WATER RIVER MORMON FERRY JULY 15   FORT LARAMIE JULY 1

[N] RIVER

DEVIL'S GATE   *Grizzly bear*

Mountain goat   *Thunderstorm*
Mule bogged JULY 21   *Mirage*   JULY 13   JULY 6   SCOTTSBLUFF JUNE 29
['S] CUT-OFF   BLACKSMITH JUNE 28
[g]ht March   *Dead oxen*   *Buffalo*   JULY 7   N E B R A[SKA]
[Y] 25   ‡ *Sage hens*   *Beaver*   "Old Jack" sick JUNE
            SOUTH PASS   JULY•4th   ASH HOLLOW   JUNE 18
            PACIFIC SPRINGS   DEER CREEK   *Frenchman*   *Wagon accident*
[A]'S FORK   JULY 23   LA PRELE CREEK   JUNE 30   JUNE 19 JUNE 16
[s]   LA BONTE CREEK   CHIMNEY ROCK   JUNE 16
[Y]   INDEPENDENCE ROCK JULY 18   COURT HOUSE ROCK   *Rattlesnakes*
        GREEN RIVER   JUNE 27   JUNE 23 *Buffalo*
                                                   Sioux *Forded Platte* FORT KEARN[EY]
                                                                              JUNE 14
                                          Wagon accident

[U]T[AH]   SOUTH PLATTE

H   C O L O R A D O

continental divide   K A[NSAS]

June 15–July 22, 1849

the mystery of our last night's failure was explained. Some careful soul to prevent the mules from starting while waiting yesterday, had locked the wheel & forgotten to loose it   the heavy sand took hold of it in a way that defied the strength of all our mules. Once unfastened & our mules hitched to over we went without trouble abt 2 P.M. I forded the Platte myself 9 times in the whole scrape getting wetter every time. Yet enjoying this more than anything on the trip except our buffaloe hunt. Here was some excitement, some variety, & far preferable to the teams jogging on over the level prairie following a wagon.

There was a large train of mule wagons from Boston crossing at the same time we did & such cheering as one wagon after another got safely over & when all were across 9 such hardy cheers woke up the hills around & made all ring again. Doct Jacksons mules also crossed with us & about 3 we made 4 gallons of coffee, got out our Brandy & the Beans, Rice, Bread, crackers &c, of our three messes invited in Doct J's men & had the merriest dinner ever eaten, & paying us well for our troubles & wetting. We took everything out of our wagon & sunned for two hours, then harnessed up & at 5 made a fresh start to find better grass which we did & camped at 1/2 past six, being in advance of everything which was in sight when we reached the ford.

Looking back from our camping place the plain looks like the camping ground of a great army. Trains of all kinds are stretched far as the eye can reach, some camped & others making preparations to do so, & the scene is quite animated & exciting. We are bound to keep ahead of them all & shall make an early start tomorrow.

Our Vicksburg friend Harlan is with us & we find him one of the pleasantest fellows in the world, full of spirits, a great hand to tell a story of which he knows any quantity & an admirable singer being almost theatrical in his voice & manners. He is now amusing our boys with his songs & yarns for which latter I will match him against Charlie Rhody[?] all day.

Distance today 5 M, up the south fork of Platte. We cross over tomorrow to North Fork & follow that to Fort Laramie. Day has been pleasant & warm. Wind West, roads very wet.

Thursday June 21. Start 1/4 before 5. Morn clear & pleasant. Breeze south. Travelled up the south fork some 5 miles when the road turned up the sand hills & after 5 M more of uneven roads we struck the North Fork of Platte & followed its bottom up some distance, once more ascend the sand hills & camp in Prairie at 6.

Saw quite a number of wolves today & several were fired at but none killed. No antelope or Buffaloe. Distance 20 miles, roads heavy.

Friday June 22. Morning clear & pleasant. Wind S.W. Start at 1/2 past 5. Roads today have been very heavy & deep sand making the traveling very hard for our animals.

About 11 o'clock one of our large wheel mules seemed to stagger in his harness for a few rods & then fell flat on his side, as for some days most of the mules had shown signs of fatigue & being nearly "done over," we supposed "Old Jack" was what is called "fagged out," & our countenance lengthened "about a foot" very suddenly at prospect of losing him, & well we might feel blue for we had not an animal to spare. However on examination by the knowing ones of our train he was pronounced out of danger & only laboring under an attack of cholic, & a dose was prescribed for him. Sp. vini, Rect, Zry, Tinct opii Zry, Agua Zry which prescription I put up "scrumndum artum."[13] Our patient however like many others at the prospect of taking an unpleasant dose suddenly grew better. After rolling & groaning for some minutes, he all at once got up on his feet & began to Eat apparently as well as ever. Harnessing in one of our horses in his place & tying the old fellow behind our wagon we moved on again.

Day has been warm Evening pleasant as usual. Camp on Platte at 6. Distance 20 m.

Saturday June 23. Start 5. Morning clear with a delightful East breeze. About 10 arrived at a pretty place known as Cedar Grove where we laid in a supply of fire wood. Here is a grove of cedar trees entirely isolated. Not another green thing to be seen but the grass of the Prairie & we were much refreshed by their shade. Trees are a perfect luxury to our Prairie sick eyes.

At 11 we arrived at Ash hollow[14] where the Main road from the South Fork crossed over to the North Fork, a pretty ravine lined with a growth of stunted ash trees & a rivulet following down it to the River. Here we had some romantic rocks & bluffs which for variety were very acceptable.

About half way down the hollow we found a fine spring of cold water & a short distance below the spring we met the advance guard of a band of Sioux & soon after the whole tribe.

[13] Perkins' experience in his father's drugstore would account for the jocular prescription given here.

[14] Ash Hollow, a well-known landmark, was a popular stopping place for trains. Edwin Bryant described his party's experiences when it arrived there, June 19, 1846. "We descended into the valley of the north fork of the Platte, through a pass, known as 'Ash Hollow.' This name is derived from a few scattering ash trees in the dry ravine, through which we wind our way to the river bottom. There is but one steep and difficult place for wagons in the pass. I saw wild currants and gooseberries near the mouth of Ash Hollow. There is here, also, a spring of pure cold water." *What I Saw in California,* 97. "Off at 5 1/2 oclock, road bad all day sandy & heavy, particularly through a deep gorge in the bluff for 2 or 3 miles, called 'Ash Hollow.'" Gray *Diary,* June 4, 1849. "Here at Ash Hollow we met, last evening the first Indians since we passed the agency 30 miles west of St. Joseph—here they came *swarming* around us—to beg—they are pitiable objects. Many of them are naked. They are of the Sioux Nation." Solomon A. Gorgas, Diary of an Overland Journey, May 21, 1849. "We came into Ash Hollow this even, plenty of good wood & water but no grass of consequence for our teams. We have ascertained that grass is all cropped short for 8 miles and as the blackness of the western horizon indicates a storm we concluded to stop in the gorge." Steuben Diary, June 5, 1849.

Men, women & children, horses & dogs such a motley mass as never greeted my wondering eyes before. The men fine muscular fellows, though short in stature riding their long maned ponies & the women leading the pack horses.

Their way of carrying their goods and children is curious enough. It has often been described. Two large poles the larger ends fastened over the horses back and the other ends dragging on the ground. Willow baskets tied to the poles about midway & in them stowed promiscously the children, Buffaloe meat, cooking utensils & all the nameless articles belonging to an Indian camp.

The almost numberless dogs were all harnessed in the same way. Some of them carrying loads which I could hardly have believed possible. These dogs are really useful animals to the poor Indian carrying his goods when moving, helping him hunt his supply of provisions & keeping a most faithful watch over his scattered property at night. The same amount of canine stock though in a civilized community would soon occasion a mayors proclamation offering 1.00 per head for all dogs running at large.

After a certain time, as much as we were all gratified & amused by this train they were unintentionally the cause of the most serious accident from our journey. "Polly" our leading mule walked along part way through the Indian procession paying them some attention to be sure but apparently not much alarmed till she met a huge Indian warrior wrapped up in a red blanket & his face & the exposed parts of his body painted most hideously. This was too much for her courage & "Woozed out at his finger ends" immediately & making a short movement to the right about she went "right astern" all of the other mules following her of course, & before they could be stopped & in fact almost before we were aware of her intentions there was a great crash & down came our omnibus with the tongue broken short off, & the off wheel under the wagon the axle also snapped. Here was a fix to cause us sorrow indeed, & we stood for some seconds mutely gazing at our misfortunes. The driver at the time came in of course for his share of the blame but the thing could not have well been prevented as we did not anticipate any difficulties and of course made no preparations to avoid it. The mule it was now easy enough to see should have been led. However all to do was to repair damages as soon as possible.

The wagon was unloaded. Wheel got out. Axle taken off & our spare axletree which two or three times we had been on the point of splitting up for firewood was commenced upon, cutting it down to fit &c. The other teams having stopped their tools were brought out & with the assistance of Adsit's wagon maker & Doct Riggs Blacksmith, after 6 hours hard work we had a new axletree in, wheels on & greased, load in Mules harnessed up & were off again. A more expeditious job I venture to say than many could have hired in the states. Our tongue was cut off where it was broken before, & made a splice of it shortening it some two feet which the very length of which we had often complained admitted of very well.

Once righted & again underway we began to inquire as to the cause of the accident &c, & it seemed to be pretty generally decided that the driver was so fascinated by the pretty face of one of the Indian girls & many jokes were passed at his expense. Several of these girls were quite pretty & appeared to belong to the aristocracy being much better dressed than the rest in finely tanned deer skins almost as white as linen, handsomely fringed with beads & colored threads.[15]

Camp some 2 miles from the place of our accident on the Platte. Distance 15 miles.

Sunday June 24. Morning clear & pleasant. W. SE. Another portion of the Sioux tribe came into our camp this morning about 7 with a half breed for an interpreter & through him I had some talk with them, learned their way of holding & drawing their strong bows, & tried to buy a bow & arrows but no, no, was all the English I could get. They were very friendly in their manners, wanted to shake hands with all, continually repeating their "how, how." Their bows & arrows are formidable weapons. The first being as much as I could conveniently draw & the arrows about 3 feet long with a steel point sharp as a knife.

The half breed was a mixed looking individual half Indian, half white man in everything, dress, manners &c. He spoke English very well & Indian better, I suppose as I was no judge of this latter accomplishment. They wanted to trade ponies for our horses, one pretty creature I should like to have, his long mane & forelock reaching over his shoulders & head entirely. As they could not trade to suit themselves the whole posse left about 8.

Owing to our accident yesterday & consequent detention we did not reach our camping ground and so were obliged to travel today to get better grass. Started at 9 oclock. Found the roads very heavy a deep sand for miles at a stretch in places

[15] Not until the emigrants met the Sioux on the move at or near the North Platte did they feel that they were really in Indian country. Almost every emigrant journal mentioned these colorful Indians along the trail below Fort Laramie. Here are three other accounts:

"Passed an encampment of the Sioux Indians, there were about 204 of them male & female. They were *Frenchmen* with *Sioux wives* & some of the children were as fair as our home productions. Several of our men bought of them leggins & moccasins made of deer and antelope skins." Gray Diary, June 4, 1849.

On July 26, 1849, James Pritchard saw what may have been the same village that Perkins saw, two miles above the crossing of the South Platte. There were, he thought, "some 2 or 300 of them of all sex ages & size were loitering around our wagons pretending to trade Mockinsons & Skins &c for something to eat. There were among them ten or a dozen Frenchmen living with squaw wives." Pritchard, *Diary*, 73.

"Near where we forded the South Platte we had the good fortune to come upon a large village of the Sioux which was squatted temporarily in the locality. These Indians struck me as being decidedly comely specimens of their race—neat, healthy, self-poised. Their dress was made chiefly of white-tanned skins, and looked very picturesque in its elaborate decorations of beadwork and other fanciful adornments peculiar to savagery." Leeper, *Argonauts of '49*, 20-22.

through which our mules would insist on stopping almost every 1/4 of a mile.

Camped at three o'clock on Platte. Day has been oppressively hot & I am confident that the thermometer would have shown nearly 100°. Distance 10 miles.

Monday June 25. Start 5. Morning cloudy. Wind E. The Sand hills which have bordered the Platte all the way are beginning to change their appearance & character somewhat being more abrupt & large with occasional rocks & bluffs. As we go on up the change will be greater till we get to the Black Hills.

Travelling today has been very heavy. The warm deep sand of yesterday & the same slow travelling. Past today two large trains of ox teams plodding slowly on their weary way, although we pass oxen so easily I'm not sure we gain on them very much. They keep steadily on without stopping up hill & down & we have been several days passing & repassing the same train before finally losing sight of it.

We frequently see the names on trees & boards by the side of the road of some of Chapins company & they have kept their distance of seven days in advance of us regularly. We were in hopes to catch up with them, but 'twill take us sometime if they go on as at present. I should like much to see them.

Camped tonight at 6. Distance today 22 M.

Tuesday June 26. Morn clear, cold wind W. Finding our load too heavy for these sandy roads, we determined to pack our spare animals & so lighten & we staid behind to do so. Riggs & Adsit leaving as usual at 1/2 past 5. Just before they started a large wolf trotted by near our camp & two or three lucky rifle balls brought him down & he was hauled in in triumph. He was a savage looking rascal with an enormous head & long fangs with which he kept both our dogs at bay before finally yielding.[16]

On over hauling our wagon to get out articles to pack I discovered my bundle of India Rubber goods containing coat cap & leggins were missing. I knew they were safe yesterday afternoon & was confident a train which passed yesterday evening after we camped must have picked them up. So saddling one of our ponies I started in pursuit & after a hard ride of six miles overtook it, & on inquiring I found that one man belonging to the train had picked up goods of the description I gave & with some difficulty I found him & recovered my valuable

16 Numerous entries in diaries commented on the terrifying experiences of hearing wolf howls, and of seeing the animals passing around camps. They were capable of doing real harm by their thievery, and by stampeding the picketed livestock at night. L. Dow Stephens appreciated the fact that not all wolves were alike. "There were," he wrote, "many different classes of wolves to be seen on the prairies; the common prairie wolf, the gray, the black and another, a large long-legged wolf, the latter being found always near the herds of buffalo and was a constant terror to the calves. While the herds were traveling the cows and calves always in the center with the bulls on the outside, affording protection against the Buffalo Rangers, as these wolves were called. The wolves were ferocious, and a band of them would attack men, if hungry." Stephens, *Sketches of a Jayhawker,* 10.

articles. Without them I should have been in an uncomfortable predicament, exposed to every storm with no protection. On my return I found the boys had packed & were nearly ready to leave started at 9.

Roads began to be hilly and we left the river about 11 & went back behind the bluffs. At 12, camped in sight of the famous Courthouse Rock, Chimney rock looking like a needle in the distance. Started P.M. 2.

Camped at 5, near a fine creek. Just as we camped a heavy rain came up from N.W. & we had some wet work to do setting Tent & picketing mules &c. The water fell in torrents for some two hours when we had a clear up & fine Evening. Distance 18 miles.

Wednesday June 27. Morn cloudy Wind W. Start at 1/2 past 5. Day quite warm. Passed within about 3 or 4 miles of Court House Rock, which presents a grand impressing appearance I should call it a castle as it is nearly round & looks as if it might be fortified.[17] In size it must be immense tho I had no time to go over & examine it. It rises abruptly out of the rolling country & stands entirely alone with the exception of a smaller square rock some hundred feet from it called "The Jail." It seems to be the vanguard of the Rocky Mts. as beyond it the rocks & irregular outline of bluffs & mountains increased in frequency with distance while this is the first large mass of rock we have seen. It is well worthy a place in a sketch book of survey on this route.

At 12, camped opposite to & about 1 mile from Chimney Rock. I had some curiosity to see this as I had noticed a plate of this scene in Frémont's work and it far exceeded my expectations.[18] The plate spoken of does not do it justice. Imagine a pyramid standing alone though surrounded by many rocky precipices some 150 feet high & from its center a column of rock 100 feet high & at its base 20 feet through & then compare with Fremont's picture & his will look like a pile of stones on a hill. No conception can be formed of the magnitude of this grand work of nature till you stand at its base & look up. If a man does not feel like an insect then I don't know when he should. The pillar itself is somewhat irregular in its formation having the appearance of strata laid here one on another. Or as one of our boys expressed it "looks like a pile of pancakes." It is said to be fast washing away & travellers 25 years hence will see nothing of it but its base. If so one of the principle attrac-

[17] Courthouse Rock and Jailhouse Rock stand just south of present-day Bridgeport, Nebraska. Legend has it that the main rock was named by St. Louis traders who thought it resembled the St. Louis courthouse. Jailhouse Rock no doubt was named by a wag who could not conceive of a courthouse without a jail nearby.

Edwin Bryant, (*What I Saw in California*, 99-101) showed more curiosity about the impressive formations than did some travelers, and went off the road with his traveling companion, Benjamin Lippincott, to get a closer view of them.

Charles Gray wrote: "Last evening just as the thunderstorm cleared up, I remarked there is the Court House, a formation of the bluff resembling a large buidling & for which I had been anxiously looking for the last day or two. . . . Today we came in front of the Court House. . . . It is not inaptly named, as it bears a strong resemblance to some monstrous building, with regular sides & dome, the sides appearing quite square and regular, but several of our men who went down to it . . . represented it as quite rough & irregular. These bluffs are quite picturesque & contrast strongly with the features of the country." Gray Diary, June 7, 1849.

[18] The plate to which Perkins referred may be found facing p. 38 of the 1845 Senate edition of Frémont's *Report*.

Charles Gray thought that Chimney Rock resembled most closely the chimney of a sugar refinery. "The Chimney Rock after all days travel is now about 4 miles ahead of us & stands up against the sky in fine relief, it is about 200 feet in height & some years ago was calculated at double that—but that is doubtful. . . . A great many persons on foot & horseback were visiting it." Gray Diary, June 7, 1849.

Other travelers such as William E. Taylor, Nicholas Carriger, and Virgil Pringle also noted passing Chimney Rock. Their diaries are given in Morgan (ed.), *Overland in 1846*, I, 118-33, 143-58, 159-88, respectively. Pringle remarked (June 19, 1846) that it reminded him of the foundries in St. Louis. Edwin Bryant was likewise struck by this formation (*What I Saw in California*, 99-101) and like Perkins believed that it would shortly crumble away.

tions of this journey will be lost. The view from its base is magnificent.

All the rocks & hills in this vicinity appear thrown up abruptly by some convulsion of nature & one can hardly divest himself of the impression that he is gazing at the ruins of some vast city. Castles with turrets, towers, square buildings with the tops gone, palaces, one in particular having a strong resemblance to the plates giving the Tuilleries, appear on every side, all standing separate & alone, looking exactly like the work of men's hands only far too vast & grand. I thought for even in such scenes sometimes thoughts of profit & loss will intrude that twould be well worth the while of some Daguerreatypist to come out here & take a variety of views from different positions of this & other positions on the route, take them home & have them engraved. Few of the prints now sold would excel such in grandeur or beauty, & the speculation could not but be profitable.

Started P.M. from our noon camp at 2, with regret leaving Chimney Rock behind us. Travelled til 6 when we came up with our other wagons & encamped. Day hot & pleasant, breeze from W. Distance 20 M.

Thursday June 28. Warm cloudy day pleasant quite warm. North breeze. Start 5. The plains here and for some days back have been covered with Prickly Pear which being in full bloom appears quite showy. The flowers are much like one of the cactus tribe cultivated at home though rather small & of a pale yellow color, some white. Came in sight of Scott's Bluff about 10, & camped opposite it at noon.[19] Another abrupt pile of rocks projecting out into the plain. More fine views of castles &c which cannot be described must be seen to be appreciated.

At 2 P.M. started left the river again & went in behind the bluffs & began the ascent of a long steep hill or a succession of ascents altogether some ten miles in length. About 1/2 way up came to a spring of water coming out of the mountain sides the like of which I never expect to see again. Colder than any well water I ever drank it had a smooth pleasant taste that made it a perfect luxury & you hardly know when to stop. I sat down by it with my cup in my hand & for half an hour took real comfort tasting the water & enjoying the cool of the grotto whence it came.[20]

Camped near top of the mountain at 6. Distance 20 miles.

19 Scott's Bluff impressed everyone. James A. Pritchard viewed the calcareous formations of sand and clay and saw in them a great variety of architectural shapes, including gothic fortifications and towers, arches, spires, domes, minarets, and temples. Pritchard, *Diary,* 79. Scott's Bluff rises to an altitude of 4,662 feet, or 750 feet above the high plains. According to Washington Irving, "The bluffs were named for a trader (Hiram Scott) who met an untimely death under the bluffs. His skeleton was discovered the next season and identified, thus his name was given to the bluffs." *Adventures of Captain Bonneville,* I, 104-105. Bryant told the story of how a party of trappers had abandoned their disabled companion Scott to die. His skeleton was found and buried near the bluffs. *What I Saw in California,* 104. Captain David DeWolfe was one of the few emigrants known to have climbed Scott's Bluff. He wrote his wife on July 7, 1849, "Next comes Scott's Bluffs higher than all the rest I went to it & after a good deal of hard work succeeded in getting on top when I stood six hundred feet above the sea level of the plain. On top & sides are a number of pine trees growing.

"Here I saw about thirty Mountain Goats & it would surprise you to see with what fleetness they climbed the sides of the rocks which I had extremely hard work to get up." DeWolfe Diary, July 7, 1849.

20 "About twenty miles from the Chimney rock we had found a very beautiful spring of excellent and cold water; but it was in such a deep ravine, and so small, that the animals could not profit by it, and we therefore halted only a few minutes, and found a resting place ten miles further on." Frémont, *Report,* 38. Solomon A. Gorgas agreed that the cold spring near the North Platte was an excellent resting place. "Camped at about half-past 10 'clock three miles west of a fine living stream gushing from the hill side to the right of the road. Here we rest for the night until tomorrow morning." Gorgas Diary, May 23, 1850. Almost everybody noted this spring. Joseph Sedgely confirmed Perkins' description. On June 25, 1849, his party broke camp and drove five miles into a hollow where "We found a spring of pure water and took lunch." *Overland to California,* 29.

21 "At this place [near Scott's Bluff] a Mr. Roubidoux has established a trading post—to trade with the Indians & to take advantage of the vast migration which invariably passes this point a number of mechanical shops (or tends rather) are here in operation during the season of the emigration." Buffum Diary, May 25, 1850.

"There is a temporary blacksmith's shop here, established for the benefit of the emigrants, but especially for that of the owner, who lives in an Indian lodge, and had erected a log shanty by the roadside, in one end of which was the blacksmith's forge, and in the other the grog-shop and sort of grocery. The stock of this establishment consisted principally of such articles as the owner had purchased from the emigrants at great sacrifice and sold to others at as great a profit." Stansbury, *Explorations*, 52.

22 Travelers grew weary of looking at the greenish-gray sage brush. R. C. Shaw, an unusually perceptive traveler, described sage less negatively: "Wild sage (*artemisia*) is a small shrub from one to six feet high. It is found from the British possessions on the north to Mexico on the south. It delights in dry, sandy plains and gravelly hillsides, but is shy of river bottoms and rich soils. In color it resembles the common garden sage, and exhales an agreeable odor. It furnished travelers with the means of cooking when no other fuel could be had. It also relieved the desert country of much of its monotony." Shaw, *Across the Plains in Forty-Nine*, 70.

23 These mountains are now known as the Laramie Mountains; the name "Black Hills" was evidently derived from the dark evergreens growing on their slopes. "The plain between Scott's bluffs and Chimney rock was almost entirely covered with drift wood, consisting principally of cedar, which, we were informed, had been supplied from the Black hills, in a flood five or six years since." Frémont, *Report*, 38-39.

24 "Today we have seen large quantities of every description of articles, which have been abandoned by the gold seekers. The road and plain is strewn with them, hats, coats, boots, broken wagons, old iron, powder, balls, lead, boxes, trunks, valises, salt, bacon, flour, everything in fact, which could give a team relief; a wasteful but necessary step." Gray Diary, June 11, 1849.

25 Perkins had before him an ascent of 1,773 feet in thirty-one miles crossing over the Laramie Mountains. Federal Writers' Project, *Oregon Trail*, 84. William E. Taylor wrote in his diary: "Here the Road Leaves the River and takes through the Black Hills

Below the spring in the ravine a Frenchman has established himself in a little cabin as a jack at all trades, Blacksmith, Shoemaker, Tailor &c, & keeps a little stock of groceries, hardware, &c. Some of his prices were amusing enough to us though not to the unfortunate traveler whose misfortunes compel him to refit here. Sugar 40 cents per lb, Flour & corn meal 8 cents per lb. Shoeing mules 2.00 per shoe, tin lantern 1.25. Yard muslin 40 cents &c &c.21 Camp at 6. Distance 20 miles.

Friday June 29. Start 1/2 p. 5. Morn clear Wind E. Completed the ascent of Scott's Bluff at 1/2 past 6, & were greeted on our arrival at the top with one of the most magnificent views imaginable. Far down below us was the plain with the River winding through it & hundred miles & miles away by a range of mountains beyond which rose the peaks of the Rocky Mts. & above all 150 miles distant was Laramies peak towering far above all. Behind us was a distant view of Chimney Rock. Court House Rock with their beautiful scenery & beyond the vision extended to indistinctness hills & plains being mixed up in one confused outline. We estimated that we could see in almost any direction 100 miles. How some of my romantic lady acquaintences would have gone into ecstasy over this prospect & much did I wish as I gazed at it that some of them could be with me to participate, without undergoing the hardships incident to the journey.

We stopped our wagons some time to take our fill of the picture spread out before us & then reluctantly commenced the descent to the plains. The distance down was nearly equal to that up. Then level road till half past six when we camped. 20 miles.

Passed by the side of the road today the skull & horns of a rocky mountain sheep. The horns were enormous, curling round spirally they must have been 2 1/2 or 3 feet long & at the roots the size of a mans arm. Either would weigh more than two such skulls & I wondered whether the animal did not tire of carrying such a load. We are in hopes to see a live specimen though it's doubtful whether we do unless we camp & go up into the Mts to hunt them as they are very shy.

Saturday June 30. Start 6. Morn clear day hot. Wind S. This morning early a half breed Frenchman trotted into camp with a little cart made of some old wagon wheels & the sides of raw hide very light. He belongs to the establishment at the Cold Spring on the mountains & was going to Fort Laramie for a stock of goods. He spoke broken English & appeared to know nothing but to shoot his rifle, drive his horse & make sharp bargains, still he was a very clever little fellow, active as a cat & helped us well out of a slough we had to cross, wading by the side of the mules & with his sharp sudden yells scaring the animals into harder pulling than all our whips could have done.

Immense quantities of the Cactus scattered over the Barren sand hills today & not much else but wild sage which also grows in abundance.[22] This country looks desolate enough everything looks dry & parched up. Our half breed travelled with us this forenoon then whipped up saying he must be at the Ft. tonight 25 miles, a pretty good drive. Camped at six. Distance 18 M.

Sunday July 1. Concluded to drive to the Fort this morning & camp started at 1/2 p. 5, roads good, some sand & hills. Nearing the Fort we began to see the preparations made by the over-loaded teams for crossing the Black Hills.[23] Piles of bacon lying by the road side some must have had 2000 or more lbs, more lbs beans, corn, all kinds of iron implements & tools stores, &c &c, one train we passed told us they threw away over a ton of bacon several barrels of bread, six dozen steel shovels, axes hose &c &c, amounting in value to nearly 1500 dollars![24] Their crackers we determined to have some of so Chesebro & myself took two animals with some empty sacks & went to their camping ground & got about 90 lbs of the finest crackers I have seen & are far superior to our own.

We arrived at Fort Laramie at 9, & crossing Laramie Creek which was over three feet deep, & where we got our wagon body wet, tho we had taken the precaution to pile our goods inside up on boxes &c &c so that they were not damaged, we camped on its bank. Thousands of animals of all descriptions were scattered over the plain & not a spear of grass was to be found.

We have determined to leave our wagon here divide our stock into companies of two & each take care of itself & shall stay two or three days to carry out this resolution. For some time we have been talking of & making up our minds to this we find our wagon so heavy & so much load that we dare not try the Black Hills as we are and though we cannot see anything to throw away, as a company yet when divided the whole weight can be much reduced.[25]

Most of us will probably pack, it will also we anticipate be much less work & trouble to cook for ourselves only and take care of our own goods than to do the same for the company

this part of the Road is not verry good tho when it is compaired with some other parts of the Road it is excellent." Morgan (ed.), *Overland in 1846,* 132-33. Dominating the mountains is Laramie Peak with an elevation of 10,272 feet. Dr. E. A. Tompkins recorded sighting this landmark in his journal: "As we continued on our weary way (having rested one day at Scott's Bluffs) we soon came in sight of Laramie Peak, one of the very highest prominences of the spurs of the Rocky Mountains which projects eastward. This point can be seen full 150 miles and perhaps much farther, it is a little over 11,000 feet above the level of the sea." Quoted in Coy, *The Great Trek,* 139.

By the time the emigrants reached Fort Laramie they had to decide whether to attempt the steep grades ahead with wagons or to use pack mules and go on more lightly equipped. Annoyance with personal conduct and mannerisms, which was exacerbated by the harsh conditions of travel, also contributed to the breakup of parties, although apparently this was not a cause of the decision by Perkins and his companions to split their forces. Mrs. Haun in her diary recorded an annoying mannerism of a member of her company. "One of our ox drivers was a great hand to whistle and in these weird surroundings I believe that he never drew breath that was not a whistle. This annoyed some of the more nervous temperaments and afterwards when we were safe in a less dangerous locality and the terrors of these horrors was forgotten in the anxiety of new and present difficulties, much fun was poked at him, charging him with having whistled to keep his courage." Haun Diary, 25-26.

Joseph Sedgely arrived at Fort Laramie on July 1. The thing that struck him the most was the great variety of articles which the emigrants were obliged to leave behind. His own party had to throw away 500 pounds of lading. *Overland to California,* 33. Impressed by the problems of overloading, the emigrants tended to travel too lightly the next year. "There is one thing Sertain that the People ware most too mutch in lightened this year they did not carry a nuff flower for there appitites we bot 340 lbs 96 lbs dry bread we weighed it out as we cooked it and found that we would be rather Scarce and de termed to buy some the first opertunity consequently 4 miles above fort Laramie we met with an opertunity of doing so from John Roberts and Fulkerson they ware leaveing there wagons and prepareing to pack on there mules we got 100 lbs at 12 cts per pound." Marshall (ed.), "Road to California," 253-54.

besides being more expeditious. I was anxious to get through in the least possible time without regard to comfort.

Tomorrow we shall talk over & arrange our affairs for a fresh start. Doct R. & company will leave us here tomorrow though they intend camping two or three days to refresh their stock & we are in hopes to see them again. I should regret not doing so as they are fine fellows & I shall always remember them with pleasure. Distance today 9 miles.

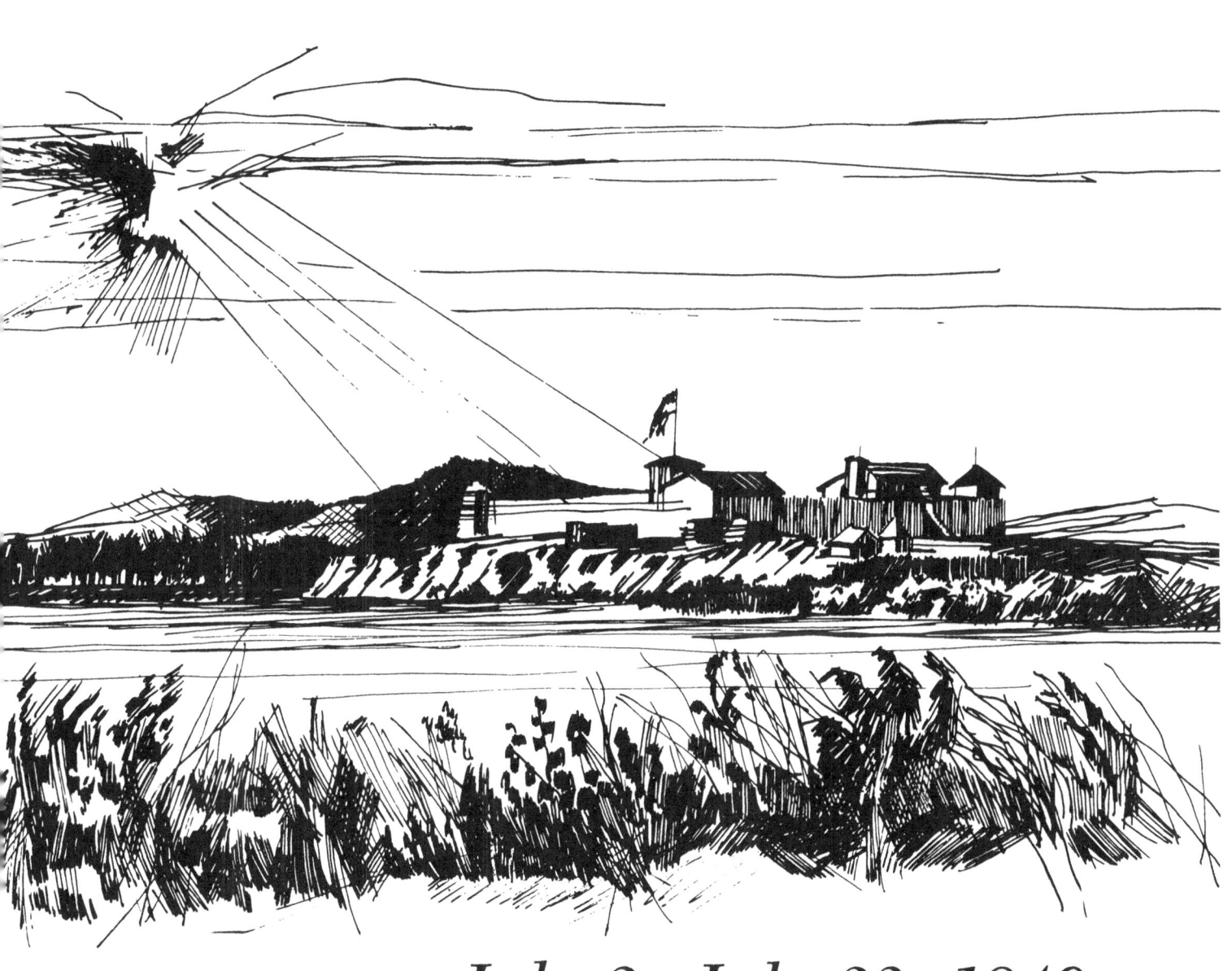

July 2–July 22, 1849

MONDAY JULY 2. Morn clear, night cool. Wind W. We called ourselves together pretty early this morning to transact business & have accomplished a good deal for one day. All being agreed, we proceeded to divide our mules. Three fell to the share of each pair, Stephens & Chesebro travel together, Cross & Cunningham & I take Johnny Huntington under my care. Our mules disposed of we proceeded to weigh out our provisions &c equally & other things were then portioned out as each could agree with the others.

The wagon we are offered $5 for! Stephens swears before he will sell it so, he will make a bonfire of it & run away by its light. I think tho I shall put it to another use. Cunningham & Cross went out to another camp this morning after the division bought a old cart for which they paid $20. I have seen a number of these French conveyances about here & they drive to so easily & are so light that I made my mind up to convert our wagon into one as I can travel much easier & more comfortably & with packs & I am told here almost as expeditiously. This afternoon I went to work at it. Took the bed off the wheels separated the hind wheels from the fore & have engaged a French wagon maker to come tomorrow morning & to put some shafts to them. Meantime I have been some 2 miles, cut a couple of cotton wood saplings some 12 feet long & five inches through at the butt, which I fastened one on each side of my mule, after the manner of cart shafts, towed them to the river then tied them together & with about 10 feet of line fastened them to the pommel of my mountain saddle. I plunged in the water up to my horses back & the logs pulling downstream in the swift current got them safely over however, fastened up again & finally deposited them by the side of my wheels ready for my carpenter.

I am rather fatigued with my days work, but shall attack the wagon body tomorrow morning & hope to get off by evening. Stephens & Chesebro have been making pack saddles all day & will be ready tomorrow also. We have need to be expeditious for our animals can hardly pick enough here to satisfy their hunger much less to recruit which they much need. Doct R. & Adsit left us today at noon promising to wait two or three days at the first good grass for us.

I went up to Fort Laramie about one mile above us today to try to get some canteens &c. It is a rectangular establishment with projecting towers at each corner. I forget the military name for them built of the stiff mud found in abundance here which baked in the sun becomes quite hard & solid. The clay or mud is cut in cakes something the shape of a brick about 1 foot long by 8 inches square laid one on another & they soon adhere quite strongly sufficiently as to be Indian proof tho a slim protection against an 18 pounder. Their walls are some 12 or 15 feet high with one large gate of heavy plank if I remember rightly pointing south. Altogether tis quite an establishment & has been a great headquarters for all the Indian trading of this region.[1]

Last Tuesday it was sold to the U S. & the troops here under command of Maj. Sanderson took possession with great parade, guns, firing &c. The Am. Fur Company's men moving out & the troops moved in. Where the Fur Company will make their headquarters next is uncertain, an express which we met has been sent over to the states for orders. Meantime Mr. Hosmer the agent with his company of trappers traders &c is encamped about 4 miles from us on the river.

About a mile & 1/2 below Fort L. is the ruins of old Fort John built on a larger place than Fort Laramie & must have been a better establishment but deserted on account of its unhealthiness so I was informed, now used for herding cattle.[2] It stands on a low plain while the other is on a higher ground near Laramie Creek. A fine plain lies between the two in the fork where Laramie Creek runs into the Platte & ordinarily it must afford the richest grazing but now is gnawed down to the very ground.

In front of us are the Black Hills, black enough they look too, being covered with a stunted growth of cedar & pine. Laramie Creek is here quite a stream being I should say at least 100 feet wide & from three to 5 deep. The road divides here, one following the Platte while the other takes the Hills. The latter is said to be the shorter & as Doct Riggs intended taking it we shall follow suit.

At Fort L. I made the acquaintance of the junior surgeon a young man educated at Yale, & a very pleasant fellow. I recognized him as an Odd Fellow before being with him half an hour & we were of course at once intimate. As I had been unable to procure a canteen he undertook to get one for me and after some difficulty made me a present of one, which he coaxed out of a soldier, a gift then which nothing could have been more acceptable.

This evening we took our last supper together as a company & had of course the best that the ingenuity of its members could

[1] The junction of the Laramie and Platte rivers was known to trappers and traders almost from the beginning of the Rocky Mountain fur trade. In May, 1834, William Sublette, founded a post named Fort William. The fort was sold in 1835 and later passed into the hands of the Chouteaus. Fort Laramie was a major trading post, Indian center, and way station for all sorts of people traveling up the Platte. Hafen and Young, *Fort Laramie,* 17-38. Fort Laramie was purchased for the government in June, 1849, by Lieutenant D. P. Woodbury. An account of his survey and recommendations is to be found in Willman, "History of Fort Kearney," 259-68. Fort Laramie became a military post on June 26, 1849, when the post was transferred by the fur company to Major Winslow F. Sanderson, who reached the fort with Company E of Mounted Riflemen. This company consisted of 5 officers and 58 men. Hafen and Young, *Fort Laramie,* 138-56. Fort William in 1841 had been reconstructed of adobe and named Fort John, evidently in honor of John Sarpey. *Ibid.,* 68-70.

[2] Perkins referred to Fort Platte, begun in 1841 and abandoned in 1845. His is one of the best descriptions of the terrain at this time.

invent & few suppers at home tasted better. Thanks to my Mother H.[3] for her cooking directions, for many a nice dish have we had from them.

Tuesday July 3. Morning sultry, no breeze. My Frenchman came as per agreement, & looked at my sticks & said they would do, & went to work striking a blow once in a while & stopping to look then sitting down to rest & poking along in such a way that I was soon satisfied he would be only a drag on my operations. However in the course of the forenoon he actually got all the bark off them, cut one end square & laid it on the axletree! I had agreed with him at six dollars for the job which at home would have cost about 1 & I told him at noon he might clear out & I would do the rest myself. He wanted his price though, complained of such hard work. "He neviare vill work so much before in de hot sun by gar!" & I gave him his money to get rid of him.

Meantime I had sawed off the body of the wagon about six 1/2 feet long cut it down, to 1 foot deep, taken off the irons & nuts & refitted them & had a neat strong & light cart body. The bows of the wagon were then fitted on with wrought nails, cover cut off ends sewed in & put on & in the afternoon I nailed the body to the axle & she was done. For binding a couple of irons to put on the shafts I paid $2.00, a 25 cent job at home or hardly that. The imposition on travelers here who are compelled to make any repairs is outrageous. Knowing they cannot help themselves the blacksmiths & carpenters take advantage of their necessity and rob the unfortunate traveler of sometimes all he has.

About 4 oclock, I loaded up & harnessed in my mules tandem, leading one for the saddle & we started, Cross & Cunningham leading off I following. Grand as a lord. At the Fort we stopped an hour & there I wrote a letter, in the quarters of my young friend the surgeon to my wife, the last we are able to send till I get through. At six we started again & travelled till 10. The night was beautiful & a bright moon lit us on our way. Stephens & Chesebro left at 2 P.M., promising to pick out a good camping ground & wait for us, we saw nothing of them though, & at 10 camped by ourselves on a plain near the hills. Distance 10 miles.

Wednesday July 4. Hurrah for fourth of July! Here we are in the Black Hills far from home & little glorification have we

3 Mrs. Samuel P. Hildreth, Perkins' mother-in-law.

4 "Our course today was over the Black Hills so called a/c of their somber appearance. They consist of a range of hills, bluffs, running parallel to the R. mountains and right in the front of them as we approach from Fort Laramie they are mostly stony Barron hills on which grow a few scattering

had this day.⁴ Though often have we thought of what was probably going on in dear old Marietta.

The sun rose fiery enough this morning, & day has been excessively hot, no breeze, started at 5 oclock, travelled on finally, at 1/2 past six came upon Stephens & Chesebro still camped, went by them some mile or more when in descending a short hill, snap went one of the shafts of Cross' cart. Here was trouble number one our tools were got out & with some pieces of boards I spliced it together again driving pins through & wrapping with rope till it was as strong as before. While we were at work our two packers came along & stopped & asked if we wanted assistance, or whether they could render us any. We let them go without moving & tho we thought how lucky twould have looked to offer their aid. 3 quarters or 1 hour saw us off once more. Travelled on till about 11 oclock, without water & almost choked. The roads & hills were nothing but sand & wild sage & the reflection of the suns rays nearly blinded us.

At 11 we came to a steep descent & Cross being a little in advance went over first, & as I came to the top I saw a sight about half way down for fourth of July! There was the old cart bottom side up all the goods & chattels spilled out & Cunningham underneath the whole! In descending the hill both shafts had again broken & the body swung under the axle capsizing the cargo animate and inanimate as I have described.⁵

Cunningham crawled out from under his sugar & flour & somewhat discomposed, but unhurt, & we began to deliberate what was best to be done & had about concluded that they must pack, & I was making up some bundles in suitable form when an ox team appeared on top of the hill which proved to belong to an old acquaintance of Cunninghams, Sargent by name from Parkersburg, Va.

Our first inquiry was for water for we were almost done over with thirst, & as I held the canteen to my lips & felt a relief I never before experienced because I never before suffered for want of water. We had been without water since last evening using up all our canteens held at our camp & when Sargent came twas 1 oclock. In this desert country a man needs drinks of some kind every hour of the day.

Sargent after looking at our difficulty said we must not give it up so as he had some pieces of iron &c in his wagon to mend with & he never left any man in trouble when he could help it. So out with our tools again & with Sargents assistance, iron &c

a distance but when you are among them serve to render them very beautiful on a/c of the novelty of seeing such trees in this region. We sometimes travel in the gorges between the hills & sometimes mount to the summit when the prospect would be enchanting. Right before us is Laramie Peak, one of the highest of the Rocky Mts. most always in sight whether you are in the valley or on the hill top. I cannot describe the Mts. they are so lofty, dark, rugged, dismal and hideous that they remind me of nature in Chaos." Steuben Diary, June 15, 1849.

"Hurra for freedom's jubilee / God bless our happy land. Independence Day, Oh, what an Independence. This morn we fired a salute at sunrise which was indicated (We being in the valley) by the reflection of its hundred rays from the Eternal snow peaks that surround us. The morn was clear & beautiful but the temperature 20. Cold almost beyond endurance. We traveled today 20 miles & camped on Little Sandy River. Its banks are full owing to the melting snow in the mountains." *Ibid.*, July 4, 1849. "The 4th of July. After dinner that night it was proposed that we celebrate the day and we all heartily join in. America West was the Goddess of Liberty, Charley Wheeler was orator and Ralph Cushing acted as master of ceremonies. We sang patriotic songs, repeated what little we could of the Declaration of Independence, fired off a gun or two, and gave three cheers for the United States and California Territory in particular!

"To the rollicking music of violin and Jew's harp we danced until midnight.

"There were Indian spectators, all bewildered by the weird (to them) war dance of the Pale Face and possibly they deemed it advisable to sharpen up their arrow heads." Haun Diary, 31.

⁵ Travelers along the trail to California were constantly threatened with breakdowns. One of the most frequent accidents reported was the breaking of wagon tongues. James A. Pritchard broke one of the hounds of his wagon during a night march on July 29. "I lashed it with ropes and it lasted me across to the river." He made a new hound out of an oak coupling pole. Pritchard, *Diary*, 127.

we patched the miserable shafts & had the cart loaded by 4. S. then invited us to his wagon where one of his men had been cooking & treated us to some coffee biscuit ham &c, & we all then started on together. At 5 came to a spring where we filled our canteens & at 6 camped on creek. Distance 15 miles.

Evening is most glorious. Seems to me I never saw the moon so brilliant or such a clear air. What a way to spend Independence day. I think for novelty our celebration will take the lead of any in the states.

If we go on as we have commenced our carts will not be very expeditious modes of conveyance. Mine does well enough but I find they are very hard on the shaft mule jerking him about as the wheels strike stones or ruts & going down hill the steeper it is the more load is thrown on his back. Though they are very comfortable & we can ride & take our ease yet I am inclined to think I shall pack yet. I must get ahead as fast as any one on the road or I shall not be content.[6]

Thursday July 5. Morn fine, wind South, start 5. Came to a clear creek about 10 oclock & a little beyond steep hill, on top of which we nooned. The roads are becoming very hilly up and down all day & in some places very steep. We have commenced the ascent of the black hills & may look out for rough travel till we get to the upper Platte ferry some 80 miles.

At our noon camp we had a grand thunder storm with heavy hail, which fell so fast as to whiten the ground like snow. We were up very high & could see the hills & mountains for a great distance & the view of the storm among them was fine & exhilirating. I thought of Byrons fine description of a storm among the Alps where "from Peak to Peak leaped the live thunder, and Jura answered from her misty shroud. Back to her sister Alps who call to her aloud." Calling Laramie's Peak which towered up not far south of us "Jura" & we had Byrons scene on a small scale.[7] The lightning was most brilliant & the thunder terrific.

The rain lasted some hour or more & at 3 we started, travelled up & down to Heber springs 2 miles beyond which rose a hill which our wagons never could have ascended in the world without double teams, & which made our little carts a heavy drag.[8] We kept going up, up, up till we turned the top on a level ridge road, & then another of those grand views where the mountain Peaks can be seen one behind the other till all becomes indistinct by distance. Laramie's Peak was south of us & so near that we could distinguish trees on its sides tho

[6] Although the Marietta party had from the start shown some sense of responsibility for the welfare of their animals, Perkins became more concerned as he started up the eastern slopes of the Rocky Mountains. Here he seems to have realized that his salvation depended largely on his mules and ponies. If they lost shoes or became lame or otherwise disabled, he and his companions faced hardships that could easily prove fatal.

All along the trail emigrants thought of their mules in terms of personalities. In many entries their draft or pack animals are called by name, and they told stories of dead and dying animals that were almost as frightening to fellow emigrants as word of cholera among the emigrants themselves.

[7] "Childe Harold's Pilgrimage," Canto III, verses lxxxvi to xcviii.

[8] Ware told in his *Guide* (p. 21), of "Heber Spring, near the timber on the right side of the road. There is a creek a few rods north from this spring. 2 1/2 miles' travel brings you to a long steep bluff; you may have to double teams to get over this place."

they looked like little shrubbery. I could not help feeling as if twas too bad that the magnificent views should be placed out here in such an inaccessible place, though the difficulty in reaching such a point of view is of course a consequence of its very grandeur. Camped on top of this hill at 8 oclock after a hard day drive of 25 m.

Friday July 6. Start 7. Morn clear cool. Wind East. At 12 after some very rough roads struck the La Bonta River & camped, a little creek some 20 feet wide very good water, but no grass, as usual entirely eaten down.[9] We begin to find a scarcity of provender for our mules & shall have to go up some of these streams 4 or 5 miles & let them pick up. Anywhere near the road every green thing is eaten off & the animals begin to fall away already.

Started at 2 camped at 4 near spring at the foot of a remarkable hill being composed entirely of stones of all sizes piled up into a ridge some 200 feet high. Distance today 15 m.

Saturday July 7. Start 6. Morning clear warm. Breeze W. Passed two large trains & drove on with my cart leaving Sargent & Cross behind some miles. I stopped behind to talk with the Capt of an ox train about 12, letting Johnny H. drive on. The last I saw of him he went over a hill down to La Prele River,[10] & when I came up there lay my little "Express" as the ox teams named her, flat on her side in the water & goods of all descriptions jumbled together inside. It seems Johnny got some man to lead the forward mule on the creek who was afraid of wetting his feet let go at the edge & the mule shied off & ran one wheel upon a bank & over she went.

Some half dozen men from the trains behind took hold with us & we lifted it on its legs again & drawing it aside to make room commenced unloading, & found nothing damaged except a few bottles from the medicine chest, among them my collodin which I regretted much. Once more we loaded up, & started after eating our lunch at 3, & before Cunningham & Cross had come up.

Truly we seem to have been unlucky since leaving the Fort, but all these accidents do not discourage me in the least. They are soon repaired & when over we feel better than ever again. One thing though does cause me some anxiety, "Old Dave" my large shaft mule shows evident signs of being tender footed & unless I can get him shod somewhere soon this may occasion me serious trouble & delay. He was neglected in the wagon til

[9] On June 18 Wood's party had made remarkable time in reaching Heber Springs and had crossed La Bonte River. Wood described this as very rough country. Joseph Warren Wood, Diary of Overland Trip, I, June 18, 1849, hereinafter cited as Wood Diary. La Bonte River was described by Ware as "a fine camping place, grass, timber and water, plenty; the river is about 30 feet wide. A branch of the La Bonte, crosses 5 miles onward; the banks are steep. In the next 13 miles the chances for camping are poor; the road is but tolerably good. The curious may here look out for toads with horns." *Emigrants' Guide,* 21. "Crossed the LaBonte some distance beyond found a little water and camped among the Wormwood. It seems very disagreeable." Buffum Diary, June 14, 1849, p. 57. "We camp on LeBone [*sic*] River, grass poor, water good. Our route still among the Black Hills, one hour we are mounting to the clouds & the next diving to the depths below. The roads are good. The hills over which the road passes are composed of gravel with a slight mixture of clay which the powerful rains help to form into cement." Steuben Diary, June 16, 1849.

[10] The name of this stream caused the emigrants a great deal of trouble. They could not understand its correct spelling from the slurred pronunciations heard along the trail. This was the La Prele River, a swift, narrow stream forty-five miles east of the Upper Platte ferry. Ware, *Emigrants' Guide,* 21. "We are camped on the A. La Pule river, a small stream lined with short timber." Wood Diary, June 19, 1849.

his hoofs were badly knocked to pieces while some of the other greater favorites of our drivers were carefully shod. He is a noble old fellow, strong as an Elephant & I must take all possible care of him.

At 6 I stopped on top of a hill to wait for the other cart to come up, not seeing anything of it I concluded to spend the night. Distance 18 m. A large ox train camped some quarter of a mile off, & I was invited down to supper with them, where I had a treat of splendid coffee with *milk,* hot biscuit & butter &c. Find this evening that my revolver is gone, the thread which fastened its case to my belt appears to have worn out & I suppose some one will pick it up & I hope it may do him more good than it ever did me. Evening cool. Strong west wind.

Sunday July 8. As I was encamped where there was no water, must go on till some is found so I harnessed up & waited till 7 for my companions when they came up & we all started together. Warm clear hot. Strong west wind. Travelled till 10 to Deer Creek[11] on which we camped. Distance 8 miles. Roads heavy.

Monday July 9. Start 6. Morn clear & warm, strong west wind which raised such a dust in our faces that we could neither see nor breathe, & so were actually obliged to halt on the Platte at 9 having travelled 8 m.

I was determined to stop somewhere soon to recruit my mules & Cross & company would do the same. Hearing that grass was very fine some 5 or six miles up Deer Creek we concluded to stay where we were two or three days & send our animals over the hills some 4 miles to the fine grazing. So this afternoon Cross & company took all the mules but one which John H. rode after Harlan & went over leaving me alone with the carts. We shall gain time by this operation as for the next 150 miles provender will be scarce. Near me is encamped a large Missouri train who have sent all their cattle over to the same place, & a blacksmith belonging to this mess has promised to shoe my mules for me, so that I shall start fresh.

I have determined to give Cross & Cunningham my cart in place of their old crazy thing & pack. I shall travel much faster & as to comfort I dont regard it at all when compared to speed, having packs with us I can soon fit out & shall be ready to start Thursday morning. Have been regulating some of my clothing &c today & find it needs it. This is the first leisure time I have

[11] Deer Creek was truly a landmark. Scarcely a traveler crossed it without making a note. On July 26, 1842, John C. Frémont's party had reached the mouth of Deer Creek, finding it "twenty feet broad, and well timbered with cottonwood," with good grass, "the largest tributary of the Platte, between the mouth of the Sweet Water and the Laramie." *Report,* 51. James Pritchard noted, on June 9, 1849, "In 4 miles this morning we struck the river. Here we found good grass & plenty of timber. In 5 miles we struck Deer Creek—a fine camping spot. The Stream is large & handsom & said to contain an abundance of Fish. The road has been fine today, by noon we had travelled 15 ms. Here we found in the bend of the river a splendid bottom of blue grass. It was from 12 to 15 inches high and heavily seeded. It resembles very much the blue grass of Ky— it is somewhat courser with very heavy heads, and of a more nutritious [character] than any grass I ever saw in any country." Pritchard, *Diary,* 87.

had since I left home, & I have plenty to do for myself. The dust & wind for some days back have been more annoying than anything I have experienced. We are almost suffocated, & our poor mules when on the road could hardly be prevented from turning their backs to the storm of sand which drove into their nostrils & eyes.

Have been regaling today on a piece of Buffaloe meat given me by my Missouri neighbors, which they killed last evening. Buffaloes are plenty in these hills though I have not yet seen any. The Buffaloe gnats are so annoying that I can hardly write getting into my eyes nose mouth & ears in swarms & when they light biting almost as bad as a mosquito. One of our mules was brot up a few mornings since so badly bitten by them as to be swelled in places large as my two fists.[12] An application of the oil which we brot along at the recommendation of my Father-in-law soon chased the torments & the swelling went down.

Tuesday July 10. Morning clear & hot, strong west wind. Was regaled in my solitary bed last night by a most melodious concert of wolves & owls. I should suppose there were 50 of the former rascals come down to the bank of the River & set up their long dismal howls & sharp whines. While right on the tree under which I lay were some owls putting in their "hoot, hoot, hoot, to hoo" by way of variety. However unmelodious this combination of sounds, there was a monotony about it which soon put me to sleep & I left them to serenade at their leisure.

I saw this morning a drove of Buffaloe on the hillside above us & started some of the Missourians after them. No success though. At 11 Cunningham came over the Bluff from Deer Creek after me & I went back to see how our animals were doing. On my way I saw a large elk standing on a ridge looking at me but I had no arms but my Bowie Knife & could not molest him. He appeared at 300 or 400 yards as large as a horse but had very small antlers. I think the Elk like most animals of the Deer kind sheds his horns every spring.

A ride of some 5 miles over bluffs & ravines brought me to the camp on the creek bottom. Found our boys in a beautiful shady grove in company with our old friend Sargent & the Missouri train & having nice times. Last evening a fiddler was found & there being a number of girls in company with our Missouri friends they "tripped the light fantastic toe" till 11. Sam C. acting as usual as "caller off." The mules were luxuriating in

[12] Diarists marked almost precisely the spots where certain animals and insects would occur. Long before Perkins reached the Rocky Mountain valleys, Edwin Bryant had had trouble there with mosquitoes and gnats, and with wolves. "Our mules as well as ourselves suffer much from the myriads of buffalo gnats and mosquitoes, which take up their [abode] near all the water courses and fertile spots. The evening is beautiful. The howling of the wolves and the low hum of insects, are the only sounds which disturb the profound solitude." Bryant, *What I Saw in California,* 121. "The mosquitoes drove us out this morning & we yoked up before daylight & drove on." "Last night for all the wind the mosquitoes were really awful. Pollard & I could not sleep." Wood Diary, July 21, 22, 1849.

a growth of timothy, red top & wild oats two feet high. The hills & valley of Deer Creek abound in grain of all kinds, Buffaloe we have seen in abundance. Also deer, antelope, some mountain sheep.[13]

Yesterday a huge grizzly Bear was killed near our camp by some Tennesseans, weighing 800 lbs & his paw measuring 7 inches across. Two balls only were fired at him both going through his heart, & dropping him almost instantly I was told by some old hunters & mountaineers of whom there are several in this vicinity that the stories of the ferocity & tenacity of life of the Grizzley Bear are mostly fabulous.[14] Though it is not safe for one man to attack one alone as when fired at they start immediately for the place where the gun cracked, yet one ball rightly "put in" will kill them as easily as a Buffaloe, & they never make fight with man unless wounded, then they are formidable antagonists.

Some two hundred yards from our camp in the creek opposite is a Beaver dam which I shall visit in the morning. Evening warm & pleasant with fine moon. I saw a splendid meteor slowly falling towards the horizon & resembling a huge ball of fire.

Wednesday July 11. Morn clear & warm, breeze East. Wind has changed at last & the disagreeable west wind, raising such clouds of dust has left us. I hope never to return.

This morning early one of the men brot into camp a large Beaver which he Killed at the dam. I never had seen one before & was much gratified by this opportunity for a close examination of this singular & sagacious animal. It resembles almost exactly the muskrat in the appearance of the body & head though the specimen I saw was over two feet long without his tail, & 1 foot broad across the back. The tail which seemed to serve as broad paddle, & stirring oar was some foot long five inches broad, flat hard & scaly like a fish or more like some parts of an alligators hide. Teeth like a squirrels only very large & strong set in jaws whose strength is indicated by the cuts they can make in hardwood. The whole animal seemed to be stronger in proportion to its size than anything I had ever conceived of, taking their works as a criterion.

I visited their dam & was astonished at the ingenuity & calculation exhibited in its construction. It was thrown across the stream where two projecting points in the bank formed butments & a bar in the middle a support. It was constructed of the willows with which the banks are lined, varying in size from

13 "Some of the men caught a couple of mountain trout & reported that they were excellent eating. Elk horns are abundant through here—some being of great size. I also saw a Mountain Goat's head. His horns were 2 1/2 ft. long & the shell was in front 1 1/2 in thick & was very heavy." Wood Diary, June 20, 1849.

14 Grizzly bears appeared along the trail from the Laramie Mountains in present-day Wyoming to the Sacramento Valley of California. The grizzly was the largest of the American carnivores, a frightening and wondrous animal. Some stood four feet in height on all fours, were eight feet in length, and weighed 2,000 pounds. It is little wonder that Perkins was excited at seeing one. The Tennesseans were fortunate that they were able to kill a grizzly with so little experience in dealing with the breed. There is a good history of this bear in Storer and Tevis, *The California Grizzly* and a detailed description of the animal in the foreword of Bancroft's *History of California*, II, 560-61.

1 to 3 inches in diameter all cut off clean & smooth & as if by the stroke of a hatchet. Three cuts of their teeth suffice to fell a bush 1 1/2 inches in thickness as indicated by the slight splits in the end where each cut had been made, & the incisions were made diagonally downward as a man would strike with an axe. These bushes were laid with great regularity parallel with the course of the stream, the bushy ends upward in layers & between each layer were heavier sticks placed crosswise thus binding the whole & forming a very strong & durable work. Above this dam they had thrown in moss & leaves which catching the sand had completely stopped the water & raised it four feet. The dam was some 40 feet long & altogether a better constructed & more durable piece of work than our state job at Marietta.

Wishing to see them work I labored hard for more than half an hour making a small breach, & sat down under concealment of the willows patiently watching for an hour or more, but probably the shooting one of their number in the morning had entirely depressed their energies for that day, & I came away disappointed. Inumerable fish of all kinds & sizes were lying on the dam & in its vicinity but whether killed by the Beaver or caught by the network of brush, I could not determine. Some of them were bitten through the back while others showed no marks of violence. I do not wonder that the superstitious Indians ascribe a soul to the beaver. The evidences in his favor are sufficient to almost over set the theory of instinct alone.[15]

When we left the frontier we were told great stories about the selfishness & want of feeling among the Emigrants that the hardships and uncertainties of the journey had soured what "milk of human kindness" they might have possessed. I wish to bear my testimony against this slander. Never have I seen so much hospitality & good feeling anywhere exhibited as since I have been on this route. Let any stranger visit a camp no matter who or where, & the best of everything is brought out, he is fed, & caressed almost universally. If at meal time the best pieces are put on his plate & if the train has any luxuries they are placed before him. Nor have I seen any man in trouble, deserted, without all the assistance they could render. There are of course individual exceptions to all this, & such men are known to almost every train following. One fellow left an old man on the road without money or provisions. He was picked up & brot along by the next train, & I have not overtaken or fallen in with any companies yet but in course of conversation had something to say about this affair. "They would like to catch the fellow out alone." Some "would give him five hun-

15 Beaver, of course, had attracted the attention of trailbreakers into the Rocky Mountains long before the 'Forty-niners started through the passes. R. C. Shaw may have overstated the case when he recalled, "Beavers along Little Sandy were quite numerous, and wherever there were trees near the banks we found traces of their work. We saw trees one foot through which had been cut down with their teeth." Shaw, *Across the Plains in 'Forty-Nine*, 83.

dred lashes." "We damn him Id go in for hanging him up to the first tree" &c.[16]

I had a great treat last evening at the camp of the Missourians, hot coffee with plenty of milk, warm biscuits & butter. Buffaloe tender loin &c, this being one among the many civilities I have received from different trains. Started at 12 to go back to camp on the River, just at having a huge Buffaloe come galloping by & the mounted men gave chase, with what success I do not know.

I walked alone over the bluffs 5 miles under the hottest sun I have seen for many a day & arrived in camp in an hour & a half pretty well heated & tired. Having determined to pack from this place. I have spent this afternoon in making pack saddles arranging provisions &c. I shall take only 90 lbs each of bread & meat, being allowance for 60 days though the mountaineers tell me I might go through in 40.[17] Once started I shall push on with all possible expedition. My cart I turn over to Cross & Co as being far superior to their own.

This evening very cool, North Wind began to blow about two hours since & I almost apprehend first snow tonight. Such things have been here at this season of the year.

Thursday July 12. Morn quite cool & cloudy. Some appearance of rain though most too cold. Some Buffaloe came down the Bluffs to the River this morning & the boys of the Missouri camp with Cross & Harlan lay in wait for them on the bank & succeeded in killing two. We however received no benefit from this exploit for the Buffaloes succeeded in crossing the River before falling, where we could not get at them. While the Missourians who crossed with all their teams afterwards procured a fine lot of fresh meat.

Have been at work all day making up my packs & getting ready to try this new & speedy mode of travel. Of course only necessaries can be taken & a limited quantity of these.[18] I have laid out for my stores after this & which I shall pack, 80 lb crackers, 40 lb flour, 60# hams & dried beef. 20# sugar with some tea choclate pepper salt &c, two tin cups, two spoons 1 plate, 1 knife & fork which with our clothing will make our load about 260 lb. I shall pack on my larger mule "Old Dave" abt 140 & on the other "Cara" 120 lb. The above allowance is for 60 days.

If nothing occurs to delay my progress I shall be able to make the necessary distance in 40 days. One of my Buffaloe robes I have cut in two to put under the saddles as it is said this will

[16] There were plenty of veterans along the trail to do the "damning." William E. Chamberlain said in his diary: "I do not think there ever was as many men ever together or on any road so shockingly blasphemous as the emigrants on this route to California—they hardly use any expression to horse, mule or ox except 'G——d'm your soul!' or 'heart to h'll,' or to 'damnation.' I think I hear it 50 times a day. Woe for California, if such is the character of the future population—this habit extends through all ages from the boy of 12 to the old men on the border of eternity." William E. Chamberlain Diary, July 5, 1849. Perkins was irked by the constant swearing even though he himself wished to damn the heartless creatures who set an old man afloat on the trail without resources to sustain life. His personal tendency to charity was further stirred by the fact that he belonged to two mutual-help fraternities, the Masons and the Odd Fellows. Another observer wrote, "Profane language is the common dialect most frequently heard of the great majority of emigrants. The poor Indians apply the [terms] most frequently heard to us. On one occasion a party was enquiring for a good camp ground of an aborgine and was assured that there was plenty of grass ahead for the 'whoa haws,' but no water for the 'G'd d'mns.' " Hulbert, *Forty-Niners,* quoting the V. A. Williams Diary, 183.

[17] Perkins was indeed optimistic. He did not arrive at Sacramento until September 27, eighty-seven days later.

[18] "here we threw away our cart and went to packin all the clos i had on war what i had on my back we had a horse a par and one single blankit and one hundred bonds of grub my comred come in to say what will we dow I ust yell him that we wod git a lon some way or rather he sad he war afraid he was agont to dy on the plans and he usto frett a good dall about dying." Anson Clark, Reminiscences, N-12.

effectually prevent the greatest difficulty packers have to contend with, sore backs on their animals. Day has been quite hot. Wind west. Buffaloe gnats very troublesome.

Friday July 13. Last night was almost freezing cold.[19] I could hardly keep myself comfortable under the buffaloe robe cloak & blanket. The wind blew strong from north all night, & was very piercing. The difference between the temperature of the days & nights in this lattitude is at times astonishing. Yesterday for instance the thermometer would have shown 80° in the shade last night lots of persons would have considered it safest to house their choicest plants. This morning quite cool till 8 or 9 when the sun began to pour down his scorching rays & we have had another hot day.

Yesterday evening two of our company went down the River some two miles to a blacksmith connected with a company camped there to get two of our mules shod & finding pleasant company & good quarters did not return till 11 A.M. today. We intended starting this morning but on account of the above delay lay still till the heat of the day was over.

At 3 commenced preparations & at 4 left our camp & traveled till six. My packs did finely not moving or requiring tightening which considering everything was new, & I not an old hand at the business was very encouraging. Camped on Platte with tolerable grass. Evening pleasant. Distance 5 miles.

Left many more articles at our last camping ground which at home are considered indispensable. Our keg of powder using only two or three pounds per use on the journey was thrown out. Also saw, some other tools, soap, herring, fire kindling &c, &c.

Saturday July 14. Another quite cool night wind north. Arose at 4, refixed our packs somewhat & were off at 1/2 past 5. Morn cloudy with appearance of rain, cleared off at 9 & day intensely hot.

When some 5 miles from camp a large solitary Buffaloe was seen standing near the road intently observing our movements, Cross & Harlan succeeded after some 4 or 5 shots & a chase on foot over the steep sand hills in flooring him, & when I got to them were engaged in a discussion as to the best parts to be taken. We took out his hump tender loin & some steak from which we have had some choice messes. Some express themselves disappointed in Buffaloe meat, not finding it so good as they had anticipated. On the contrary I think it is "hard to

[19] Bryant noted that "The Buffalo-robes (which compose a portion of our bedding) were hoary with frost, and the grass through the whole valley was stiffened and white with the congealed moisture which had been condensed upon it during the night." Bryant, *What I Saw in California*, 119.

beat" the only objection being its dryness, rendering it unfit for broiling, though some which I laid on the coals today was very sweet & not too dry to be quite eatable. I should like no better speculation than to have some half dozen of these fine fat fellows weighing some 1200 or more in Boston market. They would be worth I think 1.00 pr lb.[20]

Strong East wind again, raising clouds of dust in P.M. arrived at ferry which had been abandoned & camp for the night. Distance 15 m. Not much feed for animals. Country sandy.

Sunday July 15. Arose at six night warm. Morn clear & hot sun. Not a breath of air. Our party being determined to cross today Huntington swam the Platte to a "dug out" or shapeless canoe lying on the other side which he brought back & we went over to examine the ferry boat & on what could be done in the way of getting her over. Found her hard aground & had about abandoned our undertaking when six mule wagons from New York & Miss.[21] came up. Fastening together all their lariats they conveyed them over & tied to the old craft & all hands laying hold succeeded after an hour's hard labor in getting her across. After all nothing could be done. She had no oars or poles & the current was so swift that the line would not bear the strain of warping over, so after all we were obliged to give up trying to ferry ourselves, & at 11 started for the Mormon ferry distant 7 miles, arrived at 1/2 past one, & were compelled to wait the ferry mans pleasure till 5.[22] Meantime I had one of my mules shod at the Mormon blacksmith shop for which I paid 1.00 pr shoe!

Some 30 or 40 Mormons were camped here, having come from Salt Lake 1 month since to make money by blacksmithing, ferrying, & selling various articles necessary to Emigrants, & well had their time been put in. They had realized so their Capt. told me, over 3500$ from the ferry, 1500$ from the shop, & I dont know how much from sales of sundries, but judging by the price of whiskey, sugar &c it must have been profitable also. The former article was 50 cents per pint & a great demand for it at that, the latter 50 cents per lb.

Were ferried over at about 6 with the mule wagons spoken of above, in very neat & expeditious manner, paying for wagons 2.50 each. Cross's cart 2.00 & my packs 1.00. Mules all swimming a short distance above. These ferry boats are all constructed of trees cut some 12 or 15 feet sharpened at each end & dug out & 5 of them lashed & bound together with poles & pins, & they answer a very good purpose.

[20] Charles Gray's impression of buffalo meat does not accord with that of Perkins, for he commented that it "looked & was quite tender for bull beef & almost all of [the men] were delighted with it, but to me it tasted like the 'chef d'ouvre' *of the devils kitchen,* the most offensive meat I ever tasted & so I found it impossible to eat." Gray Diary, June 2, 1849.

[21] Edwin Bryant reported to his old friend George D. Prentice, editor of the Louisville *Journal,* that there were two Mississippi companies in the field. One of these, the Pontotoc and California Exploring Company, was under the command of T. H. Vaught. Louisville *Daily Journal,* April 12, 1849. Perkins had seen a New York company back on the Kansas plains, which may have been the one led by McNulty.

[22] In 1847 Brigham Young detailed nine men from his pioneer party to remain at the Upper Crossing of the North Platte to operate a ferry for the Mormons and others who wished to cross. The ferry was maintained for several years thereafter during the emigration season. On June 10, 1849, James Pritchard found 175 wagons waiting to be ferried across, and, like Perkins, his party attempted to cross on their own but were frustrated by the strong current of the river. Pritchard, *Diary,* 88. For an account of the Mormon Ferry see Morgan (ed.), "The Mormon Ferry on the North Platte; the Journal of William A. Empey, May 7–August 4, 1847," *Annals of Wyoming,* XXI (1949), 111-67. For a general account of the ferries on the upper Platte see Morgan, "The Ferries of the Forty-Niners," *Annals of Wyoming,* XXXI (1959) and XXXII (1960). Morgan supplies the full context of Perkins' entry at the Green River crossing in XXXII, 195-96.

Harnessed & packed up & at 7 started went abt 2 miles & camped on Platte. Distance 9m. Day has been exceedingly hot, & no breeze to cool us at all.

Monday July 16. Morn cloudy Wind W. Start at 1/2 past 5. Road left the River & went over the bluffs, & in the morning sandy P.M., hard & good. Passed today 3 large trains, slowly crawling on their way to the Gold regions. The country thro which we have passed today has been of a very different character from any we have hitherto seen. Everywhere are evidences of volcanic action in many places rocks are heaved up bearing the marks of intense heat, being many of them fuse, & crumbled. In some valleys was a strong smell of sulphur & many holes under huge masses of rock where the blackened edges & ashes show fires to have been pouring forth. All ponds & streams of water taste strongly of Alkalies & a deposit of tolerably pure carb. potash can be picked up on the banks of any of them. The water and herbage of the district which also partakes of the same nature is very poisonous to cattle though not appearing to effect mules. We counted today 15 dead oxen by the road side. In what way it affects them I cannot say. Most however appear much bloated.[23] Day has been cool & pleasant. Cloudy with a slight rain P.M. from east with much wind.

Camped at 1/2 past 7 on a bottom with luxuriant grass growing however out of almost pure salaratus, but very agreeable to the mules. Found a fine spring of cold water with a little carb. soda dissolved in it. We could hardly drink enough of it so cool & pleasant was its flavor & the quantity taken by some of us was astonishing. Distance today 26 m. Packs are firmly [set],

Tuesday July 17. Morn fine & clear. Night quite cool. Wind East. Owing to the fine pasturage we laid in camp till 8 that our mules might get the benefit of it. Start at 8. Have seen today a number of sage hens, a fine large bird of the grouse species. Much larger than the prairie hen of a mottled gray color, & so near the color of dried sage grass that it is difficult to distinguish them among the bushes of this plant. Where they are I believe exclusively found. A company which we overtook had killed several & we found them from a taste we had by invitation very sweet & juicy.[24]

For the last two hundred miles we have seen crickets of such enormous size & in such quantities that they are really worth remarking upon. There appears to be two kinds one green the

[23] Stansbury confirmed Perkins' observations. He counted numerous dead oxen along the way and reported that a quantity of goods, thrown away by emigrants, was strewn over the trail. "The country," he wrote, "all the way from the crossing of the Platte, is a dry, sterile, and dreary desert. The artemisia constitutes the whole growth, and what little grass had come up has been completely eaten off by the hundred of thousands of animals that have passed before us. Thirty-one head of dead cattle were passed on the roadside today, and on the bank of a small drain, where the efflorescence of alkaline matter was very abundant and rendered the water nauseously offensive, nine oxen lay dead in one heap. They had been poisoned, doubtless, by the water." Stansbury, *Explorations*, 63-64.

[24] "Soon two of the boys hallooed from the adjacent hill, 'Bear'! At the instant three or four of us with guns in our hands and knives and pistols belted around us, were off at full speed: fully expecting to have a rencounter with an *Old Grizzly!* The hill top gained we heard the discharge of two guns in quick succession! re-animated and spirits wrought up to the highest pitch we quickened our step and soon gained the place and to our amazement and sad disappointment found the object of their aim to be nothing more nor less than a flock of *sage hens!*" Paden (ed.), *Moorman Journal*, 40.

25 "The large crickets or grasshoppers spoken of by Palmer & Bryant are frequently seen & are very fat looking fellows considering they have such poor pasture. If one is killed he is devoured by the others in double quick time." Wood Diary, July 23, 1849. Edwin Bryant in 1846 saw and described the grasshopper catchers: "The Indians of this region, in order to capture this insect with greater facility, dig a pit in the ground. They then make what hunters, for brevity of expression, call a *surround*;—that is, they form a circle at a distance around this pit, and drive the grasshoppers or crickets into it, when they are easily secured and taken. After being killed, they are baked before the fire or dried in the sun, and then pulverized between smooth stones. Prejudice aside, I have tasted what are called delicacies, less agreeable to the palate." *What I Saw in California*, 162-63.

26 Willow Springs, still so called, is shown north of Independence Rock on Part II of T. H. Jefferson's Map of the Emigrant Road from Independence, Missouri to San Francisco, California, 1849, reproduced in Morgan (ed.), *Overland in 1846*. On July 9, 1849, Joseph Sedgely came to Willow Springs. "The water is cold but contains minerals. After filling our canteens, we went up a steep bluff, whence we obtained a view of the mountains and the surrounding landscape. Coming to a patch of grass, at seven, we camped, making a distance of forty-four miles. Our mules are all sick caused by drinking poison water." *Overland to California*, 36.

27 "We moved about seven miles up the valley, and encamped one mile below Rock Independence. This is an isolated granite rock, about six hundred and fifty yards long, and forty in height. Except in a depression of the summit, where a little soil supports a scanty growth of shrubs, with a solitary dwarf pine, it is entirely bare." Frémont, *Report*, 56-57. In 1850 Solomon Gorgas reached Independence Rock by June 6, and realized that he was in for trouble. "Grazing poor overloaded we repacked our load & threw away some articles of clothing & some medicines &c—to lighten up our load." Gorgas Diary, June 6, 1850.

28 "The rest of the company startid the sam day we did hear we gow winchind our way threw the black hills this is ruf country

other black, as large as the outer half of a man's little finger many of them, & great fat bubberly fellows that will hardly get out of your way in the road. The Indians make use of them something after the fashion of the asiatic locust, drying a bushel or two in kilns underground & eating them in their dried state or mixing them up with fat & making a kind of marmalade. The Mormon Capt told me he had seen parties of Indians sit down around a Buffalo skin on which was spread any quantity of these insects & make a meal of them with the greatest relish.[25]

Day has been hot. P.M. wind S.W. Camped at noon at Willow Spring a fine spring of pure cold water.[26] Have seen 23 dead cattle. Camped 6. 18 miles, roads heavy & sandy.

Wednesday July 18. Start at 6. Morn clear & fine, night quite cold. W. East, roads sandy & quite heavy, over plain we crossed of some 4 or 5 miles sand was so deep as to be almost impassable. Finding that by staying with Cross' cart I was losing some 8 or 10 miles per day. I determined upon leaving & going forward. With my pack mules I could pass the cart 3 or 6 times per day & leave them at a rate of nearly 1 mile per hour they traveling about two miles & I three & without any stoppages for hills or ravines. With this determination I passed Cross & Cunningham about 9 bidding them goodbye & with Johnny H. shoved ahead & the last I saw of the cart was from the top of a high hill, when it was plowing through the sand plain spoken of above.

At 11 overtook Capt Burnhams train, which had camped with us on the Platte & whom we knew to be very clever fellows. They supplied me with some 15 lb powder, spoons, forks &c, & travelling went with them some 4 miles to Independence Rock.[27] I left them & went on again 2 miles from Ind Rock passed another train, crossed the river, then left it & came to it again near Devil's Gate where we nooned started again at 4 & camped at 7 near a smaller train, having travelled 28 miles. This is encouraging & looks like getting on & as though our journey was not to last forever.

Ind. Rock is a huge mass of smooth stone oval, some 1/2 mile long or more rising abruptly out of the plain & standing entirely alone though a little beyond commences the Rocky Mountain range in full view. The greatest attraction & curiosity of Ind. Rock however is the vast number of names inscribed upon it. Being very smooth it makes a fine intelligence board & thousands upon thousands of names are thickly inscribed in large letters on the two sides round which the road winds. A man could spend all day looking for some familiar initials &

not get through them, Hardly anyone passes without leaving his mark with tar or charcoal.[28] Unfortunately I had not ambition enough & perhaps did not feel like spending the time either as I was now running away from the cart, to take the trouble to thus hand my name down to posterity.

The Sweetwater where we struck it is a pretty clear stream some two feet deep & 40 or 50 wide.[29] Its valley is very beautiful, that is, as beautiful as any prospect can be when lacking trees, the great ornaments of all scenery. The valley is wide & all stories we have heard about scarcity to the contrary, is covered with as fine grass as animals ever grazed in. The prospect from Devil's Gate for miles showed the rich bottom covered with oxen while their white-topped wagons could be seen in little patches along its borders.

Arrived at "Devil's Gate" at 12 & camped 4 hours, spent most of the time in exploring this remarkable gap through which the Sweetwater forces its way.[30] The rocky mountain chain seems to branch off into spurs & ranges, one of which crosses the valley of this river & the water by some means seems to have cut itself a passage through the solid rock where it is some 400 feet high, & through this channel of some 40 or 50 feet in width, it rushes roaring & foaming down to the plains below. The gap I should suppose to be 1/4 mile in length running nearly East & West. The wall on the south side being perpendicular while on the north it is more broken & slants away somewhat at the top. The rocks are of a reddish brown color having something the appearance of granite in their formation. Huge masses have at different times been precipitated into the bed of the River & the action of the water has worn them into various fantastic shapes & boiling pots. The channel with its rocks & water reminded me much of Bellows falls in Vermont. Near the top of the south precipice some 350 feet above the stream a solitary cedar has found root in a crevice & forms the only variety in its bald surface. Twas altogether a grand & imposing place & I spent an hour in admiration of this massive work of nature.

Left at 4 P.M. course lay up the valley, forded the River once & camped at 7. Distance 29 miles. Have seen today 38 dead oxen & two horses. No doubt hundreds lie in the ravines out of observation from the road.[31] The Raven is here very numerous & have taken the place entirely as scavengers of the Buzzard. The latter bird I have not noticed any after 100 miles or more.

Thursday July 19. Morn clear & cool, wind from E. Was soaked last night by a squall of wind & rain & as fortunately

the next place was the sweat wator hear is what the [call] independence rock this rock 25 feet high and 500 feet long 100 ft wid and a hard to be a granite rock a setin on the top of the ground and every one that came thar rot thar names on it wit tar and whar they hailed from it was all covered i saw names that war rot in 44." Clark, Reminiscences, N. 10-11.

[29] "Today we crossed over to the Sweetwater River, descending into its valley by the side of a small tributary, whose course was nearly south, and encamped on the left bank of this beautiful little stream, a mile below Independence Rock. This river is about seventy feet wide, from six to eighteen inches in depth, with a uniform and tolerably rapid current of clear, transparent water." Stansbury, *Explorations*, p. 64. The Sweetwater was a haven to emigrants going west. Here was a passway of rather easy traveling, grass, water, buffalo, and antelope. Stewart, *The California Trail*, 132.

[30] "5 miles west of Independence Rock brought us to a place called Devil's Gate. This is one of the most singular freaks of Nature that the eye of man ever witnessed. I stated above that the valley was walled in on each side by lofty palisades. They vary from 400 to 700 feet in height. At Devil's Gate the river suddenly turns from the plain and dashes through the only fissure to be found in this towering granite wall." Tompkins Diary, quoted in Coy, *The Great Trek*, 143.

"Through this range the Sweetwater passes in a narrow cleft or gorge, about two hundred yards in length, called the 'Devil's Gate.' The space between the cliff, on either side, did not in some places exceed forty feet. The height was from three to four hundred feet, very nearly perpendicular, and, on the south side, overhanging. Through this romantic pass the river brawls and frets over broken masses of rock that obstruct its passage, affording one of the most lovely, cool, and refreshing retreats from the eternal sunshine without, that the imagination could desire." Stansbury, *Explorations*, 65-66.

[31] Occasionally this road was difficult, the water poisonous in places. Perkins seems to have been alone in noting dead animals along this particular stretch. Possibly they had become so numerous and commonplace that no one else took the time to note them.

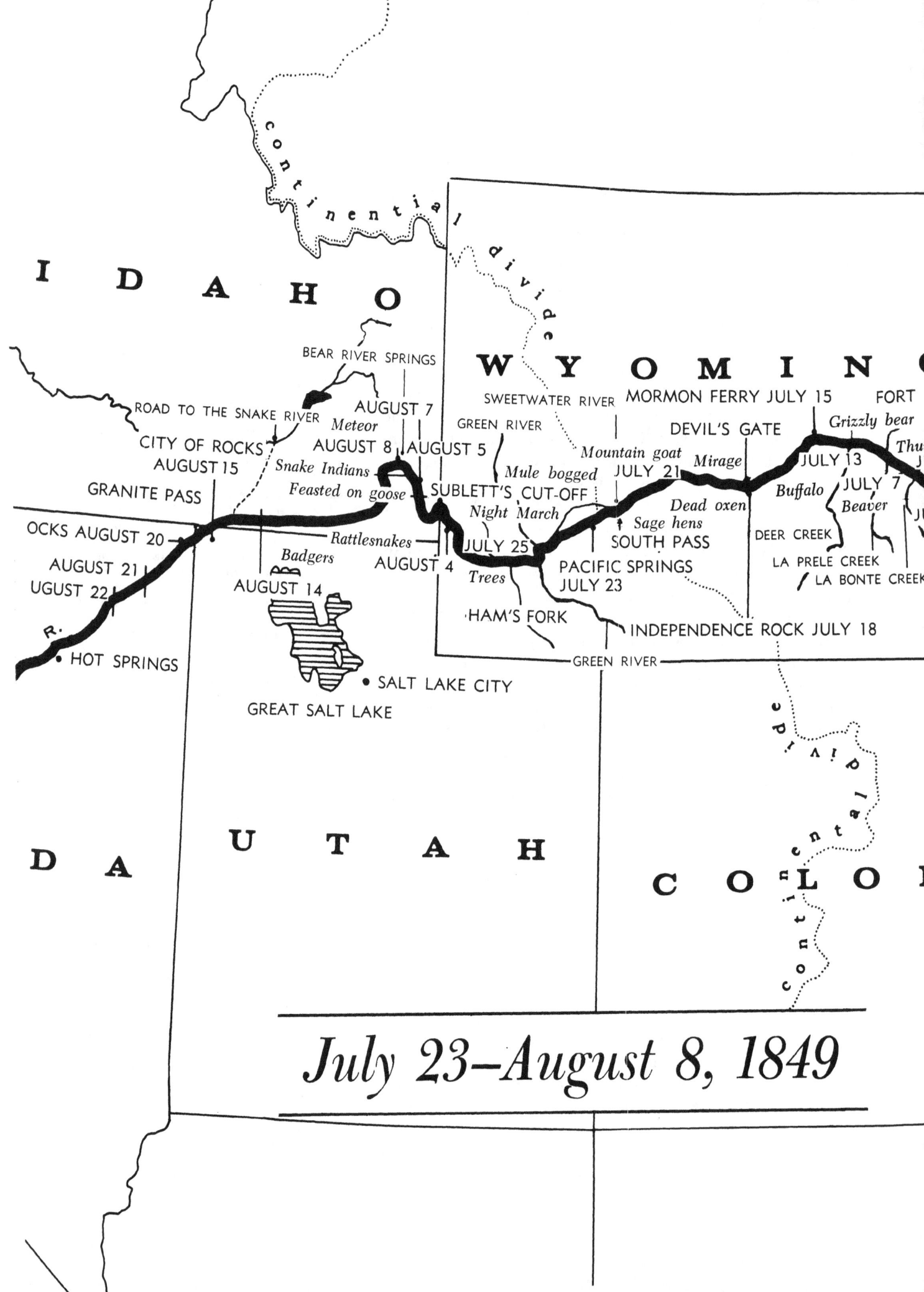

I D A H O

continential divide

BEAR RIVER SPRINGS

ROAD TO THE SNAKE RIVER

AUGUST 7

Meteor

CITY OF ROCKS
AUGUST 15

AUGUST 8   AUGUST 5

Snake Indians

Feasted on goose

GRANITE PASS

OCKS AUGUST 20

AUGUST 21

UGUST 22

R.

• HOT SPRINGS

Rattlesnakes

Badgers

AUGUST 14

SUBLETT'S CUT-OFF

Night March

JULY 25

AUGUST 4

Trees

HAM'S FORK

GREEN RIVER

Mountain goat

Mule bogged

W Y O M I N G

SWEETWATER RIVER   MORMON FERRY JULY 15   FORT

DEVIL'S GATE   *Grizzly bear*

JULY 21   *Mirage*   JULY 13   *Thu*

Dead oxen   *Buffalo*   JULY 7

Sage hens   *Beaver*

SOUTH PASS

PACIFIC SPRINGS
JULY 23

DEER CREEK

LA PRELE CREEK

LA BONTE CREEK

INDEPENDENCE ROCK JULY 18

• SALT LAKE CITY

GREAT SALT LAKE

GREEN RIVER

continental divide

D A   U T A H

D A

C O L O

continental divide

July 23–August 8, 1849

there were some three or four wagons camped near, I hastily waked John who for some time could not be made to comprehend where he was or what was to pay, we scampered off & took shelter under one of the wagons where we spent the remainder of the night, leaving our faithful dog Watch to guard the packs. Had we been camped above us as contemplated in the Evening we should probably have been somewhat dampened. Start at 1/2 past 5. Day cloudy & cool.

Our road lay up the fine valley of the Sweetwater & was very hard & good. P.M. the sun came out clear & warm. Had a fine specimen of what I have before seen on the Platte, the mirage.[32] At the horizon a wavering indistinct line of light appeared & above it & entirely separated from the ground beneath we could see the top of trees & hills which were far beneath the horizon, as the distance when we made it proved. It is a singular appearance, though to us not very attractive as it indicated a long road over dry level country.

Thousands of a bird almost exactly resembling our meadow lark, though somewhat smaller & having no resemblance in its note which is far different from the sweet clear tones of our Ohio bird. Have seen them as far back as the crossing the Platte though not in such numbers.

Crossed the Sweetwater four times in one mile this P.M. to avoid projecting points of rocks. The road here was through beautiful & grand scenery. The ranges of rocks are a great curiosity in themselves, rising abruptly out of the green plain almost perpendicular some 4 or 600 feet & as bare and smooth as a pebble stone, being in fact nothing but solid rock, sometimes rising into Peaks, sometimes round knobs one of which looked exactly like the top of an enormous human skull & the majority being sharp ridges. In places there were gaps through where a strip of green grass followed its course till connecting itself with the plains which we could see beyond stretching far away, at the place where we crossed the River so often, two ranges came together forcing the water from one side to the other like the rebound of a ball, & leaving space enough at this base for our trail.

This travelling among the mountains is less fatiguing & far more pleasant than the eternal sameness of the level Prairie. Various names are inscribed on the rocks wherever they approach near enough the road to be convenient. This afternoon saw in rather fresh tar that of "Paul Fearing, Harmar, Ohio," with no date attached thereto & I cannot form an estimate of the distance Chapin's Company are in advance of us.[33] We

32 Almost every traveler who went over the California Trail was intrigued by the mirage in this alkaline area. Edwin Bryant wrote, "The *mirage* has deceived us several times during the day's march. When thirsting for water, we could see, sometimes to the right, sometimes to the left, and at others in front, representations of lakes and streams of running water, bordered by waving timber, from which a quivering evaporation was ascending and mingling with the atmosphere. But as we advanced, they would recede or fade away entirely, leaving nothing but a barren and arid desert." Bryant, *What I Saw in California*, 135.

"I will here state a fact that has been noticed by all persons who have traveled the overland route to California. In all the mountain region through which we pass, the space between us and any distant object seems diminished in a remarkable manner. An object is really four or five times further off than you would judge, until by experience you learn to estimate distances in this singular region of the globe." "The philosophic cause of this phenomena, may be the abruptness and height of the mountains, and the thinness and transparency of the atmosphere." Phillips (ed.), *Scenery of the Plains by Franklin Langworthy*, 100.

33 The name "Paul Fearing" [see appendix] was well known to Perkins. Harmar, Ohio, adjoined Marietta. Paul Fearing perhaps was the grandson of Paul Fearing, 1762-1822, a famous lawyer of Marietta, Ohio. *Biographical Encyclopedia of Ohio of the Nineteenth Century*, 449.

have been within 5 days travel of them & I should think from appearances I must now be much nearer than this. Should like much to overtake them.

At 6 began to look round for a camping ground & spying some wagon tops a short distance down the stream went to them & found to my great pleasure my old friends Sargent & Anderson, & with them my train of coffee & milk memory. Camped with them at 1/2 past 6. Distance 25 miles.

Just as we arrived one of the train killed a Buffaloe & we soon had a glorious supper of "Buff steak" Sargents fine biscuit & coffee with milk from "that same old Cow." After supper pipes were brot out & we have spent a jolly evening more the cheerful camp fire which the coolness of the East wind renders quite comfortable. 33 dead cattle today.[34]

Friday July 20. Morn dark & cloudy. Wind cold from E. night was quite cool. After a fine breakfast of Buffaloe &c, ox teams left at 7. Having promised to camp with my kind acquaintances again tonight & knowing I could easily overtake them I staid till 9, to allow my horses to pick up some of the grass which was pretty good.

Started at 9, road dusty, terribly so, the soil being light of an ashy color & the dust from it on the least disturbance floated up and around us like clouds of smoke filling our eyes mouths & lungs. My throat & lungs are quite sore tonight from breathing so much of it. Considerable alkali is mixed with all this soil which may have contributed to produce this effect.[35]

Road lay parallel to the ranges of rocky mountains, with almost continual, though slight ascent. Passed over a sandy desert of sixteen miles without water, the only product of this waste being a few stunted sage bushes, except in places where there is an alkali lake with some green grass around it.

Overtook Sargent & Co at 12, & travelled with them two hours when went on to an appointed camping ground which I reached 1 hour in advance of the train. Distance 22 m. Camped at 5. Day has been hot. Clouds cleared off at 10 & sun came out scorching. P.M. had a squall from the N.W. which passed off with a slight sprinkle after raising clouds of dust. Have counted 27 dead cattle. Camp on Sweetwater with tolerable grass.

Saturday July 21. Morning cloudy with very heavy fog, quite a rarity in this region, & the only one I have seen since leaving the vicinity of Fort Kearney. Night cool. Stayed in camp till

[34] Whereas most travelers busily noted human deaths, Perkins observed the animals he saw along the way. The ground was highly charged with alkali at this point, and the water was especially deadly to cattle.

"We have passed timber mentioned in Ware's Guide & found 3 trees. An ox was dying near them. We have passed many dead oxen." "Dead oxen are so numerous that we are seldom bereft of their odor, but our train has not yet contributed its mite for those who succeed us." Wood Diary, June 29 and July 5, 1849.

[35] Alkali dust was a ferocious enemy of the peace and comfort of the Forty-niner. "After crossing the north fork of the Platte, we had long stretches of dry, barren plains, vast fields of wild sage, scraggly hills, deep and rocky ravines, and miles of volcanic rock and ashes. The dust from the ashes was very annoying to both man and beast. Nearly all the men had their lips covered with court-plaster, while their inflamed noses and eyes showed the effects of the vicious alkaline dust." Shaw, *Across the Plains in 'Forty-Nine*, 70. "Today, the heat has been oppressive which join'd to our long walking through the hot dust make us feel it very much. This morning we left Green River & all day it has been nothing but sand & heat & dust." Gray Diary, July 5, 1849.

8 to fix one of my pack saddles. Then harnessed up, passed our friends who started some half hour before me & bidding them goodbye pushed forward again. At 10 overtook Doct Jacksons pack mule company which had been encamped for 3 days on Sweetwater & camped for noon with them at 11.

I should have mentioned that we parted company before reaching Ft. Laramie they remaining behind to recruit their animals & passing us again while lying at Deer Creek, of course there was a great amount of shaking hands done, & we were treated for dinner to a dish of antelope & mountain Goat the latter of which was new to me & surpassed in flavor & juiciness anything this wilderness has yet produced. Even excelling my favorite Buffaloe meat.[36] It has something the flavor of fine fat mutton but is far richer. Had some drizzling rain from North while nooning.

P.M. clear & hot. Travelled with Doct J & camped with him this evening on north fork of Sweetwater at 7. Distance 23 miles. I have remarked that there has been no graves for the last two hundred miles with some half dozen exceptions, which satisfies us that we are out of the influence of the cholera. The air of these high regions is remarkably pure & bracing & we cannot but feel & be well.

This afternoon my eyes were greeted by a new sight & a strange enough one too for July, being nothing less than a view from the top of a hill we ascended of the Wind River mountains their tops covered with pure white snow.[37] I for some time did not perceive that what I saw was a range of mts, but supposed it to be a bank of clouds on the horizon, until I could see dark ravines running up in the clouds as I thought, which arrested my attention, & pointed out the curiosity to my fellow travellers, & with the help of so many eyes the matter was set at rest, twas snow & no mistake.

This evening when arrived at the branch on which we are camped I picked up a lump of hard snow which I suppose to have floated down the stream & Huntington & myself gnawed off the corners to have it to say that we had eaten snow in a hot day in July! No artificial article either but snow of natures own manufacture. The water of this stream is almost as cold as ice,[38] being the product of the melting of this bank it having its rise some 12 or 15 miles distant in these mountains. Have passed 27 oxen dead.

Sunday July 22. Morn clear cool. Night cold & heavy dew. Doct J & Co left me this morning at 7. I being determined to

36 "We remained encamped today [July 29]. I went out from camp a short distance into a small piece of timber and on my return a young deer ran before me and shot it with my Pistol through the heart. This is the first deer that has been killed by the company. Mr. Lyon also killed a mountain sheep, or Big horn." Webster, *The Gold Seekers of '49*, 64.

37 James A. Pritchard's party watched expectantly for the Wind River Mountains, a famous landmark. On June 15 Pritchard recorded in his journal, "This afternoon was the first time that we had a view of the Wind river Mountains." Pritchard, *Diary*, 91.

38 In his peculiar phonetic spelling Anson Clark wrote: "hear was a little lak had froson over and the ground had washid over it and grass was grean all over it and it look butifull we do don one foot and came to ice it war to fot dap and how much more I dont know." Clark, Reminiscences, N-11. Many emigrants mentioned ice on this part of the trail.

camp all day. At 8 hearing from a train that grass was much better on the Sweetwater 6 miles distant & my animals needing all the nourishment they can get I started over, struck the stream at a place where some dozen trains were camped & was obliged to go up farther to seek grass.

Camped at 11. Distance on the road 6 m. Am camped in a beautiful place on the banks of the River & entirely surrounded, screened & shaded again by impenetrable thicket of water willows. Had a storm of wind & some little rain this afternoon from W. blew over in 1 hour but we made up our minds for a terrible drenching from appearances at first. Of course our packs must be kept dry & our India Rubber cloths are spread over them while I wrapped in my I.R. coat, & Huntington who has none wears my cloak. Should it storm at night as there is still some appearance that it may do, as we have no friendly wagons to take shelter under, we shall have to lie quietly & count the drops trickling down our sides & backs.

Made some bread for supper this evening, or rather short cake, having no pan was obliged to mix it in a frying pan the iron of which gave it somewhat the color of rye bread succeeded however in producing some rather good cakes which at home would have been pronounced rather too full of ashes cinders &c to say nothing of the brown appearance of the inside, but which we thought quite nice. Saw today 12 dead oxen.

July 23–August 8, 1849

Monday July 23. Morn clear and cool—night quite chilly, wind W. Start 1/2 past 6. I find my packs and saddles require some changing & attention every morning which prevents my starting as early as I would wish, but hope in a few days more to get everything so systematized as to be able to get off at 1/2 past 6.

Road left Sweetwater & went over the sandy bluffs—rather heavy— At 10 made the South Pass of the Rocky Mts.,[1] a place of great anticipation to all emigrants but of any little, if remarkable in its appearance & no one would know that he had arrived at the greatest height & turned it from anything observable at the place itself. The summit is between two ridges of barren sand & granite & is about two thirds the distance through a ravine the road follows between these ridges. To the north can be seen the snow-capped mountains I observed yesterday & day before. While south on the barren hills of gravel, no appearance of mountains, rocky or otherwise.

On emerging from the ravine we came to a spring from which flows a small brook running west, a sure indication that we are now descending to the Pacific Ocean. This is called Pacific Spring.[2] Water good, and valley though covered with fine grass entirely inaccessible for grazing purposes, for some four miles being a perfect bog of soft mud. One of my mules attuned by the prospect and somewhat hungry determined to have a taste & broke from the road down to the valley. Three or four steps from its edge & down she went, to her packs in the mud, & there she was compelled to lie quietly till we could take the load from her back, when with some assistance she floundered out & looking quite muddy and chop-fallen. The rascal muddied my packs somewhat, and probably learnt some wisdom from her adventure & I hope will leave to me to choose her grazing ground after this.

Our reflections on "passing the Pass" are rather of an agreeable nature. We are now considerably more than half-way through our journey, with fine road & grass most of the remainder & down hill. From here to Sutter's is something less than 800 miles, & I hope to make it in 30 days. The first part of the trip seemed like "up hill work" while now it is down hill.[3] Still we shall have some hard times & will require some energy to go through safely & expeditiously.

With most persons the first part of the journey is expected to be very pleasant. They intend to hunt, fish or have plenty of provisions some luxuries & perhaps enjoy themselves much. From now, however, stern perseverance without regard to per-

[1] Ware described South Pass as "distant from Fort Laramie three hundred miles, or about nine hundred and fifty miles from the mouth of the Kansas. Alt. 7490 feet; Lat. 42° 27′ 15″; Lon. 109° 27′ 32″. It is difficult, from the gradual ascent of the Pass, to find the precise summit; the point of culmination is between two low hills, about sixty feet high. The Pass is about nineteen miles in width, without any gorge-like appearance." Ware, *Emigrants' Guide*, 24. Frémont wrote one of the most fascinating descriptions of South Pass, levied upon by Ware, as seen above. "About six miles from our encampment," he said, "brought us to the summit. The ascent had been so gradual, that, with all the intimate knowledge possessed by Carson, who had made this country his home for seventeen years, we were obliged to watch very closely to find the place at which we had reached the culminating point. This was between two low hills, rising on either hand fifty or sixty feet. When I looked back at them, from the foot of the immediate slope on the western plain, their summits appeared to be about one hundred and twenty feet above. From the impression on my mind at this time, and subsequently on our return, I should compare the elevation which we surmounted immediately at the Pass, to the ascent of the Capitol hill from the avenue, at Washington. It is difficult for me to fix positively the breadth of this pass. From the broken ground where it commences, at the foot of the Wind River chain, the view to the southeast is over a champaign country, broken, at the distance of nineteen miles, by the Table rock; which, with the other isolated hills in its vicinity, seems to stand on a comparative plain. This I judged to be its termination, the ridge recovering its rugged character with the Table rock." *Report*, 60.

[2] When emigrants reached Pacific Springs, many of them felt they were at last well on their way to California. On July 16, Joseph Sedgely wrote, "We climbed up the mountains, 7,490 feet above the level of the sea. This pass is about nineteen miles wide, with a good road—the best we have seen for several hundred miles. Here are the Pacific Springs, so called from being the first water on the road that runs toward the Pacific Ocean." Sedgely, *Overland to California*, 38-39. Joseph

sonal comfort is necessary. Your provisions are reduced to hard crackers & bacon. No more game of consequence to find animals pretty well worn down & yourself considerably tired of your undertaking & you have nothing cheerful to look forward to but the getting through. The temptation is great to find some cool shady place to waste time in listless idleness. Many are so dallying now & most of these with ox teams too, & I am much mistaken if they do not have to pay for their comfort now by spending the winter in the mountains. How many have provisions to keep them in case they are compelled to do so I know not but vast quantities have been thrown away.

This morning a fine flock of sage hens crossed the road before us & I made up my mind to have a nice meal—but on trying to shoot neither of our guns would budge a ball—snap- snap- went the caps but uselessly. While the fine birds stood looking at us wondering no doubt what we were waiting for. In this predicament we were compelled to go on & leave them "alone in their glory" & accomodate our disappointed palates as we best could.[4] The rain yesterday so dampened the powder in the tubes as to make our arms useless against sage hens or Indians. This morn we spent in getting them in shooting order again & satisfying ourselves of their fitness by shooting at marks.

Day has been cool, fine breeze from West. P.M. storm passed to south. Another crossing of Pacific Creek some 5 miles from Pass. We to avoid the mud & wading went some rods higher up stream where two of our mules jumped over safely, but "old Davy" stepping too short one of his hind feet sank in the mire and threw him on his side, & the weight of the pack prevented him from rising we were obliged to unload. In attempting to get up the old fellow worked himself round into the creek & finally had all four legs fast so that after two or three desperate efforts he gave up trying nor could we induce him to move again. Fortunately an ox train came up & putting some ropes under him all hands took hold & dragged him out bodily on the bank. From what I have been told of the nature of the beast I have no doubt had not help arrived in time the mule would have laid where he was & died.

The Capt of this train advised me not to try to travel much farther this afternoon with him as the straining in the mud was more than a days work, & invited me to camp with his train some three miles on. Which invitation I accepted & spending an hour in washing old Dave & arranging my things I followed on & found the wagons encamped on a green meadow with very good grass. Turning out my mules I was preparing

Warren Wood described his experience, and the novelty of the divide. "Today we have passed through the pass & are encamped on the Pacific Springs. It is a fine spring of pure cold water, it is a tributary of Green River, below called the Colorado of the West. . . . I have drunk of the waters of the Atlantic and the Pacific mingled in the same cup." Wood Diary, July 3, 1849. "We went on to miles flather an com to som good water an it tastid goog to one that war dri i thot that war the best water i ever drink in my lif." Clark, Reminiscences, N-10.

3 Perkins thought too much in terms of declivity at this point. As always he was too optimistic about his capability to travel fast. He was still sixty-five days away from Sacramento, and had the most difficult traveling before him.

4 Game appeared to be fairly plentiful in this mountain reach. Joseph Warren Wood went hunting on Sunday, August 5, 1849, with good luck: "I went down the stream to hunt & crossed over it for that purpose. I saw plenty of Ducks, Doves, Meadow Larks, Black Birds, Prairie Hens, Sage Hens, Crows, Cranes, &c &c, but killed only Sage Hens & Prairie Hen & a crane. Pollard shot a Lark. The Crane was a Sand Hill one & very large. We used our hens to make a soup for dinner & our crane forms a body to preserve." Wood Diary, August 5, 1849.

to make a cup of tea when the Capt called out to us that we must eat with them, & glorious treat they gave us. Nothing less than a real old-fashioned richly seasoned dumpling sage hen soup! and with fine white bread found the choicest meal I have eaten on this journey. Found after supper when talking over our pipes round the camp fire that three of our hospitable entertainers were brother masons & we had some pleasant conversations relating to the order. At bed time we were pressed to take shelter under their tents but being accustomed to outdoor life & the night fine declined accepting & spread our Buffs & blankets on the grass. Evening fine & clear. 28 cattle [seen dead]. Stars brilliant—wind W. Distance today 20 miles camp at 6.

Tuesday July 24. Morn clear & cold. Wind N.W. Night very cold. Awakened this morning with a sensation as of some one pinching my proboscis & found my cloak which covered me perfectly white with heavy frost! The warm air reminded me much of a winter morning at home. Sun rose bright & clear & sun sent down his rays sufficiently hot to dissipate all appearances or imaginings of winter. Owing to the delay occasioned by my mule's miring yesterday I did not make the 28 miles I intended which would have brought me to within twelve miles of the desert in Sublett's cutoff. The twelve miles I should have made by 9 this morning & after nooning today expected to cross the barren plain tonight. Shall be obliged to defer doing so until tomorrow night and as I had only twenty miles to make today & tomorrow concluded to let my mules lie most of today on the good grass of the meadow & enjoy it.[5]

Started at 1/2 3 this P.M. travel 10 miles, shall make the other 10 to the Big Sandy tomorrow, & lying there one day be ready for the thirty-five miles of desert travel in the Evening. The Missouri train with whom I camped last night left this morning to go to the Sandy today, & lay tomorrow & they promise to put up some directions by the road sides so that I may find them tomorrow morning & spend another day with them. Which I shall be glad to. The many such pleasant acquaintances which I have made & the enjoyment consequent therefrom are the brightest spots in my rather monotonous journey.

Day has been warm with fine fresh breeze from N.W. Am camped in the dry bed of a creek some mile from the road & shall be compelled to put out camp fires before going to sleep to prevent attracting the attention of any wandering Indians that may chance to be about. Have passed 15 dead cattle.

[5] Most of the teams which had made the long pull up from Fort Laramie were weakened by physical exertion and the sparseness of grass. James A. Pritchard of Kentucky noted at this point that his teams were in good condition but that "It is an alarming and fearful thing to see as we do every day teams broken down by the mismanagment of their owners in this remote wilderness. The teams that alarmed our boys so much a few days since has not been seen since—and we learn from packers as they come up that they are falling back every day." Pritchard's train stopped long enough to take advantage of grass two miles off the road. Pritchard, *Diary*, 121.

The N.W. breeze freshened up toward evening into a strong wind which was hard to pull against & made the walk of 10 miles more tiresome than 15 would as in quiet air, but little grass when we see some thin patches only.

Wednesday July 25. Morn clear, night cold. Wind from W. A train camped some mile from us had water frozen in their buckets ice made 1/4 of an inch thick![6] start at 6 road's good. At 1/2 of 9 crossed Little Sandy some 40 feet wide water very cold & at 11 made Big Sandy[7] on the edge of the desert & found my Missouri friend's encamped here on the opposite shore. Their cattle were two miles down stream & as they were about changing guard sent my mules down with the relief.

Had indescribable trouble at starting this morning with one of my mules she being always of an extremely obstinate nature & has made me trouble before. I was leading her with my rifle on my shoulder when she made a perfectly unsuspected and unaccountable surge ahead. Knocked my gun from my shoulder & broke the stock short off at the lock. This might have been repaired so as to have been serviceable but I was made so angry by the thing that I picked up the barrel instantly & without a minute's thought struck her over the head with it, bending it to a beautiful curve & ruining it entirely. As twas of no use to carry it further I stacked the pieces up by the side of the road as a warning to others against indulging in a hasty impulse. After all I do not feel very badly at the loss though the gun was a beauty, as I am relieved of the trouble of carrying it, shall have very little use for it on the remainder of the trip as game is scarce & I am not going to California to hunt. My pistol & fine bowie will answer for all purposes of self defense. I am more vexed at the way I lost my gun than at the loss itself, first that my mule should be such an unmanageable brute at times & that I so losing my temper as to finish the work she commenced. Had I given it away to someone who would have prized it & taken care of it I should have felt perfectly satisfied. One of my Missourians would have been glad to get it & I even talked of going back some 10 miles after the pieces as there is a gunsmith in the crowd who could repair it. I should have been glad to make them this return for their kindness. On our arrival this morning they set out Biscuit broiled ham & warmed up some coffee for us, & we felt just in humor to do justice to this hospitality as we had walked 15 miles instead of the 10 we supposed the distance to be.

At the forks of the regular road & where Subletts cutoff leaves

[6] Everybody suffered from cold in the vicinity of South Pass. "This morning the water in our pails near the campfire were frozen over an indication that we were on a good deal of an elevation." Staples "Journal," 137.

[7] On the banks of the Big Sandy David Jackson Staples said his party shot several sage hens and hares and made them into "soup," meaning, apparently, a kind of stew. Other diarists used the term in a similar context. Staples, "Journal," 138. This creek is shown on the map accompanying Ware, *Emigrants' Guide.* L. Dow Stephens camped on this stream the latter part of July and found heavy frost and ice. *Sketches of a Jayhawker,* 12.

by [passing] Ft Bridger I saw some 40 or 50 notes stuck up in forked sticks with directions and news &c from those in advance to acquaintances behind, none however from anyone I knew.[8] This kind of post office is very common at the different points on the road & I have never known an instance of any note or scrap of paper being disturbed or misplaced. Every person looks to see who they are from & goes on leaving all as he finds it.

The country so far this side of the Pass is desolated & barren enough. Sandy plains or basins surrounded by sand banks & giant ridges. No vegetation but a scattered growth of wild sage giving the country an ashy appearance. The road crosses a plain mainly in a straight line passes over the sand ridge & goes down into another plain & so it has been all the distance as yet. The Big Sandy is quite a river larger than the Sweetwater, but not so clear by far.[9] Water cold & good drinking, it must be 40 feet wide & from two to 3 deep.

Start from camp on Big Sandy at 4 to cross the 35 miles of desert. Found the road good & considerable Buffalo grass in spots. According to directions we travelled all night expecting to make Green River[10] about 7 A.M. today Thursday. At 11 o'clock stopped 1 hour to rest animals & have lunch. I travelled with the train of my Missouri friends thinking 'twould be better for my mules to let them travel slow. For lunch they produced peach pies, biscuit, ham &c, & a fire being kindled with the wild sage which was to be had in abundance as we shortly had some hot coffee & went into the eatables with an appetite strengthened by an unusual & rather severe travel. Started again at 12 & travelled till 7 when we halted for breakfast within as we supposed 5 miles of the river.

There was something very unique & to me exciting in our night march. The long train of wagons & oxen scarcely to be distinguished in the darkness, the indistinct forms of their drivers & men belonging to the train & all moving quietly & without noise except the occasional words of encouragement to the animals. It was to me something new & I could easily imagine our advance to be the strategy march of an army upon some slumbering town. I do not wish however to make many such travels for the sake of novelty or anything else. The dust was terrible, & fatigue & want of sleep so overcame me that I several times found myself with both arms around the neck of my horse in the act of falling from his back & several times thought I would lie down by the roadside & sleep till morning & catch up with the train afterwards.

About 11 Doct Jacksons train passed us hurrying on at a

[8] David DeWolfe saw a roadside postoffice on the Humboldt. "At the forks of the roads was a postoffice consisting of a large water cask it had a large amount of communications in it." DeWolfe Diary, 47. "Road fork'd, right to W. left round bend of River. 1 wagon camp'd there. Boards & sticks & notices—greater part taken right hand r'd. Red barrell, sqr. hole, in head, branded 'POST OFFICE,' half full of letters, notes, notices, &c." Bruff, *Journals*, I, 291.

[9] Ware described the Big Sandy region as barren, and warned that there was no feed for the next twenty miles. *Emigrants' Guide*, 27. Joseph Sedgely said that when he passed the Sandy he saw 210 wagons in the river valley, recruiting their stock, preparatory to crossing the Colorado Desert. He found the soil around this spot a curious mixture of yellow and green colors. *Overland to California*, 39.

[10] Beyond South Pass, James A. Pritchard noted that "At the distance of 7 miles this morning we reached & crossed Big Sandy at 7 A M—The starting point to cross the Green wood or Sublett's cutoff." Pritchard, *Diary*, 94. See note 12, below. Stansbury had difficulty at the Sandy crossing. On August 9, he said, "Our road today lay along the right bank of Big Sandy, until we reached Green River, which we crossed above the junction, and encamped a couple of miles below. The increased altitude, and the consequent dryness of the atmosphere, had so shrunk the woodwork of many of our wagon-wheels, that various expedients had to be resorted to, in order to prevent them from falling to pieces." *Explorations*, 72.

rapid rate trying to make River by daylight. About 2 o'clock a magnificent meteor broke nearby overhead & shot away to the horizon. The report was much like the crack of a whip, & at first appearance was that of ball of fire brilliant as the sun & lighting the road distinctly, & as it began to move the light changed to a beautiful color decreasing in brilliance till near the horizon leaving now a trail of sparks of fire. When the first explosion took place the ball appeared nearly as large as the sun. When it disappeared it seemed to have burst down to that of Venus. The changes of color were almost equal to those of the rainbow & seldom have I witnessed any fireworks to compare with the whole display of the heavens. I felt amply rewarded for my night's watch.[11]

[*Thursday July 26.*] After breakfast we started on. I leaving the teams & pushing ahead to get the packs off my mules & wait for the train instead however of seeing the creek from the top of a ridge at the place which we had camped & could see instead the road winding over hills & ravines for miles away towards the mountains & after crossing several deep & steep gullies & ascending another high ridge I could still see the road miles ahead & no appearance of any stream, & so it continued for miles & miles & then untill 12 M. I could see the valley of Green River a thousand feet below me & at 1 camped on its banks at the ferry having come since morning 22 miles instead of the expected 5, & when I learned that the distance across the desert instead of as laid down in the guides to be as measured last week by the Pioneer Train 52 miles! & now to my great discomfort of course learned that we were not on Sublette's cutoff[12] but on a road made by the Emigrants this year which left Sublette's road somewhere in the night turning down the River on account of a ravine in the cut-off impassable for teams but which I could have travelled with ease had I known where to find it which would also have reached the river in 35 miles saving me 17 miles travel.

My poor mules were almost used up having traveled since yesterday morning in some 28 hours with only 4 hours rest, 67 miles! They looked so gaunt & weak that I was frightened lest they might give out, & shall stay with them here two or three days as I found some fine grass. After swimming them over to this side of the river, some three miles from the ferry. The ox teams did not get in until 6 o'clock this evening & thus cattle have been sent to where my mules are picketed & will stay there a day or two to recover after such a forced march as they made.

Green River here is quite a stream, some 75 feet wide & not

11 No other diarist is known to have noted this heavenly phenomenon.

12 Sublette's Cutoff (formerly the Greenwood) cut across a loop of the Oregon Trail between the western mouth of South Pass and the Bear River. Although about fifty-three miles shorter than the Fort Bridger route, it included a waterless stretch of about fifty miles between the Big Sandy and the Green rivers. The cutoff was shown on T. H. Jefferson's map, but not in detail. The chart in Ware's *Guide* attempted to show the cutoff, and Ware gave instructions how emigrants could save five days' travel by following it, but his was hearsay information and did not apply to the route actually traveled. See analysis in Morgan, "Ferries of the Forty-Niners," 167-69.

A considerable number of emigrants followed the Salt Lake road south to Salt Lake City. Dale Morgan found that seventy-eight diarists went by Sublette's Cutoff and thirty-nine by way of the Salt Lake road. See Pritchard, *Diary*, 30.

fordable: a ferry crossing teams at the moderate charge of 3.00 per wagon. We are some three miles below the branch spoken of in the guides which Sublettes trail strikes it about 5 miles above this branch. About 12 miles from here this road comes into Sublettes again.

Day has been hot wind from W. blowing a disagreeable cloud of dust into our faces. Shall camp tonight with the mules & cattle leaving Johnny here with the packs and wagons. Distance today, last night & yesterday from camp on Big Sandy, 67 miles.[13]

Friday July 27. Morn clear cool. Night cold Wind N. Started last evening to guard my mule camp where several of our Missourians were also with their cattle. While I was at the wagons they had changed their position & on my return about dusk I could get no trace of them. I rode my little Victoria up & down the creek bottom in the dark for miles but without success & was finally compelled to drop myself upon my cloak which fortunately I had with me & lie down on the bare ground without supper or bed, first picketing my mule at my feet. The night was quite cold & when I awoke in the morning for I did sleep though disturbedly, my limbs were stiff with rheumatism & I felt uncomfortable enough.

Despairing of finding the cattle ground I returned to camp this morning & Johnny went back with the relief guard, & I have spent the day with the wagons. Wind changed to W & the dust blowing over from road has rendered our camp uncomfortable. Took a good 3 hour nap to make up for lost time today in one of the wagons and woke up refreshed. Have amused ourselves reading, playing cards &c.

The country round here has a curious appearance. Clay & slatey hills & bluffs 1000 feet high & washed by the action of the water into various fantastic shapes & appearances of architecture.[14] The lines of different stages of water are very apparent on the sides of the bluffs many hundred feet above the present level of the river. There is however nothing interesting in the appearance. I had been anticipating real mountain peaks vegetation except the stunted wild sage & some little brown weeds.

We have left the vicinity of the Wind River Mountains with their snow capped tops. I was much disappointed in their appearance. I had been anticipating real mountain peaks running up into the clouds as represented in the plates of Frémont's works[15] but there are none such to be seen from the

[13] The rugged stretch of road from Pacific Springs to Smith's Fork took its toll of travelers' energy. Joseph Warren Wood noted that he had a fever over this stretch, and was too sick to make entries in his journal. His party paid $4.00 to be ferried over Green River, and then struck across the mountains to Bear River. Wood Diary, I, July 17, 1849. John Edwin Banks wrote, July 10, 1849, "It is a dismal journey most of the time dust three or four inches deep and wind blowing. I was driving loose stock and toward the end of our march became very much exhausted; had to lay down to rest several times. When I would raise my head I felt faint, and my mouth burning with thirst. Within one half mile of Green River (which seemed to me to be receding as we advanced) I found a spring of pure appearance though surrounded by alkali. I drank and felt much refreshed. I had never felt so prostrated in my life. All were fatigued, and some nearly as much as myself. Two nights' loss of rest and driving made my lot the hardest. I cannot forget this dreary and desolate land." Scamehorn (ed.), *Buckeye Rovers*, 38-39.

[14] J. Goldsborough Bruff described this approximate spot. On August 5, he said that he had "Sent the mules to the Northward, about 4 miles, under a guard, where they obtained some bunch grass, and wild oats. . . . Country now becomes more irregular, sand deep, crossed several ravines and deep hollows; and over high round hills of sand and slate. . . . Late in the afternoon we arrived at a range of very high and steep clay & sand bluffs, some parts perpendicular, and even overhanging, ending in this sort of bold rugged promontory, the base of which rests in the bottom of Green River valley." Bruff, *Journals*, 71-72.

[15] See plates facing pages 67 and 71 in Frémont's *Report.*

trail now followed & the largest of the ranges are no more inspiring in their appearance than the Alleghanies where the stage route crosses them. Excepting the snow the tops these W.R. Mts are far above those of the Alleghanies, from the level of the sea tis true, but above the level of the surrounding country I think the Alleghanies are the highest.

I saw today a number of Snake Indians galloping across the bluffs on the other side of the river & they rode their fine ponies splendidly.[16] Several wigwams of them are camped some two miles up the river.

A French mountaineer arrived yesterday from California with considerable gold. I judge though he was too fond of the lazy mountain life to labor to dig much. He says any man can get gold in abundance in Cal. & also reports that England had objected to Gov. Smiths proclamation[17] excluding foreigners from the gold regions & would try to force her subjects in. If this be correct we may have a chance to back the Gov with our rifles, & the hardy bold men who form the majority of the Emigration would compose an army formidable Enough & sufficient to resist any force England could send us. Not a man I have seen but seems to rejoice at the Frenchman's news & at the prospect of a brush with their natural enemy. Gold digging would be entirely par in the excitement of an invasion of their rights by the overbearing John Bull.

This Evening after supper had a jolly time. A young man from a neighboring camp came over to ours with his violin which he played very well & flute was produced for my benefit & between us we made very fair music while the men of the two companies forming themselves into squads shuffled away at a grand rate & many were fantastic pigeon wings cut & considerable good dancing mixed in also. Kept it up till 11. Night clear, cool, W. W.

Saturday July 28. Morn clear Warm—Wind S.W. concluded by the seniors of the camp that 'twould be best to start from a creek 12 miles from here & spend Sunday so the cattle & my mules were driven in at 9 & at 11 broke up camp and left. To allow my mules to rest I shall not attempt to travel further than the ox teams in a day & shall remain with my kind friends. Started with them but soon left them far behind. Promising to look up Good grass & wait for them to come up. At 1/2 2 made the creek & following it up a short distance camped & turned out my animals. At 6 ox teams came up & camped by me. I gained 2 1/2 hours on them in 12 miles travel. The road today

[16] "We were further regaled by seeing a large band of the Sho-sho-ne or Snake Indians. These, too, were an interesting type of the Aborigine. They were migrating nomad-fashion, being generally mounted and carrying with them their families, many ponies, and all the equipments of the camp, chase, and the warpath. The mounted braves; the fantastic trappings; the squaws with their burdens; the motley households; the pack ponies; the platform or litter here and there erected on these poles to convey the sick, disabled and infirm; the whooping vaqueros driving the loose ponies,—all combined to form a most interesting panorama, and one the like of which is never again to be witnessed in the wilds of this country." Leeper, *Argonauts of '49*, 44-45.

[17] Here Perkins refers to General F. Persife(o)r Smith who was military governor of California for only a short time, February 28 to April 12, 1849. He had been out of office nearly a month before the diarist left Marietta. General Smith had superseded Colonel Mason in command in California. Acting as United States consular agent on the South American Coasts, he wrote the Secretary of War while en route to California from Panama that he was "partly inclined to think it would be right for me to prohibit foreigners from taking the gold unless they intend to become citizens." Since "the laws of the United States forbade trespassing on the public lands," foreigners who entered the land and dug for gold would consequently be trespassers. Bancroft, *History of California 1848-1859*, VI, 272-73.

has been very disagreeable, dusty, sandy, hilly, & crooked. The dust very annoying being blown into our eyes by a strong S.W. wind & inflaming them so that now although I have washed them in cold water several times I am obliged to keep them partially shaded from the light.[18] I have seen several persons completely blinded by it so that they had to be carried along for several days. Not seeing enough to take care of themselves. If I have yet "seen the Elephant" it is in this shape, for certainly nothing has annoyed and troubled me so much as the dusty days & sore eyes consequent therefrom.[19]

The road followed Green River valley some three miles down, & then turned to the right up a ravine between two steep hills to the left of the road I saw a path looking much like a cut-off winding over one of the ridges but would not take it not being ready to believe that the trail would turn back & go Easterly but on following the road some two miles up the ravine it ascended the hill to the left turned square East & two miles more travel brought us right opposite where we were an hour & a half below & only a few hundred yards of steep ascent intervening. I was vexed to think that I had lost the principal advantages of reaching this place, that of shortening the distance by cut offs. The road made another bend down the hill to the right & in a minute turned west up the creek, making by its crookedness a distance of 7 miles into 12.

This evening three men with two Indian ponies each passed us on a trot, with light pouches & one of them stopped to talk a short time. He proved to be one of Frémont's men—was a mountaineer who had lived & hunted in this vicinity for years & had just started with his two companions for California. He was dressed in half Indian costume with leggings & beads round his neck. Was apparently about 30 years old active and well built with an intelligent & pleasant countenance—His name I think was Burney.[20] He says he shall go to California in 20 days. Taking a cut off below Fort Hall by which he will save 200 miles. A mountain track known only to the hunters, which he says "Old Bill Williams" took him through once.[21] I wish I could go with him. Distance today 12 m. Evening pleasant cool. W. west.

Sunday July 29. Morn clear cold. Wind S.E., night very cold. Ice made on a blanket by the side of my bed 1/4 inch in thickness. We have great extremes of cold & heat in 24 hours in this altitude & elevation, & it requires considerable labour to accomodate your clothing to the changes of temperature.

18 "The alkali dust off this territory was suffocating, irritating our throats and clouds of it often blinded us. The mirages tantalized us; the water was unfit to drink, or to use in any way; animals often perished or were overcome by heat and exhaustion that they had to be abandoned, or in case of human hunger, the poor jaded creatures were killed and eaten." Haun Diary, 35.

19 Travelers were constantly complaining of the effects of alkali water and dust. Even though he was then just west of Deer Creek, Bruff wrote in July, 1849, "This barren, arid march—wither'd hether, & few dusty sage bushes—Road generally smooth, & firm—sand stone or indurated clay. Fortunately cool. Strong breeze N.W. 'Rock Avenue' very remarkable. Animals suffer'd much. No water —little grass. Streams, from all authorities, now dry sand beds. Fear of effects of alkali prevented a drink, which might have revived, but likewise injured them—probably fatally. . . . Oxen lin'd road, kill'd from drinking the water— 1 entire team of 5 yoke, in line as fell, on left of road, kill'd by bolt of lightning. . . . Dead oxen distributed along road at greater distances." Bruff, *Journals*, I, 124-25. This story could be repeated of many areas between the Sweetwater and the Truckee rivers.

20 Perhaps Baptiste Bernier, who accompanied the 1842 and 1843 expeditions. Frémont, *Report*, 9 and 105.

21 Bernier was indebted to William Sherley Williams, born January 3, 1787, on Horse Creek, a branch of the Pacolet River in Rutherford County, North Carolina. "Old Bill" was killed by Ute Indians on the Upper Rio Grande in New Mexico, March 14, 1849. He was an almost legendary figure in the Rocky Mountains, having been a trader, scout, and mountain man. Favour, *Old Bill Williams*, 18. This reference to the early history of what became known as the Hudspeth Cutoff is very interesting. Here Perkins was again attempting to save time and distance. Hudspeth's Cutoff came off the Emigrant Trail near Soda Springs and rejoined the trail on the Raft River.

Unfortunately I sent all of my woolen goods around by the capes & in consequence of wearing cotton night & day have a little rheumatic. My cloak is my only protection, & it is a great comfort too, thanks to my lady dear who insisted on my taking it with me. Day pleasant, cool, spent in camp on a fine creek of cold water. Grass tolerably good. Day pleasant, Wind W.

Monday July 30. Morn clear cool Wind E. Another cold night made ice 1/2 inch thick. The train left before at 7. Cross, Huntington and myself made some little repairs & started at 9.

Yesterday Evening while eating our supper a solitary packer came up in whom I was much surprised to recognize Cross. His mule unpacked he told me his adventures since we left him with the cart. Finding they were traveling so much slower than they could do Cross & Cunningham decided to pack at once & so camped the day we left & went to work carrying out their resolutions. The next day they bid farewell to the cart & started with their mules rather heavily packed. Cunningham was soon done over by this new life & they were obliged to camp. A mule train from Mississippi[22] coming up Cunningham made an arrangement with them to carry him through giving them one mule as pay for so doing & Cross & Harlan, who had remained with the cart, went ahead. Crossing the desert Cross & Harlan were separated by some mistake & Sam had since travelled on alone until he came up to us. We were glad to see him & shall travel in company henceforth.

Roads today dusty—sandy & hilly—almost constant up & down. Pleasant, wind shifted to west blowing the fine ashy dust into my eyes. Camped at six on the side of a mountain said to be the highest we cross between South Pass & the Sierra Nevada.[23]

Our present camp is by far the most beautiful I have seen since leaving St. Joe. The mountains tower above us with its dark ravines & banks of snow while below is the valley through which runs a fine creek of pure cold water fringed with water willows—ash trees &c, & a strip of green grass on either bank. Near us is a grove of magnificent Canada firs twining up to heaven with their cone shaped tops, & forming the first thing in the shape of trees I have seen since leaving the frontier.[24] Excepting some few stunted cottonwoods &c on the Platte. I do not think there are 50 trees near the road between here & St. Joe, & the sight of such as these was very refreshing. After camping Cross & myself took a good nap under one of the largest. It must have been 80 or 90 feet in height, some two

22 This was Vaught's "Pontotoc and California Mining Company."

23 Perkins was traversing the rugged country that lay between the Green and Bear rivers and that included the Bannock range. The conformation of the mountains here is clearly defined in Erwin Raisz, *Landforms of the United States,* and A. K. Lobeck, *A Physiographic Diagram of the United States.*

24 David Rhorer Leeper joined Perkins and Frémont in the enjoyment of this grove on the Hams Fork Plateau. "Between Green River and Bear River we crossed a divide nearly a thousand feet higher than the South Pass. This is the watershed separating the waters of the Pacific from those of the Great Basin. We were now so far above the sea-level that the humid atmosphere afforded sustenance to some of the higher forms of plant life. Our road led directly through a small grove of tamarack, alder, and aspen which crowned one of the more f .vored elevations. This grove was truly an enchanting spot; at least it so appeared to us after our thousand miles of timberless monotony. Comely trees and shrubs; bright foliage; refreshing shade; fragrant flowers; pure, cold springs; sparkling rivulets; luxuriant grasses, the chirp and chatter of many birds,—such was the scene as my memory now recalls it. It seemed indeed like a precious gem plucked from fairy land. No weary, parched and sand-beaten traveler of Sahara could have been more enraptured upon sighting this oasis than were we upon entering this cheery, sylvan spot." *Argonauts of '49,* 42.

feet in diameter at the base. The limbs running from within six feet of the ground, but horizontally the largest at the bottom and tapering away to a spire forming a regular sugar loaf cone & covered with thick dark green foliage which formed an impenetrable shade. From incisions in the limbs by insects or birds exudent the genuine Bals Fir.[25]

How gloriously we enjoyed ourselves for an hour, under the shade of a tree once more, & such a grand time too. I would give half my interest in California to have a couple of them in my Fathers front yard at Marietta.

Finding excellent grass on the banks of the creek we are determined to spend tomorrow under the shade of the firs that our mules may recruit. They will need rest & good grass as often as we can afford to give it them. Having been so hardly used in the wagon they are far from being in good condition to make rapid march & until we reach Fort Hall we shall travel slow & rest often & try to get them in condition to push from there onward.[26]

Day has been warm. P.M. wind changed west again. Distance today 15 miles. This evening to our surprise Harlan was seen from our elevated position winding his way in pursuit of Cross, & soon came up & was of course astonished to find us all together again. He had crossed Green River at our ford—Cross at another & had laid waiting for Sam all day. Our Missouri friends whom we overtook this P.M. received my friends as theirs & invited us all to a good supper to which ample justice was done.

Thursday July 31. Morn clear cool—breeze N.W. The air clear & bracing & the beauty of the scenery over which the rising sun cast his rays continually changing the shadows over the landscape as he advanced upward, rendered this the most delightful morning almost we ever witnessed. Stay under our fir tree & luxuriated for an hour admiring the varying & extensive view & enjoying the refreshing breeze.[27] Rare indeed have been such days & places on our tedious journey & we can appreciate them when they do come. We lay in our camp till 4. The teams leaving in the morning.

Harlan finding Cross was to trail with us in future & would not hurry as much as he wished, concluded to go ahead & lost no more time an arrangement to which I did not regret particularly though he was a very agreeable companion. So left at 12, & the last we saw of him was winding his way over the mountain top. We shall I think come up to him again before

25 Perkins was here once more enjoying trees after traveling over considerable stretches of arid country. Emigrants often noted clumps of trees along the way. Where there was moisture in the valleys and ravines, there were trees. Frémont noted in this area that the country was "timbered principally with willow, and here and there an occasional large tree." Frémont, *Report*, 130. J. Goldsborough Bruff wrote, of approximately the same spot, "After quenching our thirst in the cool, leaping, & babbling, brook, and threw ourselves down on a sloping moss-covered bank, between the roots of a large pine, with the sparkling rill at our feet, and rested, while I smoked a pipe,—one of the times when it was a great luxury." Bruff, *Journals*, I, 80.

26 Perkins had just crossed a rugged section of the trail. Frémont said that it "is a connecting ridge between the Utah or Bear river mountains and the Wind river chain of the Rocky mountains, separating the waters of the gulf of California on the east, and those on the west belonging more directly to the Pacific, from a vast interior basin whose rivers are collected into numerous lakes having no outlet to the ocean. From the summit of this pass, the highest which the road crosses between the Mississippi and the Western ocean, our view was over a very mountainous region, whose rugged appearance was greatly increased by the smoky weather, through which the broken ridges were dark and dimly seen." Frémont, *Report*, 132.

27 Frémont was much impressed with what he saw in this area. "Among all the strange places on which we had occasion to encamp during our long journey, none have left so vivid an impression on my mind as the camp of this evening. . . . the dark pines where we slept; and the rocks lit up with the glow of our fires, made a night picture of very wild beauty. . . . There were some fine *asters* in bloom, but all the flowering plants appeared to seek the shelter of the rocks. . . . It is not by the splendor of far-off views . . . that these [mountains] impress the mind; but by a gigantic disorder of enormous masses, and a savage sublimity of naked rock, in wonderful contrast with innumerable green spots of a rich floral beauty, shut up in their stern recesses." Frémont, *Report*, 65-66.

long. He will find some pleasant fellows & camp with them perhaps two or three days.

Started at 4 stopping in the valley at its base. The ascent & descent worn very steep so that in places our mules actually slid for yards at a time down the road. How teams cross I cannot imagine. The river from the top was magnificent, showing on the whole country round for a hundred miles. The Wind River Mts. with their snowy tops the low place in the range through which the road follows into South Pass.[28] The road itself uses its serpentine course up to our feet & trains of wagons in the distance looking like spiders. Camp tonight in a valley not much grass but fine water as usual with these mountain streams. Distance this P.M. 7 M.

Wednesday August 1. Morn clear. Very cold night. Ice made 1/4 inch thick. Wind N.W. Start 7. Morning so cold that we preferred sitting around our camp fire to packing up.

Road winds over hilly country. At 10 came to the camp of our old guide again & turned our mules out in the splendid grass to fill themselves. The men of the train were all hard at work cutting off & down their wagons, reducing them considerably in both size & weight. They also were throwing out some of their load in the shape of bacon, iron tools &c. & will be quite light when they start again.[29] Have spent the day with them very pleasantly. Day warm, distance 7m.

Thursday Aug. 2. Night cold. Morn clearing. W. W. Day clear warm. Lay in camp till 2 that our mules might get all possible benefit from the splendid grass. At 10 2 mule wagons came up. Proving to be the Mississippians with whom Cunningham had associated himself, & he came to my camp to pay us a visit, ask questions, & want something as usual. Left us in about an hour & other teams having arrived also.

Had this morning a magnificent sage hen soup prepared at the "Sacramento House" by Messrs Smith, McGinnis & Co. of Missouri. We also eat for breakfast no less than 13 dogs and pups; none of your common cuss but nice little prairie dogs & delicious eating they are too. Some are so fat that it is necessary to pull off a part before cooking to prevent the meat being too oily.[30] The country here is covered with them & they may be seen running or easing scampering in all directions bent on pleasure or business.

Started at 2. Crossed another mountain from whose top we

[28] Perkins was descending from the Hams Fork Plateau.

[29] One could read the tale of frustration, disappointment, and determination to get through no matter the cost in the discarded supplies that littered the California trail. Some travelers began to reduce their loads early, but even those who had not were forced to do so when they attempted to surmount the Rocky Mountain slope with failing teams. R. C. Shaw said, "On our march through the foothills we passed many abandoned wagons, also chains, ropes, saddles, shovels, spades, picks, gold-washers, crowbars, and a complete outfit for a saw-mill." *Across the Plains in 'Forty-Nine*, 70-71. L. Dow Stephens wrote that "the cattle could not stand the strain, and grew weaker day by day. So hundreds of pounds of the finest bacon, beans, flour and sugar were left on the wayside. The bacon was piled like cordwood, and some of the men poured turpentine on the provisions and set fire to them, so the Indians couldn't eat them." *Sketches of a Jayhawker*, 12. "We saw castaway articles strewn by the roadside one after another in increasing profusion till we could have taken our choice of the best wagons entire with much of their lading, had we been provided with the extra teams to draw them." Leeper, *Argonauts of '49*, 16.

[30] Joseph Price, in his colorful spelling, described for his wife the wonders beheld along the road to California. On June 8, 1850, he wrote her, "We have had plenty of buffalow and it makes the best stake that I have ever seen there is also plenty of antilope and you have heard tell of the Prairie dog we have seen any quantity of them they are near as large agane as a fox Squirl and resemble them very mutch especialy in culler they burow in the ground and feed on grass I have seen thousands of acres ground eat interely bare with them they make a noise something like a Squirl." Marshall (ed.), "The Road to California," 235-57.

had another extensive view & descended into a valley at 6 when we camped.[31] Distance 8 m. Roads sandy & dusty. Wind W.

Friday August 3. Morn cold. Heavy dew lay on our beds & the packs. Night quite cold. W.E. Started at 6. Roads fine. Camped last night near Cunningham's teams & left them this morning getting breakfast. The character & appearance of the country is beginning rapidly to change for better—instead of the barren granite knobs & sand plains we now have mountains with green sides occasional trees & valleys with fine brooks of water.

Our route today has been through a beautiful series of hills & vales—with some three or four real patches of woods, & many kinds of shrubbery besides the eternal wild sage & grass. I also noticed plantain leaves—many varieties of flowers one resembling bachelors button—also the buttercup, & I think the dandelion although now were near enough the road for us to speak positively as to the latter. For trees there were the aspen in abundance, firs, some wild plums & birch. The road passed directly through a grove of the various kinds, mostly firs, however, the only place since we left St. Joe of the kind, & the effect upon us of the dense shade & thick forest was gratifying in the extreme. We turned our animals loose & lay down for a while to enjoy the unusual pleasure & to think of home which was recalled vividly to our minds & of the fine forests of our beautiful state. The firs are particularly fine, tall thick & green, running away up to a spire. Much would I give could I transplant some of them into warm soil.[32] There is no pleasure however without its drawbacks & thus the mosquitoes are so excessively annoying as to compel us to leave in a much shorter time than we could have wished.

We have seen more trees and vegetation today, excepting always the grass, than in the whole of our journey previous. This grove was some two hundred yards through & when the road again emerged from its shade into the open sun we felt almost as though starting anew on our trip. While resting under a fine fir saw a beautiful bird which I never had seen before alighted upon one of its branches. It happened to be of the humming bird species, had a brilliant scarlet red head, body of a bright yellow with outer half of wings scarlet. Twas a beautiful creature & had my circumstances admitted of packing it, I should have killed and stuffed its skin to carry home.

Saw today a number of Snake warriors galloping over the hills on their fine horses, with their red blankets & head dresses

31 Occasionally it is difficult to locate Perkins and his companions on the trail exactly. They were now approaching Smith's Fork of the Bear River.

32 Frémont spoke of woods in this area. This was a favorite spot with emigrants. They stopped along this part of the trail to graze their teams. Frémont, *Report,* 133-34.

they made a stunning appearance. Their horses appear to be better quality than any Indian ponies we have seen.[33]

Counted seventeen dead oxen today. Being out of the alkaline region pretty much, I have neglected keeping account of dead cattle, we however pass usually from 5 to 12 every day. How much further this will continue I cannot tell. Probably however more or less will be seen all the way to California "given out" for some reason or other.[34]

At 5 had a fine view of Bear River & its magnificent valley, & descending the mountain some thousands of feet or more, as I should judge. Camped at 7 on its banks with fine grass. Distance 22 m. Evening cloudy. Moon was nearly full & very brilliant but was soon obscured. Some appearance of rain. The rays of the setting sun upon the clouds made a splendid variety of colors beyond description & surpassing the painter's art to copy. The hues of the rainbow were as nothing in comparison.

Saturday Aug. 4. Morn clear & most beautiful air so free & fresh. Night cold, water froze in our canteens. Wind N.W.

Our beds last night were spread near a thicket of water willows & our mules picketed around us. About 11 o'clock we were awakened by a great alarm among the animals which broke their pickets & were running off, & a rushing among the willows at our head. We "turned out" in a great hurry as may be imagined, not knowing but the Indians might be right upon us, & endeavored to explore the cause of the stampede. Could discern nothing in the thicket, however, & catching & repicketing our mules lay down once more & slept undisturbed till morning. Most likely a grizzly bear had contemplated a meal of mule flesh but was frightened away by our unexpected appearance, or perhaps a wolfe may have been prowling around.

We were serenaded this morning by the cries & quacks of a heretogenous assembalage of wild geese, cranes, ducks, &c which it appears abound in this stream. No where since leaving Florida have I seen thus congregated in such numbers during the day we have passed many swamps & marshy spots in the valley out of which any or all of these birds rose as we disturbed their vocations, & from the flocks of young ducks, it is evident that they resort here to raise their broods.[35] Had we time to spare & *guns* we might provide a mess without much trouble.

I was up by daylight this morning & in addition to the music mentioned was also regaled by that of various little feathered songsters. More than I have met with on the previous part of

[33] One of the most vivid descriptions of Indians in this area was that of J. Goldsborough Bruff. "After crossing this branch, and proceeding a short distance, we perceived an indian village on the right and another on the left, 2 or 3 miles apart, and horses grazing near them. The village on the right, when abreast of it, was about 1/2 mile off. Two indians from this, rode at full speed towards us, and were mounted on beautiful fat horses.—Asking the usual question, 'Schwap?' and pointing over to the band of horses." Bruff gave a graphic description of the clothing the Indians wore. Bruff, *Journals*, I, 97. "July 4/5 encamped on fork of Bear River near a large encampment of snake Indians—had 30 or 40 at our tent peacably disposed but rather troublesome in the way of begging—dressed in all manner of costumes from Breech Cloth to shirt, coat trousers and cap which they got off emigrants —squaws riding with papooses hanging to their saddles." Chamberlain Diary, 17.

[34] "But the last straw, so to speak, broke the camel's back. We still had eight or ten miles to water and a camping place. Several of the oxen became exhausted, and one after another sank down in the yoke. We had no recourse but to abandon them where they lay, and reconstruct our teams as best we could." Leeper, *Argonauts of '49*, 74.

[35] Like the buffalo, ducks, geese, and other birds were certain to be seen on particular stretches of the mountainous road. Bruff wrote, "I shot several ducks, and lost 2 of them,—they dived & holding on to the grass at the bottom, in pretty deep water, I had to leave them." At this time he was on the Bear River in what is now Idaho. Bruff, *Journals*, I, 90.

36 Ware described Smith's Fork as being twenty-nine miles below where the old trail via Fort Bridger reached Bear River. He said camping facilities were good at this place. Beyond it the road was quite winding. Ware, *Emigrants' Guide*, 28. Leander V. Loomis crossed this stream on June 25, 1850. He said it was composed of "4 or 5 streams within a few rods of each other, all of which nearly ran in our wagons." Ledyard (ed.), *A Journal of the Birmingham Emigrating Company by Leander V. Loomis, 1850*, 60, hereinafter cited as Loomis, *Journal*. Frémont and his party reached Smith's Fork on August 21, 1843. He wrote, "The next morning, in about three miles from our encampment, we reached Smith's fork, a stream of clear water, about 50 feet in breadth. It is timbered with cottonwood, willow, and aspen, and makes a beautiful debouchement through a pass about 600 yards wide, between remarkable mountain hills, rising abruptly on either side, and forming gigantic columns to the gate by which it enters Bear river valley." Frémont, *Report*, 133.

37 Service (or sarvice) berries were also found along the trail. "Near the Indian camp was a large grove of service-berry bushes, and quite a quantity of the ripe berries had been gathered and dried by the squaws, who were very anxious to exchange them for such articles as our boys had to trade. They offered to give a half-bushel or more of the dried fruit for a darning needle; but when it was learned that the fruit had been dried on the flesh side of green deer skins spread on the ground, with flies, dirt and dogs around, there was a sharp decline in the berry market." Shaw, *Across the Plains in Forty-Nine*, 116.

38 Banks and Armstrong both described this area as they saw it on July 18, just seventeen days ahead of Perkins. Banks wrote, "Our March this day very similar to yesterday's, only not quite so bad. The soil is of a limestone quality. Saw a large flax today, as large as I have ever seen, though I have seen what is considered uncommon. Nooned in sight of Bear River. Traveled up it some miles, crossed Thomas' Fork; it is some ninety feet wide, three feet deep, current like a mill-tail. It plainly tells its origins." Armstrong observed, "Left camp

the trip. Here I saw robins, blue birds, catbirds, &c. The varieties of singing birds, however, does not compare with those at home in numbers or song.

Started at 7, & after getting tangled up in a marsh trying to cut across the meadow & being mostly mired several times, & having at last to go back to our camp to take a fresh start, we struck the road, & at 9 crossed Smith's Fork of Bear River, a swift cold stream some 40 feet wide & two deep.[36] Here we overtook our Miss. friends, who met us as usual with pleasant faces and open hands.

On the bank of Smith's Fork I found an abundance of wild gooseberries & currants. The former are much smaller than the cultivated fruit & inferior in taste, being sour & astringent. The latter are very fine when ripe, being sweet juicy & having much of the flavor of raspberries. There appeared to be two kinds, the red & the bright golden yellow, the last being the best by far.[37]

Travelled till 4 o'clock & camped on River, distance 15 m. Day hot, roads hard level & equal to any turnpike.

Sunday Aug 5. Morn cloudy night cool, heavy dew, Wind S.E. Start to find better camping at 7, & camp at 9 on Thomas Fork just below where Sublettes trail winding over the hills meets the main road.[38] Distance 5 m.

Two of "our men" remained behind yesterday to fish, & came last night with a fine mess of beautiful speckled trout which we fried for breakfast this morning, & was delicious. While sitting on the bank watching their lines towards evening a rustling was heard in the willows on the opposite side of the River & an enormous panther made his appearance. Beaufort, one of our fishers seized his rifle & let him have its contents, but so agitated was he by the sudden apparition that he missed him entirely & the animal went off making tremendous leaps over the gullies and brushes. Very likely this may have been the same rascal that disturbed our slumbers & frightened our mules out of their united equaniminty the night previous. Two ducks were killed this morning by some of the Missourians who were not very sure of the particular day of the week.

Today has been hot & sultry, Wind W. In forenoon many clouds were floating about, but disappeared entirely in P.M.

Monday Aug 6. Mostly cloudy, night mild & cloudy wind N.E. Considerable appearance of rain till 9 A.M., when clouds blew over. Start 8. The ox teams usually start between 6 & 7 ever knowing we can easily overtake them before noon take our time because of our later starting hours.

Crossed Thomas Fork this morning & a stream about like Smith's do, & following it for some two miles the road leaves the river and crosses a steep ridge. The river bottom ever since we came to it is covered with a most luxuriant growth of grass on which our animals improve wonderfully, & which makes ridiculous the big bear stories about scarcity we heard before leaving & at intervals all the way to the upper Platte ferry, we have seen fine grass & more of it since leaving the South Pass than anywhere previous.[39]

Crossing the ridge mentioned above the road descended into a beautiful little narrow valley through which it wound some 4 miles & then began the ascent of a mountain—at an angle of about 35 degrees, & continued so to the top, a distance of 1 1/4 miles being the longest and steepest hill I have yet ascended. Once on top you were well repaid for your labor by the extended view of hill & valley & mountains. I lay down while waiting for the wagons to come up. 1 hour & never tired of watching the windings of Bear River which is extremely serpentine in its course, & of gazing at the dark cedar sided mountains which border its valley. The highest mountain then seen judging by appearances only, is at the junction of Thomas Fork & Bear River, & towers far above the one from which I had the view while traveling at its base yesterday I did not appreciate its great height, & only when seen from the top of its sister mountain did I observe the difference between the two.[40]

The valley of Bear River varies from 1 to 3 miles in width, is perfectly level abounds in luxuriant grass, & with its marshy spots & green fringed river as seen from the top of the mountain where I saw it, is indeed most beautiful. The River is much more of a stream than I had supposed being about 100 to 120 feet wide at this place & varying from 3 to 10 feet deep, quite as large as the Muskingum above Zanesville.

Morn cloudy & about 4 thunder storms with wind from NW. Indian Rubber goods being in one of the packs & rather unhandy to come at I determined to take the rain as it came especially as appearances were that it would soon blow over. For otherwise, however, it poured down rain & fine hail for 1 hour & I was most thoroughly drenched. My boots were as full of water as if I had waded the River in them & my clothing clung to me with the greatest pertinacy. For while felt anything but comfortable but by jogging on I kept my blood in circulation till the sun came out again, which assisted by a mild

about six o'clock. Drove over a tremendous hill which was about four miles over in about six miles from where we started. We took dinner near a fine spring. We drove about four miles farther and we came to Thomas' Fork of Bear Creek. We crossed over. It was considered one of the swiftest streams we have crossed yet. It empties into the main river a few rods from where we crossed. We are camped here now." Scamehorn (ed.), *Buckeye Rovers*, 41-42.

[39] Jasper M. Hixson recorded in his diary that "The boys had gone into camp on the bank of Bear river, just below the mouth of Smith's Fork, only traveling seventeen miles. The grass was fine, and it was a great relief to get down from the high altitude we had been in several days. Hixson Diary, quoted in Coy, *The Great Trek*, 152.

[40] Frémont recorded the valley as being 6,400 feet above sea level; he did not, however, comment on the height of the ridges. Frémont, *Report*, 133. J. Goldsborough Bruff was more impressed by the rough crossing at Thomas Fork than by the hills. Bruff, *Journals*, I, 90.

breeze completely dried me before we reached camp. Which we did at 8 o'clock on Bear River. Distance 18 m. Day has been hot. Road on mountains steep enough till descending into the valley in the afternoon.[41]

Tuesday Aug 7. Morn clear. Night mild W. S. Start 8. Found the horseflies most annoying in this mornings march attacking both men and animals without discrimination & most ferociously. The vicinity of the River seems to swarm with them & with large savage mosquitoes. Thanks to unusually cold nights the latter dont have half a chance

Camped at noon on a small creek of fine cold water, a tributary of Bear River & soon after stopping, three of the men who had remained behind to hunt & fish came in with their mules loaded down with fine young wild geese—having killed no less than 19 & bringing in 13 of them. They were of fullsize or nearly so but their wing feathers not grown enough to enable them to fly & getting a flock of some hundred or more in the river with one man above and another below there they drove the poor things backwards and forward giving each hunter a shot everytime & they only desisted because they had killed enough leaving the rest as they found them. Such a treat as we have had this evening on their tender meat. First a fine soup with flour dumplings etc et. Goose fried & goose boiled till we are about surfeited.[42] We found them remarkably sweet and delicate.

P.M. wind N.W. with another shower though this time quite slight but being very agreeable. Cooling the air & laying the dust. Camp at 7 on Bear River. Distance 18 m. Today has been hot, roads good, following the valley.

Wednesday Aug 8. Morn clear, warm, night mild, Wind S.E. Start 8. Travelled 15 miles without water this morning & very hot & roads extremely dusty. At 1 camped on small branch of River near a village of Snake Indians who soon came swarming around us begging for everything they saw & wanting to "swap" mocassins, worn out ponies for guns, powder, & caps. They are an inferior race of beings as to stature and appearance not comparing at all with the manly Sioux. Faces broad without much expression except in some plenty of savage ferocity, hair long black & cut square across the forehead. Exactly like the pictures we see of the ancient Egyptians, in fact the general appearance of this tribe reminded me of that people. They are filthy in the extreme & their women, those that I saw were very

41 On August 16, 1849, the Marietta *Intelligencer* carried a notice of the death of Perkins' son: "Died on Monday, the 6th inst. Albert Douglas, son of Elisha Douglas and Harriett E. Perkins, aged four months. 'And is this death? / To shut as closing flowers last night / Then open in celestial light.'" Perkins apparently never learned of his son's death; in fact, he may never have received any word from home after his departure (see below the entries for October 6 and Christmas Day, 1849).

42 James A. Pritchard, on the Bannack River beyond Fort Hall, said, "Our Fourth of July was spent in travelling in the dust and fighting off Musketoes. Their attacks were more fearce and determined, and more numerous, along this river, than any of the kind I ever witnessed." Pritchard, *Diary*, 108. J. Goldsborough Bruff found this area about Bear River a good one for hunting and fishing. His party caught a number of suckers and trout. They killed some "sweet Black mice" to make a pie. Bruff said, "I strolled down banks of the stream, and shot a large hawk, and wounded 2 ducks, but lost them. My men shot red-shoulder'd hawks, bluewing & shoveller ducks, mud-hens, curlews, plovers, grouse, &c." Bruff, *Journals*, I, 89.

homely & dirty, being a perfect burlesque on the sex. We avoided them as much as possible both on account of their thievish propensities & also on account of the swarms of vermin in their persons. I saw some of the ladies engaged in a singular amusement. One would squat down with her head near the ground & another rapidly running her fingers thru her coarse hair & skillfully extracted the ped! var! & safely deposited them where we naturally put bread & meat coffee &c &c! We remained only 1 hour in our new camp & left to put some distance between us & our neighbors as in the night they might prove troublesome.⁴³ Should like much to have visited their camp as 1 was invited by an old chief to do but had not sufficient time as it was situated in a ravine in the mountains about 1 mile off.

About 3 this PM we arrived at the famous Bear & Steamboat Springs & real curiosities they are.⁴⁴ The water came up in little basins effervescing exactly like carbonized water when exposed to the air & gurgled over the edges leaving a sediment wherever it ran of a red iron rust color. Its taste was to me unpleasant being that of soda water without any syrup or flavoring, slightly acid, also a very distinct metallic taste, & a foetid old swamp like flavor combined. It was very clear & pure temperature quite warm. The St Bt Spring is one of the same character, but the water is projected up through a hole in the solid rock of an elliptical form about 8 in. in diameter—in gusts like throwing pails of water in the air, & comes up all in a foam with excess of carbonic acid gas. The water is thrown up from one to two feet & occasionally recedes down its tunnel 1 foot or more only to reappear with renewed force. These springs are all within a few yards of the River & a few feet above its surface. The appearance of the country in the immediate vicinity is decidedly volcanic. Two distinct craters exist within 50 yards of the springs are some 5 feet in diameter & perfectly round, the other about 12 feet do & more irregular & layers of lava deposited on the edges of both. At the bottom some 6 or 8 feet were currant bushes growing in the decayed stems & loose Earth. I picked up numerous specimen of fused metal sulphates &c also petrifactions & lava, while I can get much the limited weight of my packs will not admit of my carrying home with me.

We have crossed this afternoon huge fissures in the rocks which came up to the surface of the plain in many places. Extending a mile or more & evidently caused by Earthquakes. In fact the whole country shows evidences of having been at one time in a state of intense heat & commotion. I was exceed-

⁴³ Bruff saw these Indians on August 17, 1849, just ten days later. "On opposite side of the creek," he wrote, "are 2 indian lodges, about 200 yards off. The indians are continually wading over. Want to trade, have service berrys, dressed skins, &c Children, squaws, ponys, dogs, &c. These Shoshones are riding and walking around thro' camp, trading and begging." Bruff, *Journals*, I, 92. In contrast, Jasper M. Hixson wrote, "Around the fort (Fort Hall) were several lodges of Snake Indians, and a shirt was their only dress. The honesty of the Indians was so proverbial, that in traveling through their country we had relaxed in our discipline, and did not consider it necessary to keep a night guard." Hixson Diary, quoted in Coy, *The Great Trek*, 159.

⁴⁴ The names Soda, Bear, and Steamboat springs were used almost interchangeably. This was a favorite camping site for emigrants and later became noted as a bathing place for early settlers in the region. The springs, located about two miles from the town of Soda City, Idaho, are now covered by a reservoir. William Steuben wrote, "came to a series of Springs called by various names such as Soda, Beer, Boiling Gas & Steamboat Springs & in some respects very appropriate to each. The waters are very effervescent & clear as crystal & contain large quantities of Gas almost equal to Soda & Beer. Some of them boil like a pot, others form a jet from 2 to 4 feet." Steuben Diary, July 17, 1849. Loomis said of Steamboat Springs: "These springs seem to boil like a pot of water; but there is no heat in them, except one, that is just on the bank of the river, which is built in the form of a craw fish hole, about three feet high, formed of sediment thrown up by the water, which spouts about three feet high, every quarter of a minute. There is an air hole near it that makes a noise like a steamboat, but not so loud." Loomis, *Journal*, 15.

ingly interested by everything I saw & wished my geological knowledge was such as to enable me to explore further into the mysteries of nature as exhibited in this place.

Crossing the plain some 4 miles from the spring we camped in the bend of Bear River where it turns to the South. We were on an Elevated plain, on the opposite side of the River was a huge precipitous mountain & the River ran between its base & a ledge of rocks forming the boundary of the plain some 75 or 100 feet high. Our animals could not get to the water at all & we were obliged to scramble up the ledge with a bucket apiece for our own use. The scenery opposite us is very fine. The cedar covered mountain round the base of which the River changes its course, & the fine valley run either side of it. The bend of the River is about two miles across only & from a due north its course thenceforth becomes nearly South.

Camped at 7. Distance 20 miles. Had another wild goose soup tonight, & have been living on wild goose in various forms all day. From here or soon after leaving here I shall probably part with my Kind Missouri friends & push on. The Indians told us of a cut off which leaves the Fort Hall route about a mile or two from where we are camped & strikes near the head of Humbolt River being 140 miles shorter than the Oregon Trail. If practicable I shall certainly take it & thus save nearly a week's travel.[45]

45 Perkins and his companions ignored the instructions of the guidebooks by taking the new Hudspeth's Cutoff. Ware warned, "We would earnestly advise you to oppose any experiments in your party, in leaving the regular route of travel to try roads said to be shorter. You will get to California in good season if you keep straight ahead. If not, you may lose a month or so of time, and experience the fate of the Donner's party. By trying a new road they lost nearly sixty days, and were overtaken by the snow, and spent the winter in the mountains, where nearly forty of them perished. Lose no time foolishly on the road, that can be spent with profit to yourself and teams." Ware, *Emigrants' Guide*, 31.

August 9–August 19, 1849

THURSDAY AUG. 9. Morn clear, W. E., night cool & clear with a brilliant display of the heavenly bodies. Saw as I lay on my bed before falling asleep several fine meteors, course of most was south. Start at 8, & after half an hour travel came to the fork in the road the left hand being the cut-off of which we had been told. Found that it had been considerably travelled by wagons & so our Missouri Company decided to risk it too & we all started on together. At the "Forks" we found notes & cards for the benefit of various companies behind several which we read stated that "by this cut-off it is only 100 miles to the Humboldt River."[1] If this is so we shall save 175 miles. & we were told that this road struck across from the bend of Bear River to the head of Raft River where it united with the Fort Hall route but it must take a much more southerly course to reach the Humboldt in 100 miles, & I think must be the trail spoken of by Frémont's man Burns except probably varying from that sufficiently to strike various watering places. At any rate we are fairly on it though our first days travel is not very flattering, having come 16 miles without a drop of water.

We arrived about 1 at a fine stream of clear cold water which has its rise in a spring not half a mile above our present camp. The basin from which the water issues is about 10 feet in diam. & a body of water of considerable size here first sees the light coming out from under a huge mountain & cold as though off of a bed of ice. We were so fatigued & overdone by the heat & our thirst that we determined not to go any farther today & so camped on the creek at 1. Distance 16 m. at 4 our ox teams came up & we shall spend another night with them. We wish much that their rate of travel would admit of our keeping them company to California.

Friday August 10. Morning cloudy Wind west, night warm. Start 6 1/2. Crossing the brook over which we take a last drink of its delicious liquid, the road began the ascent of the dividing ridge between the great basin & the country surrounding it. The mts. were high & the road in many places almost precipitous both up & down, & will occasion wagons no little delay & trouble, though unminded by our mules we turned the ridge about 11 & found a small stream running down the mountain side into the Basin. Descending the mountain & winding along a ravine we encamped at 5 in a valley with pretty fair water. Distance 18 m.

On the road today we lost one of our tin cups, & as "cups are cups," on this route I turned around and went back in search

[1] At the junction of the Fort Hall Road and the Hudspeth Cutoff emigrants left their notes. Sometimes there were letters to people back home, sometimes warnings to friends coming on behind. Joseph Sedgely not only saw these notes, but left one himself on July 17, 1849: "Eight miles brought us to the junction of the two roads. The right [left] hand one being Sublett's [Hudspeth's] Cut-off, we took it. At this point, we found a stick driven in the ground, split in the end, and in it about thirty cards, from as many different companies. We added ours to the number and journeyed on our way." *Overland to California*, 39.

till I met the wagons coming on, & not exactly coming on either. In descending one of the steep hills by some mismanagement or other one of the largest was overturned into the brook spoken of above & I found all hands wading in the mud fishing out the various articles of its cargo. I staid to render what assistance I could & till they were underway again.

Just as I mounted my mule to leave there I heard the buzz of a rattlesnake under his feet in the grass & jumping off I succeeded in killing a fellow of 13 rattlers & nearly as large around as my wrist. Mounting on a run I turned my animals head towards the road when "buzz buzz" went another right under him again & I feared much lest he should be bitten. Dismounting again I beat about till I drove this fellow from his cover & killed him also. He was a larger snake than the other but had only 10 rattlers. Some of the Missourians came up at this time & told me they had killed three in the mornings travel. Judging by our experience they must abound here.[2] Taking the scalps of my slain foes or their rattlers instead I left those parts well satisfied with my days work. Found my cup in a wash by the road side & rejoined my party just before camping.

Saw several new plants & trees today that is new to this journey. Among them the service berry of the states full of ripe fruit the swamp maple though of small size, the genuine little snow drop looking too pretty for this savage country & plenty of large rose bushes without the flowers however. The sight of these familiar things was quite cheering. Evening cool. Clouds floating about. Day hot & dusty beyond parallel.

Saturday Aug. 11. Morn clear cool. W. W. Night cool with much wind, & clouds flying. Start 6 1/2 road over an uneven country & cheerless in its aspects. Nothing worthy of note occurred in todays travel. Camp 6 1/2 on creek of fine water. The mountains of this vicinity tho high are uninteresting in their appearance being nearly regular piles of granite and sand with occasionally sandstone and slate projecting. Night clear. Today has been hot & dusty as usual. Distance 24 miles.

Sunday Aug. 12. Morn cloudy cool, night do, wind south. Start 6 1/2. As we had lost so much time of late my travelling companions did not feel willing to lay by today, & I am half persuaded that we might as well be on the road as spending the day as our circumstances & situation usually compel us to spend it. Our course lay over the mountains this morning & we traveled fast & well till noon without seeing water & were

2 "Copperheads and rattlesnakes are numerous." Sedgely, *Overland to California,* July 28, 1849. He was just beyond Steamboat Springs at the time he noted this fact. Frequently travelers mentioned killing rattlesnakes. Joseph Warren Wood wrote, "Rattlesnakes abound here & I have today seen the body of a huge one." Wood Diary, June 18, 1849. The next day he wrote, "Rattlesnakes abound here & visited us in camp. They were suddenly dismissed."

compelled to take our noon rest on top of a hill with barely enough water left in our canteens to quench our thirst.

Started in P.M. confident of finding water in the first ravine but travelled on over hill and dale through dust & sun almost choked until at 4 we descended into a valley & found a fine spring of pure cold water. What quanities of it we poured down our parched throats, & how deliciously it tasted! Talk of fancy drinks to a thirsty man! I never enjoyed the flavor of anything liquid as I have fine cold water after a long dry march. Camped on a small brook from the spring at 5. Distance 24 m. P.M. wind west blowing clouds of the fine disagreeable dust in our faces. This dust is one of the greatest if not the greatest annoyances of this journey. It not only arises around you filling your nose eyes lungs &c, but there seldom being any rain here to lay it, it becomes very deep so as to render walking through it very tiresome. Imagine yourself ploughing through an ash heap ankle deep all day & a few miles before you kicking it up to be driven into your face by a west wind & you have a tolerable idea of our dust.

To avoid the dusty road there are paths along parallel with it made by foot pads & horsemen which as they soon become as bad as the road itself are directed farther back.[3] These paths going through the sage & savine bushes frequently run right across the mouths of a badger or wolverine hole making it dangerous for a horse or mule without great care. Two of my mules fell today in consequence of stepping into them & one I was sure had broken his legs as he appeared unable to rise. Fortunately my fears were groundless however, but I learned a lesson of caution from what might have been.

These Badgers are very numerous & make their holes in settlements, or villages like the prairie dog—the hole is about 9 to 12 in. in diameter, running down perpendicularly for three feet & then at an angle to the bottom. As they are very shy I have never seen one alive, but one of the Missouri train killed one answering the description of the Badger as given in Natural History. They are a powerful & rather savage animal & a more wolfish looking head or formidable set of teeth cannot be found.[4] Night clear. Distance 24 m.

Monday Aug. 13. Morn clear night cool wind S. Start 6 1/2. Road led us this morning through a singular pass in the mountains. A ravine some 4 miles long, slightly ascending between precipitous mountains in many places the gulley was scarce wide enough to admit of the passage of a wagon. Winding around first north then NW, W, SW & south we finally

[3] "Traveling south we passed many who were on foot. Some had a hores or pack mule or even an ox with a pack on his back or maybe the lone man carried his own pack. Often men who started out with companies, after they were in California [*i.e.,* beyond South Pass] grew impatient and left their companions behind while they hurried on alone to the gold fields." Haun Diary, 39.

"At the Raft River camp, Good, having become restive at our slow progress, joined James Doane, a home acquaintance, who opportunely overtook us at this point, having come the most of the distance from the frontier on foot; and the two made the balance of the journey in that manner, packing their meagre outfits on their backs. Much better headway was possible traveling in this manner than in any other then available, while little additional discomfort or inconvenience was suffered; since the emigrants that year were supplied with abundance of provisions, and were so thickly strung along the route, probably at that stage of the season all the way from the Missouri River to the Sacramento, that accommodations could generally be obtained when needed." Leeper, *Argonauts of '49*, 54.

[4] James A. Pritchard mentioned seeing badgers on the South Platte, but few other travelers noted them. Pritchard, *Diary*, 71.

emerged upon a hill overlooking a valley with creek of fine water & plenty of grass & like the ravine we had just ascended bounded by high bluff mountain sides. Descending into this valley we followed its beautiful winding course for some miles & camped at 7 on its creek. Distance 22 m. Day has been hot winds west. Evening clear. Have passed 40 wagons of different trains today.

Tuesday Aug. 14. Morn chilly, raw wind from E. Night cool. Start 6 1/2, road dusty & heavy as usual. We had been told & had seen notes & cards stating the same thing that it was 100 miles to the Humboldt River from the commencement of this cut off.[5] As we had travelled that distance on it we were in hopes today to see that famous river & camp on its banks to-night. When therefore this morning we emerged from the valley where we passed the night upon a broad plain with a line of trees through its center we were sure we saw the Humboldt Valley & consoled ourselves with the reflection that twas only 3 or 4 miles now to the Sierra Nevada & made our calculations how we could reach them in 12 days &c. Judge of our disappointment & heartsinking to learn after a hard morning's travelling on reaching "the row of trees" that we were on the head of Raft River, & that twas 130 miles yet to the Humboldt!

It took us sometime to reconcile ourselves to being thus "set back" in our calculations & we begin to believe that this is indeed a long road & almost endless. There was nothing for it however but to make the best of our case so travelling up the river we camped on its banks at 7 & directly opposite where the Fort Hall road coming down the hills unites with the cut off.[6] Distance 20 m. Evening clear, wind west.

Our information as to our wherabouts was had from a mountaineer whom we fell in with & he also told us we had saved by the cut off 140 miles as we had been told before. How this can be I cant imagine. By the guide book I can only make to be about 124 miles.[7]

Wednesday Aug. 15. Morn cloudy mild. Night clear warm. Wind West. Start 6 1/2 passed by the road side near our camp a bed of real Wethersfield onions covering perhaps 20 feet square of ground. They probably are the products of some seed thrown out as spoiled by some settler, or they may be indigenous. I have repeatedly picked and eaten wild garlic of fine flavor. Heard the short sweet notes of a solitary brown thrush, at least so I judged it to be, I could not see the singer. His notes resembled exactly our brown thrush & like everything

[5] Perkins came back onto the main California Trail some distance beyond where it branched off the Oregon Trail, possibly a journey of a day and a half or two days. This point is somewhat north of the present Nevada-Idaho border. "After we had crossed the desert plain," wrote Kimball Webster, "we found a small stream of clear, cool water at which we halted two hours and became refreshed. We traveled six miles to Raft River and camped. Here was intersected the old trail from Fort Hall to California. The trail through the Cutoff—a distance of about 120 miles—is good with the exception of being considerably uneven." Webster, *Gold Seekers of '49*, p. 75.

[6] The complex of roads and trails through this section of Perkins' route was outlined in 1857-1858 by W. H. Wagner, engineer for the Department of Interior in an exploration of routes to the Pacific. This map, which bears the title "Preliminary Map of the Central Division Ft. Kearney South Pass & Honey Lake Wagon-Road surveyed and worked under the direction of F. W. Lander, Supt. by W. H. Wagner, Engr." is included in the report of this survey. Exec. Doc. 108, H.R., 35 Cong., 2 Sess.

[7] The map accompanying Ware's *Guide* gave the route from Fort Hall to the junction of the South Trail (Hastings' Cutoff) in the Humboldt Valley. Perkins appears to have been following this guide.

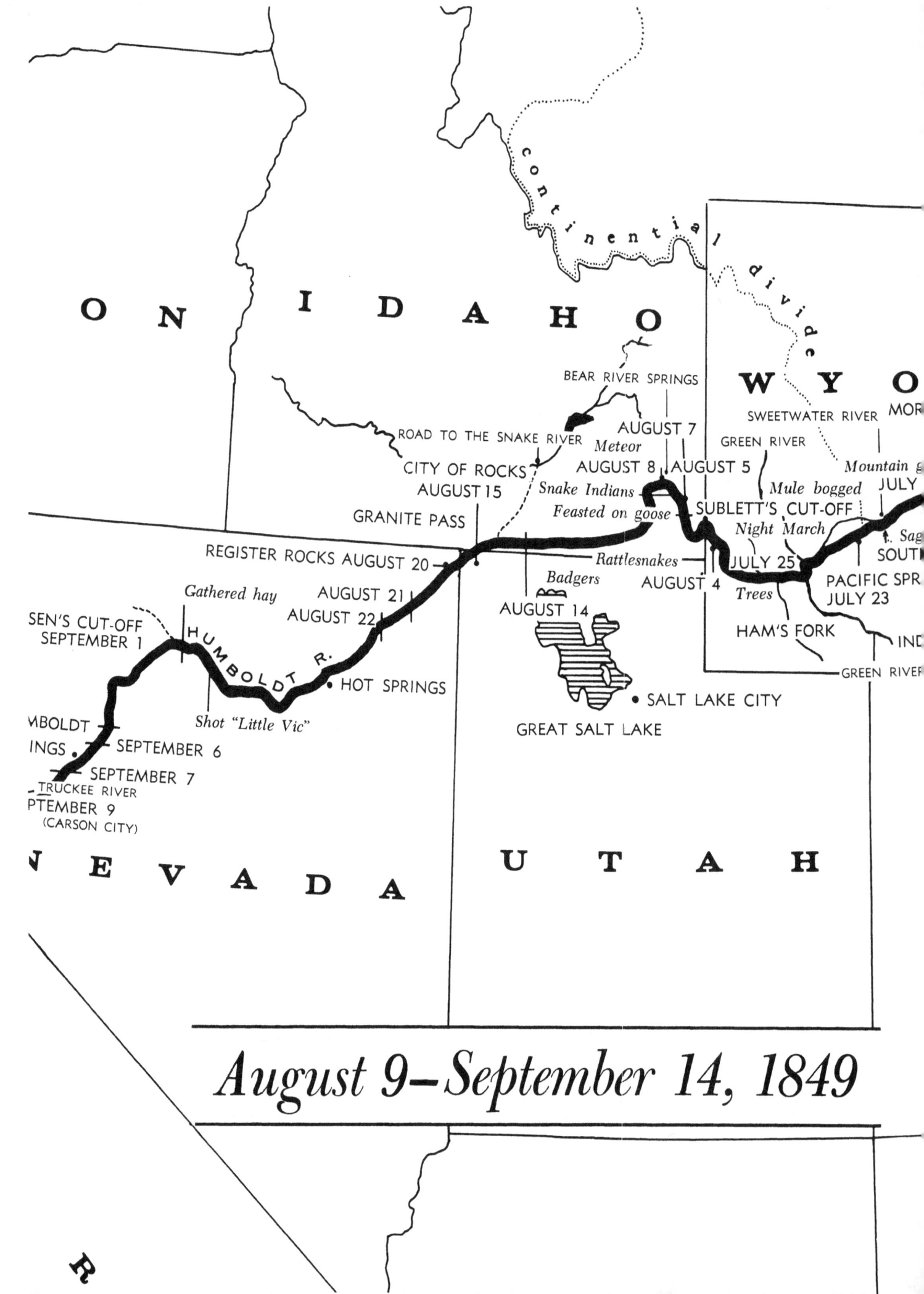

ON   I   D   A   H   O

                                    continental divide

                              BEAR RIVER SPRINGS

                                                    W Y O
                                            SWEETWATER RIVER   MOR
        ROAD TO THE SNAKE RIVER   GREEN RIVER
                              Meteor   AUGUST 7
    CITY OF ROCKS   AUGUST 8 AUGUST 5   *Mountain g*
    AUGUST 15   *Snake Indians*   *Mule bogged*   JULY
    GRANITE PASS   *Feasted on goose*   SUBLETT'S CUT-OFF   Sag
                                            Night March   SOUT
    REGISTER ROCKS AUGUST 20   *Rattlesnakes* JULY 25
                                    AUGUST 4   PACIFIC SPR
    Gathered hay   AUGUST 21   *Badgers*   JULY 23
                            AUGUST 22 AUGUST 14   *Trees*
    SEN'S CUT-OFF   HAM'S FORK   IN
    SEPTEMBER 1   HUMBOLDT R.   GREEN RIVER
                              • HOT SPRINGS
    MBOLDT   *Shot "Little Vic"*   SALT LAKE CITY
    INGS •   SEPTEMBER 6
           • SEPTEMBER 7   GREAT SALT LAKE
        TRUCKEE RIVER
    PTEMBER 9
    (CARSON CITY)

    N E V A D A   U T A H

August 9–September 14, 1849

R

I see which is rare here, but abounds with us called up recollections of the pleasant home & dear friends I have left behind me & when, & whom I would almost give my interest in California to be with again. Even the bed of old fashioned onions nearly made the tears come. They looked so like old acquaintances!

Observed today also numbers of a very pretty bird which I have seen occasionally since leaving Laramie. It is about the size of a blue jay & resembles that bird in its shape and manner of flight &c, has a black glossy head and tail, the latter quite long, body white on breast & black back & white mingled, & wings the outer half pure white, that next the body also black. They have no note that I have observed but a short chirp.

P.M. a thunder storm passed within sight to the south but only giving us a grateful sprinkle. Our road today leaving the head of Raft River passed through the ravines & over the mountains to a broad valley crossing which it entered another deep ravine & brought us among piles of white & brown rocks of all shapes and sizes some of them fantastic enough.[8] Camped at 6 against a huge rock standing entirely isolated from its fellows, & being nearly perpendicular for 100 feet in height. We are in a valley filled with just such immense detached pieces. Some are conical in their shape looking like huge loaves of white sugar others are composed of 3 or 4 pieces on top of the other & looking as though a child could push them over and send their huge masses thundering below. Near & just above us is another great Syrian mass on top of which are several hawk's nests & the birds have been whistling at us all the Evening. This again must have been once in a state of great excitement & commotion far exceeding even that usually attending the eve of our presidential elections. How these masses became broken off & sent into the valley below or piled one upon another to the height of 100 to 200 feet I leave others to determine. The whole Mts. though through which we have lately passed exhibit the effects of intense volcanic action in a remarkable degree.

Just before camping this Evening had another touch of rain accompanied with much heavy thunder. Day has been hot except during the storms, which came from the West & S.W. Distance today 20 miles.

I picked up this evening several pieces of quite pure marble white mostly, some clouded or streaked with yellow. Found also around the rock at which we built our campfire quantities of mica in small scales, some quartz crystals also a little horn-

8 This is the City of Rocks, near the present town of Almo, Idaho. On August 29 Bruff wrote, "An entire range on our left, of volcanic hills, for about 15 miles: and on our right, similar formations for about 10 ms. when we entered a very extraordinary valley, called the 'City of Castles.'—(*City of Castles*) A couple of miles long, and probably 1/2 mile broad, A light grey descripitating granite, (probably altered by fire) in blocks of every size, from that of a barrel to the dimensions of a large dwelling-house; groups, Masses on Masses, and Cliffs; and worn, by the action of ages of elementary affluences, into strange and romantic forms." Bruff, *Journals*, I, 156.

Kimball Webster, just seven days behind, confirmed Perkins' observations on the condition of the road. "Road very dry and extremely dusty. Raft River is a tributary of Snake River, which is one of the principal forks of the Columbia.

"Finding good grass three or four feet high in this valley, and our stock being in poor condition for traveling, we concluded to remain here until Monday morning. A few of our mules and horses have been left behind, they having become completely worn out or exhausted." *Gold Seekers of '49*, p. 75.

blende some little & many black hard glassy fragments that I cannot recall the name of though remember having seen them before. I think this immediate region would yield richly to the searches of the geologist.

Thursday Aug. 16. Morn clear and cool. Night cool. Wind N.W. Start 6 1/2. Road today has taken us near the highest point of the dividing ridge & descending some very long & extremely steep & difficult hills—difficult for wagons & teams. We are again encamped within the Great Basin, on a beautiful creek running westerly. Passed numbers of teams among them that of Doct B. B. Brown of St Louis, which I had a proposition to join when I first contemplated this Expedition. Evening clear & fine, quite mild. Day has been cool, P.M. wind from W. camped at 7. Distance 24 miles. Nothing remarkable in todays travel Except the hills mentioned above.

Friday Aug. 17. Morn cloudy. Wind S. Night quite warm. Start 6 1/2. Day hot & dusty. At our noon camp were visited by a storm of rain & hail with some thunder, from west. Lasted but a short time however & was not sufficient to put down entirely our old torment the dust. At 3 arrived at a creek from which it is 13 miles to water & concluded to stop for the night. Distance 14 miles. After camping for night had another drizzling rain for some hour or more. Considerable rumbling of thunder in the distance. I have seen no such thunder or lightning Either at night among these mountains as we "get up" at home.

Travelled a few miles today with an old acquaintance & college chum of Charles Rhodes. Kerr by name from Columbus, Ohio. He was riding in advance of his train & we had considerable conversations. Found he knew brother Charley of Zanesville, the Grangers, and others, &c, & we could talk of mutual acquaintances which made our short intercourse very agreeable. He wished me to present his regards to Rhodes which of course I shall take pains to do. Day has been warm Except during rains. Evening cool, cloudy,

Kerr was also acquainted with the Waters family. Knew Ava when he was working in a tin shop in Columbus & took Mrs Gill from Sewickly to Col. the first time she was ever in the latter place.

Saturday Aug. 18. Morn clear chilly wind from west Night chilly. Start 7 1/2. Travelled over a stretch of 19 miles with-

9 The Snake Indians were much in evidence to the emigrants on the Rocky Mountain stretches of the trail. They cut their trails around the cliffs, through the sage brush, and made fording places over the streams. Bryant wrote on July 14, 1846, [in present-day Wyoming] "During our afternoon's march we fell in with a party of some sixty or eighty Soshonee or Snake Indians, who were returning from a buffalo-hunt to the east of the South Pass. The chiefs and active hunters of the party were riding good horses. The others, among whom were some women, were mounted generally upon animals that appeared to have been nearly exhausted by fatigue. These, besides carrying their riders, were freighted with dried buffalo-meat, suspended in equal divisions of weight and bulk from straps across the back. Several pack-animals were loaded entirely with meat, and were driven along as we drive our pack-mules." Bryant, *What I Saw in California*, 136.

John Edwin Banks wrote graphically of the woes of traveling in arid country infested with thieving Indians. "Started before daylight. When we reached the river no grass, but all was excitement. The Indians have aroused a storm that may fall heavily on them. They stole eleven head of cattle here last night. One man who had a wife and seven children robbed of horses; his situation is deplorable. This caused some twenty men to go well armed in pursuit of the Indians. Four of the party saw the same number of Indians; killed one and wounded one. The main body saw some one hundred and fifty savages, but thought best to retreat." Scamehorn (ed.), *Buckeye Rovers*, 66.

A little farther along the trail J. Elza Armstrong's party had an experience with

out water & came to a small stream of pure water. At the spring were some Shoshone Indians, looking much like Snakes & only more cheerful countenances & more honest faces.[9] They had 3 mountain goats which they had just killed for sale to Emigrants. On my touching the animals & pointing all round over the country & asking where, one pointed to a ravine just above us & said "heap, heap" meaning that there were a great many to be had. They are a pretty animal, about the color of a deer, tho not standing quite so high but something of the same form, with horns much like a fish hook with a long shank projecting forward from directly over their eyes. I bought quite a large piece of the meat for a bar of lead & found it very sweet & fine eating much like mutton but more juicy & with more taste. Have had no fresh meat for some time so eat enormously of it, in a manner that at home would most likely have occasioned us much inconvenience, but here was nothing out of the way.

I saw also today the skin of a Mexican wild hog. Something of a curiosity. The fur or hair was tolerably fine & long wavy, of a reddish yellow color, & made a very fine saddle seat, to which use it was applied. From appearance was far different from anything of the hog kind to which we have been accustomed, being really quite a handsome affair. Camp at 7 near a well, the only water in this vicinity. Distance 25 m. Day exceedingly dusty, more so than usually if possible.

Sunday Aug 19. Morn clear cold, night very cold. Ice made 1 inch thick in our well & our canteens froze hard.[10] Ground above was frozen hard around the water & everything reminded me of a December morning at home. Clear & sharp.[11] Wind N.W. Start 7. Day cool clear.

At 9 came up to the rear of a long train of wagons which were just leaving a fine valley we had entered & saw on the cover of one written "Harmar & Co. No 4," & soon was among the boys of Chapin's Company, Hovey, Creighbaum, Fearing Soule Roe & Chapin himself came out of a wagon on hearing our voices.[12] Our reception was of course very flattering & we found plenty to keep us all talking for the rest of the day. On their insisting upon it we camped with them at 1, distance 18 m. The boys all look well & hearty but lost 4 of their men before being two weeks out, Drown, Hewlett, Cutler & Clark. Their cattle are in better order than any I have seen on the road.

The valley where we are Encamped is a remarkably fine one abounding in rich grass, & presents quite a lively appearance

the Indians. "Early this morning we were aroused by the guards. They came in and told us that one of our oxen was shot with an arrow and one missing. The guard went to sleep and did not wake up until nearly daylight. The one they had drove off was the best ox we had. He was the largest ox I have seen on the road. He was quite thin. Eight of us rigged our guns and started in a few minutes. We struck the trail and evidence three hundred yards from camp. They had killed him and taken the best of the meat and the hide with them. There had been four Indians and drove him out from among the cattle. We found two arrows. They stole two horses and one mule and, I suppose, they packed part of the meat on the horses they stole. After we struck the bluffs we seen Indians' tracks. We went twelve or fifteen miles back into the mountains. We could not see any of them. We seen a great deal of sign. We trailed the fellows to where they had laid down to rest, and there lost the trail." *Ibid.*, 80-81.

"Snake Indians to our camp every day, are very free & friendly. I gave them 2 shirts for a buffalo robe." Steuben Diary, July 18, 1849.

[10] Joseph Warren Wood said about the temperature in this area that, "The night [July 3, 1849] was very cold, & I did not suffer as much with cold any day last winter as I did at this time. I was light clothed." Wood Diary, July 3, 1849.

[11] On August 29, 1849, Bruff wrote, "Patches of snow on the adjacent mountains. We were all white this morning on awakening, with frost, and my hair being very long, the ends were froze to the saddle and ground, so that I had to pull it loose, but had to leave some, as a memento for the wolves to examine.—I thought of the picture in Gulliver's travels, where the Liliputians had *picketed* his hair down while he slept. We had to look out sharp last night about the indians." Bruff, *Journals*, I, 156.

[12] This company had started from St. Joseph about the time Perkins and his company set out for the gold fields. The Harmar Company was from the village of Harmar next to Marietta.

just now, there being wagons & camps up & down as far as we can see. We have passed in our short days travel 107 wagons! About noon we came to the hot spring which gives the valley its name of "hot spring Valley," there was a spring of water hotter than could be borne, & a stream running from it of considerable size which two hundred yards from its source was hot enough for Thomsonian uses. I amused myself for some time by trying to account for the heat & where it originated. I dont wonder at the old Dutch Emigrant who on stopping at this spring & hastily tasting it ran back to his wagon crying "drive on boys hell ish not far from dis place." Certainly there is something very mysterious in the appearance of nearly boiling water at the surface of the ground from off some great subterranean cooking stove. It was considerably impregnated with iron & carb lime as was the Bear Spring water. What is a little singular is that a spring of clear cold water issues from the ground not 1/4 of a mile from the other on the stream flowing from which we are camped this Evening.

August 20–September 14, 1849

MONDAY AUG 20. Morn clear cool. Wind North, Night cold. Could see nothing of our mules this morning on awaking. Having neglected picketing them as usual depending on their staying with the animals of our Harmar friends. I surveyed the country round with my telescope but without success & we were of course considerably alarmed. It being quite possible that the Shoshones might have taken a fancy to them. The Harmar Co. left about 7 & we scattered in all directions in search of our lost animals & about 8 found them at the foot of the hills some 4 miles from camp, luxuriating in some fine grass they had no doubt wandered off to procure.

Packed up & left abt 1/2 past 9. Soon overtook the Harmar men & travelled with them till 3 when we camped on the Humboldt Spring,[1] so called from being at the very head of the first creek which following the valley united with another some 27 miles below to form Humboldt River. This fountain head of the great Humboldt is a pool of clear cool water situated in the middle of a valley surrounded by grass & small willows & from it flows a stream of some two feet wide & 4 to 6 inches deep—increasing in size & volume as it continues onward till at its junction with its fellow stream it is quite a creek. "Larger streams from little fountains flow" &c is well illustrated by this instance.

Day very hot & dusty, distance travelled, 14 m. Chapin has calculated & put on his wagon a neat road meter & by it we find that our estimates of one days travel & speed generally, is pretty correct. My notes of distances corresponding almost exactly with his own taken from his measurement.

Tuesday Aug 21. Morn clear & warm. Wind N.W. Night cool as usual. Start from Humboldt Spring at 8. The wagons being somehow in advance of us.

After traveling some 4 miles we came to the most remarkable canon I have yet seen where thes creek forces itself through a range of mountains, forming a narrow ravine passable for wagons only by crossing the bed of the stream 9 times in the course of 6 miles the length of the pass. When we entered the gorge we found it completely choked up with ox teams slowly forcing their way through. Our friends being in the rear we pushed ahead to camp & wait for them to come up. In passing the wagons so narrow was the road that we were compelled to climb over the rocks far above the road & follow narrow paths hardly sufficient for the passage of a man. One of my mules was knocked off the path at one place by his pack striking a

[1] This spring as such was not located on any of the contemporary maps. It was situated at the head of Bishop's Creek, the canyon of which Perkins described next day.

projecting rock. Fortunately the hillside was not so steep at this particular point as at some others & he found foothold & scrambled up again.

At the mouth of the canon we had passed all the teams some 80 or 100 in number & emerged upon the beautiful valley of the Humboldt wide open & covered with most luxuriant grasses of various kinds, blue grass, herd grass, red top &c which would have been considered superb mowing at home. On leaving the road to search for a camp on the bank of the creek we waded through these fine fields of hay two feet high, & turned our mules loose among it at 2. The trains coming up at 6. Distance 15 miles. Day warmed. Camped on creek.

Wednesday Aug 22. Morn clear & warm. Night cool Wind north. Start 1/2 past 6. After a good breakfast we shook hands with & bid farewell to our friendly Harmar acquaintances & "rolled on" once more on our "own hook." Came down the valley & at 11 we arrived on the Humboldt itself, about two miles below the junction of the streams forming it. At this place it is a slow flowing rather deep creek, some 15 or 20 feet wide & two or three deep, lined with willows bulrushes, & long swamp grass, & abounding in small frogs & I am told in fine trout also.[2] The valley has widened somewhat though nothing near as broad as the Platte Valley, but presenting a much richer but so monotonous appearance. The mountains bordering the valley are high barren being covered with dry stunted sage & vary irregular in their outline, being a succession of sharp ridges narrow ravines & abrupt points. Camped at 1/2 past 5 on River. Distance 25 m. fine grass. Day has been fine.

Thursday Aug 23. Morning cloudy, mild. Wind E. Night mild. Start 6. Day hot. at 12 m. a thunder storm passed to south & we were sprinkled somewhat & about 2 another passed over us. Rain fell slowly in fine drops, very different from the usual driving storms of this region & continued to fall the rest of the P.M. & this Evening a regular drizzling rain of one of our fall or spring months. This I take to be very unusual here, judging from the parched appearance of the country & vegetation & such being the statements of various travellers.

Camped at 6 all wet & chilled, a cold breeze blowing from S.W. Have built a "roaring" camp fire however from dead willows & here we sit squatting around it. With no covering but the wet heavens over us, trying to imagine ourselves comfortable & forming pictures for friends at home to contemplate,

[2] On August 2, 1852, John Hawkins Clark wrote, "Passed over mountains to the head waters of the Humboldt river. This river, I believe, was discovered by the German explorer Humboldt. The old philosopher left his name in a wild country, but to him it may have been an interesting one; good grass, plenty of water and the wild sage for fuel. We camp at the spring from which issues this river of the desert; the spring is six feet in diameter and six feet deep. At the bottom lies a big sheet iron stove thrown there by someone who had got tired of hauling it, I suppose. We are now encamped at the headwaters of the dreadful Humboldt of which such hard stories have been circulated on the road." Barry (ed.), "Overland to the Gold Fields of California in 1852," *Kansas Historical Quarterly*, XI (August, 1942), 279. The Humboldt had actually received its name from Frémont in 1845. The stream was variously called Ogden's or Mary's river.

could they take a view of us from the windows of their comfortable houses. Distance today 25 m.

Friday Aug 24. Morn cloudy & warm. W. S.E. The drizzling rain continued nearly all night & a jolly time we had of sleeping to be sure. Our bedding was all covered up under gum cloth with the packs & we disliked to spread it out & have it all wet to get into so we sat around our fire till about 10, hoping it would hold up. Giving up the idea we finally pulled out our Buffo &c, & wrapping a robe around me I selected a ridge near the fire & lay down in the wet grass. Throwing my cloth over me to shed some of the water if it would, & slept tolerably well notwithstanding the pattering of the drops on my ears & cheeks. Woke up tolerably dry considering the circumstances & when I built a good fire & began to feel its warmth in my veins felt pretty comfortable once more. Our bedding—Buffaloes &c being all wet we stayed in camp till today to dry our goods & finally started not much worse off than before our nights adventure.

Passed this morning a real boiling spring, the water being at a temperature of 212° & answering well for cooking purposes. A fish put into this spring by an Emigrant was well-cooked in two minutes, & the steam rising from it can be seen at considerable distance.

Passed today the Rushville, Illinois, train who told us of quite a mishap which occurred to them a few nights since. Their cattle were turned out to graze at the pass of the mountain & no guard set over them. In the morning twas found that the Diggers had carried off 22 head! Some of the men started up into the mountains in pursuit of & followed the trail some 30 miles, clear up among the snow & finally found their oxen, some killed, some hamstrung, & the rest jumped off of a high bank into a kind of pen from which it was impossible to get them out without ropes and pullies, while the naked rascals who put them there could be seen dancing upon the rocks & hill tops & making all kind of jeering gestures, but taking care to keep out of rifle shot.[3] The party returned without recovering one of their cattle. These Diggers are very sly and dangerous fellows to strike. Another train a few nights since had their cattle shot with arrows while the herdsmen were driving them into corral about 9 P.M.

This afternoon the road followed the river through a canon some five miles, crossing it 4 times. The hillsides were worth noticing, being composed of various kinds of rock, sometimes

[3] Along the Humboldt, on October 7, 1849, Bruff wrote that he had seen a "Notice written to a man in the rear:—dated 2 days since, informing him, *that the dead horse & mule, in the bottom, had been shot by the indians, and belonged to Mr.* (obliterated). As the murdered animals were close by, I examined them, satisfied of the fact." Bruff, *Journals,* I, 209. "Traveled down the valley 19 miles today, making 120 miles this week. We are now among a race of Indians called Diggers, said to be very troublesome to the emigrants. They are a wretched class of beings living on insects, reptiles, etc. Are famous for their thieving propensities. Our course will now be down the Humboldt Valley for 250 miles." Steuben Diary, August 4, 1849. "200 teams passed us today, Almost every night in this region the emigrants lose cattle, Horses or Mules, by the Indians. We hear them every night imitating wolves, their barking & howling echoing from the Mountains, hills and valleys. We guard our stock & yet have lost naught." *Ibid.,* August 27, 1849. "May God help those who are far behind, otherwise they must perish. We passed a wagon on the 24th all stock belonging to it having been killed by Indians. At night camped near the train that had lost animals—4 horses & mules stolen & 3 oxen shot Wagon was left behind abandoned." Chamberlain Diary, July 26, 1849. The Diggers gave everybody trouble. Captain De-Wolfe noted on September 17 on the banks of the Humboldt that, "the Arkansas Company had four horses stolen by the Indians last night we are now amongst the Root Diggers they are a poor miserable set of wretches they subsist principally upon roots they are the most thievish set of Devils we have been amongst." DeWolfe Diary, September 17, 1849. David R. Leeper and his Hoosier company suffered from these Indians. "The diggers," he said, "raided our stock, taking six of our best oxen; one from each of our six teams, as it happened. The theft was discovered early the next morning, and a detail from our camp at once pushed out upon the trail of the thieves. The course taken by them was found to lay over a region covered with scrubby cedars showing a surface so compact that the trail could be distinguished only by the marks made by the

a solid precipice, & at this broken masses & ridges. Many places the strata of rock were mainly perpendicular resembling much that I noticed on Bear River, & having the same direction, from N.W. to S.E.

We passed company in this canon with two very pleasant fellows from Virginia, Norfolk, with whom we had some previous acquaintance, & shall endeavor to travel together the rest of our tedious journey. They are packing like ourselves & have been near us for several days. One named Coleman a lawyer, the other Gilbert who has been in business in Cin. Camped together at the mouth of the pass, at 1/2 past 5. Distance 16 m. Evening clear & mild. Day warm.

Saturday Aug 25. Morn clear cold. W. S.E. Night cold. Considerable ice made, nearly 1/2 in. in thickness. Start 6. The road left the river a short distance from our camping place to avoid a canon & took a course on the hills for 18 m. without water 1/3 of the way, we took a path down a ravine said to be a cut off from the regular road but found on getting through it after some 10 or 12 miles travel over a wild country that it was a regular "cut on" as it brought us to the river some 2 or 3 miles above where the road struck it & on traveling down to the latter place most of the ox teams which were on the road when we left it had arrived before us, making us several miles of trail for nothing. Followed down the river bank some distance to find grass which has all been grazed off in the vicinity of the End of the "stretch" & camped at 7 on River. Distance 25 miles. Day hot roads dusty, & hard travelling.[4]

What a luxury it is after such a days tramp as this has been when on arriving at camp in the evening all covered with dust so that your best friends could hardly recognize you, & your face looking as if covered with ashes, none of the features distinguishable but the mouth, to go down to the waters Edge, roll down your collar take off your hat & douse head & face under the cool elements, & how refreshed one feels after this operation.[5]

Sunday Aug 26. Morn cool & pleasant nights cool. W. E. Start 6 1/2. About one mile from our last nights camp the road takes over the hills to cutoff a bend in the river which it strikes again some 4 or 5 miles distant. We started to go over this cutoff intending to camp today on the fine grass to be found at its termination. I sent the mules forward with John while I staid behind some half hour to saddle and wash my mule. On the

oxen's hoofs in displacing the sharp, rock fragments. After the marauders had been thus tracked about twenty miles, the attention of our party was suddenly aroused by a loud shriek from behind a ledge or rocks, and, at about the same instant, a number of redskins were seen, down in a ravine walled in by volcanic bowlders, betaking themselves to their heels as fast as their legs could carry them. But the cowardly flight of the savages was of little avail to us, as every oxen had already been put to the knife." Leeper, *Argonauts of '49*, 68-69.

4 The alkali ˙ dust of this territory [the Humboldt] was suffocating, irritating, our throats and clouds of it blinded us. The mirages tantalized us; the water was unfit to drink or to use in any way; animals often perished or were so overcome by heat and exhaustion that they had to be abandoned, or in case of human hunger, the poor jaded creatures were killed and eaten." Haun Diary, 35.

5 "Started in good season this morning [September 4, 1849] and in seven miles of travel we crossed a fork of Mary's River—coming from the north—after which we traveled 16 miles and camped on the river, where we found good grass.

"We passed over some sand hills today where the road was extremely dusty. After traveling a few miles over this dusty road on a warm day a white man will be equally as black as a negro. The dust is different from anything of the kind that I have ever before seen. It bears a strong resemblance to ashes in some respects, both in appearance and smell, and seems to contain quite a large percentage of alkali." Webster, *Gold Seekers of '49*, 78-79.

road over the sand hills & at its end however I could see nothing of my company & though I travelled down some two miles making inquiries of ox teams camped at various places, then crossed over & up the same distance on the other side, making inquiries again at all points. I could hear nothing of them nor had any one seen them. So about 12 o'clock ransacking every place I supposed they could be in, I determined to go on concluding they must have done so before me.

Travelled till 3 o'clock when I overtook an Ohio train from Zanesville & Cambridge commanded by Gen. Moore of the latter place & on making my inquiries of them & telling my story they insisted that I should if I heard nothing of my company, camp with them tonight. Which I promised to do & so after a very fine supper here I am under a tent making acquaintances of whom I find a number also many who know those I know John & Charley Rhodes, Harry Grainger &c. Gen Moore is very much of a gentleman has been representative in the Ohio legislature & auditor of Cambridge County, knows Charley well & says he brought him out of a serious sickness some time since.

I saw in this camp two black tailed deer, the first of this animal I have had an opportunity to examine. They were killed by a young man of the train about a mile back on the banks of the River. They resemble our common deer very much, horns & all Except that the tail is tipped with jet black hairs.[6]

By far the most comfortable way of traveling this road is with oxen. They can carry more comforts & luxuries, cooking apparatus, &c than any other means of conveyance, & hunters of the party can mount their horses in the morning & run to the hills at their leisure without any fear of being left behind, & without being compelled to hurry to catch up, & they always have game in abundance, Antelope, blk tailed deer, wild geese & ducks sage hens prairie dogs &c, can generally be found frying or broiling at the camp fires of most any ox train, & many a good mess have I been treated to at their boards.

Were I coming this road again I would get me 2 good mules & a surry, have my packs all made up ready to put on with 30 or 40 days provisions in them, then make an arrangement with some small ox train of clever fellows to travel with them & board by the week they carrying my packs & I would go in their company as far as Bear River, when my mules being in prime order my packs could be put on & I would go the rest of the way flying, & make the trip quite as quick if not quicker than we

[6] "One of the men gave me some venison of the Black tailed deer, common in the mountains. It was excellent." Wood Diary, June 20, 1849.

shall do now, besides enjoying myself vastly more seeing the country, hunting &c. This kind of arrangement I could easily have effected when at St. Joe & had I been alone as I originally contemplated.[7] Should have done so & had a far more agreeable time than any part of this journey has proved. Distance today 15 m. Wind East. Day excessively hot. Have seen number of fine flocks of wild geese ducks &c in the River & flying over.

Monday Aug 27. Morn clear. Night cold, ice made. Wind S.E. after treating me to a sumptuous breakfast, Gen. Moore & his train left at 7 wishing me a speedy union with my company & exacting a promise that I would stop with them another night when I came up. Made me a comfortable nest in the dense shade of some willows. Picketed "Carry" where she would be seen from the road & sat down to wait for my men to come up. I having learned from some packers this morning that they were camped some 8 or 10 miles behind. Amused myself while waiting by making an Indian whip of a short willow stick & some thongs of leather.

About 11 I heard Johnny's voice call out "here's Carry & Doug must be near" & soon they all came up. Learned that they had camped in the willows above the road & had hung out a towel to indicate to me where they might be found, but though I had been within 1/4 of a mile of them, It was 1/4 of a mile higher up the river than I went, so I saw nothing of their sign. Saddled up & we went on together once more, & camped at 2 to give me an opportunity this P.M. to do some little things for myself, I intended doing yesterday. Distance 10 m. Day hot, Wind S.E.

A very fair hoax was played off this afternoon opposite our camp by the Capt of the Cherokee train which was some two or three miles back. He had been forward to look for grass & was returning when he met at our camp several trains which had stopped for water & was Being dressed in half Indian costumes, with buckskin pants, moccasins &c, & rather dark complexion, he was taken for a French mountaineer, of who there are numbers through this country, & was immediately beset by a crowd of Emigrants anxious to learn the latest news from the "diggings." In answer to their inquiries he told them that he was just from California having left Sutters just three weeks ago yesterday, that Gold was plenty, and one fellow asking him how much a man could dig in a day he replied that "a man not used to very hard work but pretty industrious could easily get 500$ per day," "Ah," says the questioner rub-

[7] At this point Perkins was dreaming. It was quite possible to make such a journey expeditiously, as was demonstrated by Edwin Bryant in 1849. He crossed the continent by way of the Overland Trail in sixty-five days. His was one of the first parties to reach the diggings in the early summer of 1849. "To-day many of our most worthy citizens leave for California. Mr. Bryant's company goes on *Pike No. 9* and Mr. Huie's on *Meteor 3.* Their departure will no doubt draw hundreds to the landing. A safe journey to you all, gentlemen; may your golden dreams be realized to the fullest extent. This will make three companies which have left this city for California—the first company, Captain Maury's started two weeks ago." Louisville *Daily Journal,* March 26, 1849. "The numerous friends and acquaintances of this gentleman will learn with pleasure his safe arrival in California. A party consisting of thirty or forty were conducted by Mr. Bryant to this country and were among the first to arrive overland this season, Mr. B. left San Francisco early in the spring of 1847, to cross the continent with the party commanded by Gen. Kearney. His notes of travel, history of events, and general observations published sometime after his arrival in the United States, form a truthful and entertaining volume, and may be properly considered pre-eminent among the works written upon the country." San Francisco *Alta Californian,* August 30, 1849. Robert Dulaney, a member of the Bryant Party wrote his brother Charles back in Louisville (Oct. 16, 1849) that "I got to Johnson's Ranch on the 3rd inst. after a very long & fatiguing journey of about 85 days." Dulaney Papers, Bancroft Library.

bing his hands with a great deal of satisfaction, "thank God I've been brot up to hard work all my life." After a number more such yarns he left them as he was in a great hurry to reach Green River. To discuss the wealth in prospect & knots of them could be heard for two hours talking over the good news & disputing about what "the mountaineer" said & what he did not say.[8] Everything of course becoming more & more exaggerated the more it was repeated. We happen to know the Capt as we had passed his train & the whole thing was exceedingly rich & afforded us great amusement to watch the excited looks & gestures of the eager dupes, & hear their different versions of "the news" to new comers.

Tuesday Aug 28. We were greeted this morn by one of the splendid soft Indian summer airs of an Ohio November morning, delicately mild & balmy. The finest by far that I have seen on the road, & quite unusual as the sun in this latitude generally rises perfectly clear & in half an hour after a cold freezing night is intolerably hot. Wind East. Night mild, Start 7.

At noon arrived at an alkaline plain where the mountains bordering the valley branched apart one range running south, the other N.W. & leaving a perfect level plain to look over which in the smokey atmosphere was like looking out to sea. Left the river & crossed the plain about 10 miles to the stream again. Wind blowing strong from the N.W. & plain being very deep with alkaline dust which made our eyes smart somewhat as it blew in them.

My mule "Carry" had been for sometime failing for some unknown reason, & had become rather poor & quite weak. In crossing the plains this afternoon she fell down under me & would not rise for some time, though I assisted her all I could, & some hour or two afterwards as I was driving her before me she fell again & I could not get her up. A train coming up just behind, I determined to sell her to them as at our rate of travel she could never get well & I found a fellow at last who agreed to take her for 5$! & I parted with her rather reluctantly I must confess, but the thing was unavoidable as we could not afford two weeks to recruit her.[9] I have seen many mules & horses left by the side of the road to take care of themselves in quite as good order as my "Carry" & I consider myself fortunate in selling her at all as I should doubtless have lost her had I undertaken to drive her two days farther. When an animal gives out on this road it breaks down all at once, & without giving you much notice. Hereafter I shall have to walk most

[8] "We had run upon a party direct from the promised land—straight from the enchanting gold fields. The party proved to be Mormons with their families en route to Brigham land. Their clothing eclipsed any we had seen for tatters and patches; but their oxen, in striking contrast with ours, were rolling fat and sleek, and thus excited our envy. The members of the party were quite communicative, and gave us a flaming account of the diggings, backing up their words with a liberal display of the shining nuggets. This was the first, real tangible proof we had of the existence of gold in California." Leeper, *Argonauts of '49*, 55. All along the trail the Argonauts listened for any story they could hear about success in the gold diggings. Many a rumor floated along the trail on this subject. J. Elza Armstrong said, "Shortly after we started we seen a man just from the mines. He belonged to Walker's train from Missouri. He sent this man on to see what prospects there was. He has twenty men hired. This man reports favorably. He says any man is sure a ten dollars per day. He says a man can make one ounce per day without much difficulty. He thinks we can get there in three weeks." Scamehorn (ed.), *Buckeye Rovers*, 73.

[9] After they had fed their oxen for about three days on willows, bulrushes, cat tails, and other low-grade roughage, the emigrants traveling with Joseph Warren Wood had reason to be disturbed. "Our cattle," said Wood, "consequently look bad! bad! bad! & our prospect is poor for getting through unless we obtain something more suitable for them to eat." Wood Diary, August 16, 1849.

of the way, but this I can easily do. I can walk now with anybody & for any time, hardly knowing what fatigue is,

while I was disposing of my mule Cross & Coleman went forward & Gilbert remained behind with John & I. At evening when we reached the river we could see nothing of them & finally camped without them at 6. Distance 25 M. This is separation no. 2 & I regret it as it may occasion us considerable delay. We not knowing whether to go on or wait, & the same being the case with our companions. A stop must be put to this going ahead out of sight of each other   if practiced much we should lose much time that we ought to be travelling.

Wednesday Aug 29. Morn clear, wind E. night quite cold, ice made 1/2 inch thick. Start 6 1/2 having decided to go on till noon & wait if we saw nothing of our companions, for them to come up. Camped at 11 & lay till 3 1/2 when they not making their appearance we started forward again   some time after we overhauled an·ox train which described them exactly in answer to our inquiries, said they camped for noon close by when Cross & Coleman were waiting for us, some 7 or 8 miles back, that they also inquired if the train had seen anything of us &c. Selecting a comfortable place in the thick willows for our camp & sticking up a notice on a stick stating where we could be found, with a saddle cloth attached to attract their attention we camped at 5 1/2 & shall lay here till they come up, being Satisfied that they are behind. Distance today 20 m. Day hot, W. N.

Thursday Aug 30. Morn clear. Night very cold, ice made in considerable quantity   was nearly stiffened by sleeping cold, having given some of my bedding to Gilbert. Sun rose hot however & soon thawed me out. Wind East.

Had a glorious supper last night,   Gilbert bought 12 hot light biscuits of a train camped near for which he paid 5.00! & I broiled some of the best pieces of my last ham, made a pot of tea & having bought 5 lb of sugar a day or two since of which I had been out for a week or more, paying for it, 50 pr lb we sat down to a meal which it is seldom the good fortune of packers to enjoy, & by contrast to our usual fat bacon & hard crackers was delicious. Our living is anything but tempting, the crackers being hard tastless & some what musty from often wetting & drying, & fat bacon toasted on a stick being a dish that at home would have spoiled my appetite to look at. However we are glad to get anything to live on. Everybody is short of provisions at this stage of the journey the result of so much throwing away in the first instance. For some poor bacon which I succeeded in procuring I paid 37¢ pr lb & they now ask 50. Crackers 25¢ pr lb, only one train has sugar to sell, which is held as stated above at 50¢ & flour not to be had at all.

Hundreds of foolish men not content with slow progress of their oxen have left their trains & taking four or five days provision on their backs & 25¢ in their pocket start off for California, calculating to speed their way through. How they are to get along I cant conceive, as Gen Moore told me that hardly a day passed but he had application from more or less of these footpads for meals, & sometimes though not often wanting to pay for food but was compelled to refuse all alike as he had no more than sufficient for his own consumption.[10] I think some of them must suffer considerably as they reach the land of promise.

We heard enthusiastic news from California yesterday by a Mormon train returning to the Salt Lake. They represent Gold as very abundant & provisions plenty & cheap comparatively. This is gratifying & the general impression is reliable information.[11] We are now within 15 days travel of Sutters & look forward to termination of our trip with the greatest anxiety & pleasure. Should nothing occur to delay me I hope to take "Doc" by the hand the 15 of Sept.

At 12 today went out to the road to ascertain whether our notice was still extant & found that some rascal had torn it down & carried off the saddle cloth, a rare instance on this road, & a rare scamp whoever he may be. Learned from a passing team however that our two lost packers had passed thru not 10 minutes before & could not be a mile in advance. Going back to our camp with the information we immediately packed up & started in pursuit at 1 & overtook the strays in their noon camp at 2 or a little after. Started all together again at 3 1/2 & at 4 was compelled to camp with Gilbert who had suddenly been seized with an attack much resembling the first stage of cholera—cholic &c. Made him some hot tea & gave him an opium pill & this evening he is much relieved.

Camp at 4 Distance 4 m. I hope in future we shall be compelled to make no more such days travel as this one on account of separation &c such delays at this stage of our journey are very annoying. I am very anxious to get letters from home which I am confident will be waiting my arrival in San Francisco, & it seems almost as if they were within my reach when I can say "Next week or week after we shall be through"

10 R. C. Shaw gave a graphic description of his company's predicament at this point in its overland journey. "We subsequently learned that our message [for aid] was properly delivered and that the Mormons sold our company four head of beef cattle, though at many times their actual value. Being about to resume the journey, our Mormon friends presented us with four or five pounds of flour, a pint of rice, and a very small quantity of salt." Shaw, *Across the Plains in Forty-Nine*, 134. Kimball Webster said, "Our stock is again in poor condition for traveling, and we have yet a long road before us before we reach the Sacramento Valley. Our cattle, which are our principal dependence for our food, are almost destitute of fat or suet, and are composed chiefly of hide, horns, cords and gristle and lean, flabby meat. It is not very nutritious living. Our appetites are wonderfully good. We have sometimes boiled the hide when we had plenty of time." *Gold Seekers of '49*, 79-80.

11 Though emigrants to California were always ready to listen to glowing reports of returning miners who assured them that staggering amounts of gold awaited them, there were those returning from the country with doleful tales indeed to tell. On September 22 in the valley of the Humboldt, Gray's party met with a man named Lawson who gave them a realistic story. "Last night," wrote Charles Gray, "Lawson whome we saw about a week ago arrived having given up his place in the exploring expedition & returns home on account of his bad health, he interested us much by his accounts of California, *even at his account we shall have a rough time of it & he no doubt puts as favorable a view of it as he can & at the same time have regarded for the truth.* He says there is much sickness in the mines & that our late arrival will be detrimental to us on that score." Gray Diary, September 22, 1849.

with what eagerness shall I have the luck to learn how it fares with all those I love so well & from whom I have not heard a word for 4 long months. God grant that nothing may have occurred in my abscence to make me curse the day in which I was induced to part with all that makes life dear, & undertake this hazardous & tedious journey. Day very hot. Wind N.E.

Friday Aug 31. Morn clear warm. Night cold ice 1/2 in. thick made in our coffee pot & continued. Wind east. Start 8, being delayed in morning to buy some flour of a team which Coleman was so fortunate to meet with. We bought 50 lb for which we paid 6.12 or 12 1/2 pr lb, & shortly afterwards exchanged 20 lb for crackers pound for pound. The price asked for crackers in money 25 so that we made quite a "spec." We shall keep the rest of our flour as Coleman has taught us a way of making a cake that is really very palatable, & far preferable to crackers forever. The only cooking apparatus we need under this new order of things is a pan to mix up dough in, & our sheet iron frying pan answers this purpose quite well.

Crossed the Humboldt today & took the Mormon trail down its south side. Met at noon a Mormon train just from California & going to Salt Lake & from them gathered considerable concerning the land whither we are bound, among other things that gold was found in great abundance & provisions plenty & cheap, flour 6.00 per hundred &c. After leaving the Mormon train we followed a path through a canon & some half way through I noticed that the pack of one of my mules, little "Vic" was loose & I stopped her, tightened it & whistled to her to go on again when she passed me she gave me a severe kick on the hip with both her hind feet, laming me considerably. I caught her again after some trouble, & thrashed her well & from her frightened appearance & trembling after the castigation I thought she was pretty well cured of her propensity. T'is only within a week that she has shown any disposition to act thus, & has kicked me slightly twice before in that time, which offences I passed by, but determined to punish her surely if repeated.

Not more than an hour afterward I punished her, I stopped her again to arrange her pack which seemed to be unusually troublesome, & as before had turned her loose to go on when she suddenly wheeled so as to bring her artillery to bear upon me, & let fly with all her strength both hind feet striking me full in the breast one of them breaking my watch crystal into a hundred pieces & injuring the watch itself, & the force of the blow knocking me backwards down the hill side heels over head. I scramble up feeling considerable soreness in my chest, & determined to punish her in a way she would remember & that instantly. Drawing my revolver from my belt I leveled it, took good aim & put a ball as I intended through the fleshy part of one of her thighs. The sharp stinging pain made her dance & squirm around lively for some minutes, & the soreness consequent lasted for the rest of the day at least & served to have the desired effect too, for though I walked close behind her & putting my hand upon her back trod on her heels &c, daring her to strike again she was very meek & evidently wished to be excused.[12] Had she repeated her trick I should repeated mine, & severe punishment following instantly the offense is understood by the most stupid of animals in time. I feel much less lameness & soreness than I feared I should this Evening, & hope no ill effects may follow the really very severe blow she gave me.

Camped at 6 o'clock on Humboldt as usual. Distance 20 m. Day warm, sultry, W. E.

Saturday Sept. 1. Another beautiful Indian summer hazy morning with its soft air, & deathlike stillness. How I have enjoyed such mornings at home, under the shade of some trees to lay & meditate & dream. But here it is nought but push on, push on, & ever must be astir. Night warm, with strong breeze from N.W. Start 6 1/2, crossed the plain of the valley from our point across a bend to another & observed unusual barren spots among the grass, thickly strewn with small sea shells, spiral & flat, colored & differing entirely from anything now found in the bed of the stream. I preserved several for further examination by some one competent to decide as to their character. At 12 the whole valley as far as the eye could reach through the smoky atmosphere was a vast marshy plain level as a prairie & covered with bulrushes long grass &c & full of wild fowl of various kinds. Being impassable along the river bank we were obliged to wind along the base of the hills some 10 miles before coming to water. I saw this P.M. many of the horned frogs or lizards, of a small size however. Have noticed them before but now to be compared for size to the Mexican animal. Just below the marshy plain the grass was remarkably fine & valley wide & hundreds of wagons were here encamped recruiting their stock & making hay, at one place the numbers of camps close together with the smoke of their fires &c seen through hazy atmosphere so closely resembled a city as to be noticed at once by us all. The haymaking was the most interesting

12 Mules were as ornery as the human beings who drove them. Sometimes the gentlest animal would become temperamental. James A. Pritchard said that C. K. Snyder's mule kicked its owner in the "brest" with both feet. Snyder was unconscious for some time, and disabled for several days. Pritchard, *Diary,* 86.

operations to me. Everything was done as at home, some mowing & others tossing & drying, while some of the wagons which had been emptied for the purpose were being loaded &c.

The cause of all this preparation is a cut off which has been discovered to the mines on Feather River some sixty miles above Sutters & those who have an object in going to the Fort can save some 60 miles.[13] There is one stretch on the road through of 50 or 60 miles without a spear of grass & hence the haymaking. They being obliged to carry provender enough for 3 days travel. We have now passed considerably the longest proportion of ox teams which left the frontier last spring, & but few ox wagons will beat us to Sutters. Still there are many ahead, but immense numbers behind. I regret that I did not from the first keeping the number of teams passed on the road.

Camped at 5 1/2. Distance 20 m. Have had a real delightful day. Soft & mild & so hazy that we could see nothing beyond the valley. Occasionally some huge mountain would loom indistinctly from out of the mist as we approached it, appearing twice its real size & seeming to tower above our heads & then fade away again as we advanced till quite lost, like the beautiful dissolving views of Dodge. Bought of a train this evening some warm biscuits & pies for supper for which we paid, pies 25¢ each miserable things too. & biscuits 3¢ each. Things eatable are held at enormous prices now & will be till inside of the Sierra Nevada.

Sunday Sept 2. Morn fine, mild & hazy. Wind S.E. Night cold, ice of considerable thickness formed. Strong wind from N.W. in Evening which was very frustrating, felt as if off of fields of snow. Start at 6, road along river bank. Saw in course of our mornings travel a drive of some 30 or 40 deer of the black tailed species also hundreds of sage hens & mule rabbits. Both which seemed to abound. Though what they can find to attract their appetites or comfort in this desolate looking region I cannot possibly divine. Camped at 4 at a place where the road, according to our Mormon guide leaves the river for 14 miles.[14] Grass grows poorer & less of it every day. Distance 19 m. Day hot.

Monday September 3. Morn clear cool. Wind N.E. Night cold, breeze from N.W. Start 6, passed a Missouri train today & found on a few minutes conversation that a number of their family lived in Washington and Gallia counties, Ohio, & know many of my acquaintances. They inquired if among others I

[13] Lassen's cutoff left the main Humboldt trail just west of present Winnemucca, Nevada. It cut northwestward across the Black Rock Desert to Black Rock Spring, Mud Lake, and Goose Lake on the western slope of the Sierra Nevadas on to the headwaters of Pit River thence through the mountain ridges southwesterly to the Sacramento. Adams (ed.), *Atlas of American History*, plate 116. Emigrants accused Lassen and others of misleading them both as to the condition of travel on the route and the distance. Kimball Webster wrote, "We travelled 24 miles down the river and crossed Feather River and camped on the southern bank. We are now very near Feather River mines, which we were told we could reach in 180 miles from the forks of the roads on the Humboldt at Winnemucca. Instead, as per my account, which I believe is very nearly correct, the distance is 514 miles as we have traveled it. We left the Humboldt River, September 14, and reached here October 17, being 33 days on the 'Greenhorn's Cutoff,' as it is now commonly known. Probably nearly one-half the immigrants came by this route." Webster, *Gold Seekers of '49*, p. 95.

[14] Many emigrants used the Mormon Guide. William T. Coleman noted: "We were provided with a Mormon guidebook, published by one who had become familiar with the overland routes during the Mexican War, and later by a trip to and from California." Paden (ed.), "Willis Guide to the Gold Mines," *California Historical Society Quarterly*, XXII, 200. The editor gives a good background account of how Willis' guidebook was produced. She also reproduced the contents. (Pages 193-208.)

knew, N.W. & on my replying in the affirmative, without ascertaining whether their language was agreeable to me or not & not knowing but I might be his son-in-law, they commenced such a tirade of abuse & enumerated such a catalogue of crimes that I was quite edified truly. Many of their stories were new ones to me & if true were sufficient apology for all their hatred, but enough now old & familiar to make me feel almost as though abt home again & seemed quite natural. To hear old tales of the kind reiterated here, nearly 2000 miles from civilization in a sandy desert, by men I had never seen before was novel enough. How a man can be happy & enjoy life with such a reputation attached to his character & curses following him thus everywhere I cannot conceive, but I should suppose the proceeds of such conduct would hardly compensate for the loss of peace of mind & conscience.

Crossed the 14 miles of desert this morning & had a hard walk through the heavy sand & in the hot sun. Some 2/3 of the way Cross overtook & passed the famous "Pioneer Train" which left the 14 of May & with which I thought at first of taking passage.[15] They have travelled no faster than ox wagons, in fact many of the latter have passed them. Their mules were overdone by fast driving the first two weeks & have now picked up again   since taking only provisions for sixty days they have been several times on the verge of starvation & have been relieved by ox trains, making purchases or borrowing flour, crackers, bacon, &c. Turner who is with this train went back to Fort Kearney some 250 miles with wagons & bought all the provisions of all descriptions that could be had then overtook his train at Fort Laramie. They number some 35 wagons & make quite a show with their mules & spring coaches. The latter are a very neat & comfortable article & had their mules held out the passengers might have travelled very comfortably. Turner & Allen must lose considerably by the operation   40 mules have died or were left from inability to proceed & in consequence some 8 or 10 carriages have also been left. These losses in addition to the outlay for provisions at the enormous prices at which all such articles are held on the plains must entirely eat up the profits & draw on them considerable besides. I can but congratulate myself on my fortunate escape in not paying them 200$ for such passage.

At 4 this afternoon we made the cutoff to Feather River[16] where were put up innumerable notices & letters speaking pro & con of the right hand or left hand roads   while reading or trying to read the half of them the Pioneers came up to discuss

[15] The Pioneer Line was organized by Turner, Allen, and Company. The proprietors planned to conduct emigrants overland across the plains in the spring and summer of 1849. They proposed to carry a passenger and 100 pounds of baggage for $200. The company offices were located at 32 Second Street, St. Louis. *The Diary of a Pioneer*, edited by Robert M. Searks (1940), is a firsthand record by a man with the first train of the Pioneer Line.

[16] The map attached to the second volume of Charles Glass Gray's manuscript journal outlines the author's route. He took the Lassen Cutoff. This itinerary was sketched on a more modern base, and the route obviously was superimposed on it years later by Gray. He indicated that he departed from the main trail just below the modern town of Winnemucca, Nevada, and went by the later Fort Bidwell on the east shore of Goose Lake. He then went southwestward by way of Alturas, Tehama, and Marysville to Sacramento. See also map by Bruff, reproduction in Pritchard, or Paden, *Wake of the Prairie Schooner*.

future action. On making a calculation twas found that though
the cutoff was some 30 or 40 miles nearer to where Gold could
be found it was 160 miles farther to Sutters than the Humboldt
route, beside having 70 miles without water or grass. Turner
was bound to the Fort & of course took the left at once.
While staying at the Forks I found that Doct [Elijah] White &
family were with the Pioneers & soon sought out the Doct
himself to inquire about Joseph Stone. I learned that he had
packed & gone on some two weeks before in fine health &
spirits, so that my letters for him will be delivered in Cal.

I saw another Doctors family also in a large spring carriage
& looking quite comfortable, consisting of his wife 3 daughters
& a little one. Miss Fanny his oldest was quite a belle in St.
Louis & is really very pretty, but looked Entirely out of the
proper place amid these rough scenes though she appears to
enjoy herself much. She has her guitar with her & sings & plays
very finely & seems to care little whether she is in her parlour
at home or sitting on a sand bank on the Humboldt, so long
as she can make music. Camped this Evening opposite Pioneers
they having crossed Humboldt to find grass for their Enormous
herd of mules. Distance 18 m. Day hot, dust exceedingly disagreeable & annoying.

Tuesday Sept 4. Morn clear cool. Wind N.W. night cold, start
5 1/2. Road has been nearly all day over on the barren plains
from 2 to 6 miles from the River & travelling has been very
heavy & disagreeable. Side roads branch from the main on
down to the river occasionally & one of these we followed about
noon, to camp. Found the River running down in a narrow
ravine some 100 feet below the level of the plain above, with
no grass & but few willows. Spent a couple of hours under the
scanty shade of an Osage orange bush, the first of the Kind I
have met with & travelled on till 6, following a path down the
desolate looking valley sometimes being obliged to climb the
sides of the South bank to avoid bends of the River. Camped
in a barren place at 6 with nothing for our animals but a few
scraggly willows or sage brushes on the River bank. Distance
20 M. Day hot & most disagreeable in every respect. Evening
clear with splendid full moon.

Wednesday Sept 5. Morn clear wind N.W. Night mild.
Coleman's horse was quite sick this morning being probably
affected by the unwholesome food of last night & we were
detained sometime in consequence. Becoming somewhat better

we started at 6 1/2. Left the River again as Yesterday &
travelled across the plains the dreary desolate appearance of
which were enough to Entirely dishearten one were they at
the commencement of the journey. About 15 miles above the
Sink on some slough where a little grass may be had and some
bad water. Leaving the main road which winds a long ways
around we struck across the plains to explore a path for ourselves & after a hard march through fields of sand black with
deep ashes & some savine bushes came unexpectedly upon one
of the sloughs, a deep precipitous ravine with some 30 or 40
feet of marshy bottom bearing reeds, rushes & some coarse grass,
& where was also a small well dug probably by some adventur-
ous packers like ourselves who had preceded us. Descending
the steep banks we unpacked turned our animals among whatever eatables they might be able to find in the rank vegetation
& lay down by the well & surfitted ourselves upon its unwholesome liquid we having been entirely without water since the
Early part of our days jaunt & were almost famished. This we
now had is a sulphurous brackish but far better than no water,
so we drank largely. Concluded to spend the rest of the day &
night here as is uncertain how far it may be to the next water
that of the river being quite as unpalatable as this in the slough.
Distance 20 m. Day excessively hot wind, N.W. Evening fine
clear & bright morn.

Heard today that three men were attacked & killed day
before yesterday within half mile of our present location by a
party of Indians. This may be only a "yarn" however though it
seems to come pretty direct but we do not apprehend any
danger & so do not trouble ourselves much about anything
except something good to eat! Our hard fare is getting very
tiresome & our dreams by day & night dwell much upon the
"flesh pots of Egypt." Oh for a slice of bread & butter or a
piece of roast turkey, or both.

Our mules were alarmed this evening by a pack of wolves
coming near about dusk & started off over the plains in great
terror. All went in pursuit but myself I staying to prepare our
simple supper. While I was there alone the rascally wolves
came down in a gang into the ravine within 30 yards of me
& treated me to a most unmusical serenade of howls yelps barks
& cries which can only be executed by a wolf & this defies
description. They no doubt were deliberating whether best
to attack me or not. I laid my pistol Knife &c by me & let
them take their own time & they finally came to the conclusion
to leave for the present & so climbed the banks of the ravine

& cleared occasionally one or two turning to give me a howl or bark by way of farewell probably.

Our boys came back in some two hours having found the mules crowded into a thicket of briar bushes so close by as they could pack themselves.[17] Antelope & deer quite abundant is probably the cause of such a procession of wolves of all colors & sizes. We saw in our today's march quite a herd of antelope playing & capering over the plains, but giving us no chance to taste the flavor of their steaks.

Thursday Sept. 6. Morn clear, Wind S.W. Night mild. Start 6. Road first part of morning after leaving the slough was over sand & ashy plains as yesterday crossing several more sloughs towards noon we began to notice the swampy ground which characterizes the country above the Sink. Road through it solid smooth & very fine, Equal to the best at home & without dust. The marshes are covered with a small species of cane, bulrushes reeds coarse grass &c & ground miry & in some spots wet. At 12 M reached "the Wells"[18] the last place where grass can be had this side of the Salmon Trout & camped to procure some hay to pack over the stretch of sixty miles ensuing. Distance to Sink 10 miles mostly marsh of saline & sulphurous nature. The wells which have been sunk here are strongly impregnated with the above substances & I cannot consider the water as healthy by far as the place where we are now camped on.

Numbers of trains & wagons making hay & grazing their cattle preparing them for the long desert stretch before them.[19] Of one of them we have procured each a bundle of hay to lash to our packs & by economizing it ought to keep our mules from starvation. As does everyone else we shall travel by night, the want of water being much less felt by both animals & men. With one of these trains I found my travelling companion of last winter & also up to St Joe, Doct Chapman. He left some 10 days before us having nothing to wait at St Joe for as had we. The old gentleman looks quite well, & pretty as ever. Very few that started abreast of us with whom we have any acquaintance before have passed. I do not remember any one now but Doct Jackson & his party must be several days in advance.

Travelling this road is far from being as rapid as most persons, myself included, calculated upon at the start. Well appointed pack parties cannot average more than 22 to 25 m per day including stoppages & this is by far the most expeditious mode of travel. By a large train camped near us have received

[17] When the emigrant companies moved into the Humboldt Valley, they could expect to be bedeviled by wolves. J. Goldsborough Bruff described an experience of September 29, 1849. "Being a mile in advance of my train, early this morning, walking, with a double-barrel'd gun on my shoulder, just on emerging from the cañon, I descried ahead of me, some 200 yds. a large grey wolf, and hastened up and discharged a charge of buckshot at him, when within 100 yds. sending him off, to the right, up a hill, on 3 legs, and his tail between his legs, like a whipped dog. I stopped to watch him; and when he had reached an elevated point, some 300 yds. off, he turned around, and barked very furiously." Bruff, *Journals*, I, 198.

[18] The "wells" at the Humboldt Sink were a dubious blessing for the emigrants. Travelers were glad to reach this spot, but many did not use proper caution for the good of their stock. "We then went three miles further," wrote Captain DeWolfe, "and stopped at the sulphur wells, those wells consists of a number of holes dug in the ground by Emigrants to get water for their stock the water is strongly impregnated with salt & sulphur. Our oxen drank it very well near those wells is a slough the water of which will kill animals if they drink it." DeWolfe Diary, September 29, 1849.

[19] This was standard practice along this stretch of road, and tons of hay were cut in this area by the emigrants. Word went back to the trains that they faced barren country without prospect of feed. By this time the teams were too thin and weak to take a chance on their surviving any extended period of starvation. Joseph Warren Wood commented on this fact in his diary on August 13, 1849, when his party was all but panic-stricken over the condition of their cattle. Captain DeWolfe's company set to work to gather hay. He said, "we here mowed some grass & made hay for our stock having a long drive to make without grass." DeWolfe Diary, September 20, 1849.

the first news from Cal. They sent one of their men in advance with 3 good mules packed to examine the road water &c. He left them at Bear River went through to Sutters & returning met his train a day or two back. He brings great account of the "Diggings" quantities of gold prices of provisions &c & we received some reliable information with regard to our road, grass &c. Distance today 10 m. Very hot, W.W.

Friday Sept. 7. Morn clear hot, wind S.E. Night mild. Lay in camp till 1 & then left to make the Sink before night & be ready for the desert at moon rise. Road fine, though through marshy grounds, there being moisture enough to prevent any dust & make the travelling solid. Arrived at the upper end of the sink at 4 & camped at its foot at 7. Day extremely hot. Thermometer I should suppose would range from 90° to 100°.

Here we see our last of the famous Humboldt & I must agree with the majority of the Emigrants in nicknaming it "Humbug River." The stream itself does not deserve the name of river being only a good sized creek, about like our Duck Creek only longer & running through a more level country. For the first two day's travel in its valley the grass is splendid, then the valley begins to narrow & feed to get poorer & less of it all the rest of its course, till for the last 80 miles except in special spots we could hardly get enough for our mules to eat, & water barely drinkable from saline & sulphurous impregnation & having a milky color. I think Baron Humboldt would feel but little honored by his name being affixed to a stream of so little pretension. It is far inferior in every respect to either Bear River, Green River, or Big Sandy. We leave it without any feelings for it at the foot of the Sink.[20] In some wells of most wretched water with which we filled our canteens though twas like taking a dose of salts to drink it, & lay down to take a few minutes rest before the moon should rise.

At 10 packed up & started for the hot springs—distance 25 miles. Road first part of travel good, last part sandy & heavy.[21] Stopped at 3 o'clock & lay down to rest being fully well fatigued, & slept till day light, pack up & pushed on to springs where we arrived at 10 A.M. almost done over by the heat of morning thirst & want of sleep.

The Hot Springs are quite a curiosity, bubbling up by themselves in this desert, & boiling hot. The Main Spring is a clear pool with a rocky bottom some 3 feet deep & 10 or 12 across. The hottest one is like a great kettle nearly round some 3 feet across & boiling evidently. As now it is drinkable, we put

20 "On arriving at the sink of the Humboldt, a great disappointment awaited us. We had known nothing of the nature of that great wonder except what we had been told by those who knew no more about it than ourselves. In place of a great rent in the earth, into which the waters of the river plunged with a terrible roar (as pictured in our imagination), there was found a mud lake ten miles long and four or five miles wide, a veritable sea of slime, a 'slough of despond,' an ocean of ooze, a bottomless bed of alkaline poison, which emitted a nauseous odor and presented the appearance of utter desolation." Shaw, *Across the Plains in 'Forty-Nine,* 172.

21 Ware potrayed the whole route west by both Fort Hall and Salt Lake City. He gave a fairly clear outline of the route along the Humboldt River, showing the position of passage around the Great Sink, and the position of the Hot Springs. Folding map attached to Ware, *Emigrants' Guide.*

22 Edwin Bryant and his party passed these springs, now called Brady's Hot Springs, in mid-August, 1846. He, Jacob, and Nuttall inspected them. "There are some ten or twelve of these springs, the basins of the largest of which are ten feet in diameter. The temperature of the water is boiling heat. To test it by the best method within our power, (our thermometer having been broken,) we procured from camp a small piece of bacon, which, being placed on the end of a

coffee in our boilers dipped them full out of the spring & our coffee was ready for use immediately & strong enough too. We drank four boilers before feeling at all satisfied. The springs vary somewhat, some being sulphurous some saline, & more pure.[22]

Camped at the springs till evening & started for the rest of the desert route at 7 1/2. Road very fine—down a valley somewhat sandy, at times crossing immense fields of hard sand, clear & white & looking in the moon light exactly like snow. Took 2 hours rest about midnight & arrived to our great joy at the Salmon Trout River at 7 A.M. on Sunday Sept. 9, & once more had a refreshing draught of pure water & was gladdened by the sight of large majestic trees.[23] The Salmon Trout being lined with the finest cotton woods I ever saw. No one can imagine how delightful the sight of a tree is after such long stretches of desert, until they have tried it, we have seen very few of any Kind since leaving the Platte, & what a luxury after our mules were taken care of, to lay down in their Shade & make up our two nights loss of sleep, & hear the wind rustling their leaves & whistling among their branches.

For me no country can have any attractions unless it can offer a heavy growth of fine forest trees. Talk as you may of the beauty of Bear River Valley, Humboldt Valley &c, I would not exchange an acre of our Ohio wooded bottoms for the whole of this—Salmon Trout is a most beautiful Stream, rushing & roaring over the rocks & its stony bed like a New England mountain river & the water as clear as crystal. At the crossing it is about 60 feet wide & 2 deep, & is every way a finer, larger & preferable stream of the famous but overrated Humboldt. Camped this morning under some pine trees having made 6 miles since Friday evening & needing rest much we have spent the day here. Mules doing pretty well. We feel now as if our troubles were about over, we having no more bad country to pass through but plenty of trees, water &c, & only 180 miles before us.

On the last desert as on others we saw great numbers of dead cattle mules, horse &c. We passed in our two nights march somewheres near 130 or 140 oxen. Killed mostly by the sulphur water & hunger. 15 or 20 were lying about the hot springs, some still alive & dying by inches, poor things I could but pity their hard fate.[24] No possible chance for them to escape. Nothing to eat or drink. The country in the vicinity of the springs & in fact all the way for the last 100 or 150 miles is certainlly the most dreary—desolate, God forsaken region that

stick and thrust into the boiling basin, was well cooked within fifteen minutes. The water is slightly impregnated with salt and sulphur. Immediately around these basins, the ground is whitened with a crust of the carbonate of soda, beneath which is a stratum or shell of reddish rock, which appears to have been formed by a deposite from the springs." Bryant, *What I Saw in California*, 204.

[23] Now called the Truckee River. Frémont called it the Salmon Trout. He and his party were on this stream on January 16, 1844. He wrote, "This morning we continued our journey along this beautiful stream, which we naturally called the Salmon Trout river. Large trails led up on either side; the stream was handsomely timbered with large cottonwoods; and the waters were very clear and pure. We were travelling along the mountains of the great Sierra, which rose on our right, covered with snow; but below the temperature was mild and pleasant." Frémont, *Report*, 316.

Ware described the Salmon Trout or "Truckie's" River as sixty feet wide and two feet deep at the usual emigrant crossing place. At its mouth the elevation was 4,890 feet. He said that fish in the river were large and of "superior quality." He advised emigrants to camp along this stream a few days to prepare teams for the ascent of the mountains. Ware, *Emigrants' Guide*, 37-39.

[24] The going was rough on any road that emigrants took across Nevada. John Edwin Banks described their problems eloquently. "Some think," he said, "they see the elephant. If fatigue, weariness, constant excitement, and awful distress among cattle make the sight, he is surely here. We scarcely made a halt until today at noon we reached Salmon Trout River. To us it seems not less than twenty-five miles. This day we scarcely were out of the smell of putridity. Not less than 200 carcasses of oxen and horses are strewn along the road within thirty miles. Some moping about waiting for death, no possibility of other relief. Upwards of fifty log chains stretch along like lengthy serpents, their owners having no further use for them. Seven of our cattle failed; some we expect to get in." Scamehorn (ed.), *Buckeye Rovers*, 78.

can be imagined—sand—sand from our hill mountains & valleys nothing but sand, with a scattered growth of miserable brown stunted sage or savine giving the landscape an ashy hue. May I be delivered from ever seeing such another tract. It is enough to shrivel a man's soul into assimilation with the products of the soil. Soil? there is no soil of pure sands of araby.

Today Sunday we have enjoyed ourselves to satiety drinking pure water & sleeping under these noble trees & Enjoying their shade, the more as the day has been quite hot. We lay & pitied the poor devils who were still toiling over the desert, & congratulating ourselves that we had no more dry stretches to think of. Wind strong from west this P. M.

Monday Sept 10. Morn clear warm—wind west. Lay over today in our pleasant camp that our mules might be thoroughly refreshed & fitted for the toilsome ascent of the Salmon Trout Valley & Sierra Nevada. P.M. I mounted one of my animals & rode down the valley & discovered some miles or more below our camp a fine thicket of what I took to be the osage orange bushes & loaded down with rich golden & red berries—much resembling currants in size at home. Riding back to camp for a pail I gathered about a gallon out of which we made a most delicious and acceptable stew, fully satisfying for once my excessive craving for something acid. These bushes would be well worth transplanting to the states & no doubt unless improved much in the flavor of their fruit by cultivation. There appears to be two varieties, the red & yellow, the latter being much the most palatable. The berries grow in clusters around the leaf stems, & at the base of the long thorns, & in such abundance as to completely envelope their supporting twigs, giving the uppermost branches of the bush a yellow color at a distance.

Day warm & pleasant wind West. Evening mild fine breeze from N.W. Tried hard this Evening to shoot a hawk which was perched in a neighboring tree, but could not coax the rifle to go off, snapped some dozen caps & thus gave up the attempt. Had we succeeded we should most surely have had "Hawk soup" for supper, so much did we hanker for fresh meat.

Our travelling companion Coleman met with quite a loss this P.M. He went down to the ford with his horse & packs & leaving the animal a short time while visiting some of the numerous camps in the vicinity it slipped over a steep bank into the rapid stream below & when Coleman got to the scene of action, the horse had scrambled out but more than half the pack was gone, comprising two valuable broadcloth cloaks, one of which cost 70$, several blankets, an India Rubber spread &c. The fastenings of the pack giving way the articles floated off down the stream, & though we traveled its banks for some miles nothing was rescued but the least valuable cloak, all the poor fellows bedding was lost. He bore it though like a major, & could hardly persuade him to accept of my blanket for the remainder of the trip, seemed to think he could get along well enough with his one cloak.

Tuesday Sept 11. Morn cool pleasant. Wind S.E. Night cool, with some clouds. I start up the Salmon Trout Valley at 7 1/2. About 4 miles from where we first struck the river we came to the first ford, & into it we plunged leading our mules slowly & carefully over its rough & stony bottom. The water was very cold & mostly waist deep, & current so strong as to require considerable exertion to keep a footing, & as may be supposed the bath was not even agreeable.

The stream crossed we pushed ahead at a rapid rate for exercise & warmth, & in hopes to dry our soaking garments, but only 1 mile & the River crossed our path again & into it again must we go & thus we slowly travelled all day crossing the S.T. 11 times & all the fords like the first, deep, swift cold, & very rough footing. Distance up River 12 miles. Soon after commencing our march today we were in the midst of the first spurs of the Sierra Nevada between which the Salmon Trout finds a passage of some 70 miles through a narrow rocky gorge or canon.

These mts. are high steep, bald, & destitute of vegetation being composed of slate stone colored clays, & some trap or coarse granite. In some places when the winter torrents had washed deep ravines down the sides of the steeps the variety of colors exhibited was surprising & beautiful almost equalling those of the rainbow. Blue and yellow & white predominated, with occasional blending of these colors producing a variety of others.

Saw in our course along the banks of the stream, numbers of those beautiful fish from which the River takes its name, but had no time to spend in catching them, indeed others who had tried it say they will not bite at this season of the year. Some of these fish were two feet long, beautifully spotted like the New England trout. Camped at 5 1/2 very little grass. Evening cloudy.

Wednesday Sept 12. Morn clear warm, Wind East. Start 7 1/2. Continued our course up the valley of the River, crossing frequently as yesterday & having occasionally small patches of good grass. Bank of River lined with cottonwood trees & willows & vast bald mountains on either side overhanging our winding way. Were it not for its serpentine course the Salmon Trout would be a perfect mill race impossible to cross, so rapid is its descent down the mountains. As it is, the force of the current is much broken by friction against the rocks & banks. I have seldom seen so beautiful a stream as this, water clear as crystal, so that standing on an overhanging rock the fish & other objects are easily discernable at the depth of 8 or 10 feet, & being made up of beautiful pools, roaring rapids. For miles in some places it rushes & foams over the rocks & boulders which form its bed, & thus subsides into a still deep pool surrounded by perpendicular walls of rock & evergreen bushes, vine, willows &c. Most lovely spots, & full of fine fish, & thus another rushing rapid, & soon to its termination at Pyramid Lake.[25] This latter curiosity by the way, I was much disappointed in observing understanding from the guide books & histories that we passed near it. The road may have been changed by Emigrants but now strikes the river some 10 miles above where it empties into the Lake.

At 4 P.M. we emerged from the canon into what is called the Mist Valley, a beautiful level plain covered with fine grass, some 10 miles across & formed by the widening of the mountain ranges. Through this valley the river winds after leaving the gorge on the other side, its course marked by a line of cotton woods & willows. Soon after entering the valley we took a trail leaving the road to the right & supposing it to be a "cutoff" as the road wound round a belt of marsh which crosses the valley nearby at right angles with the river. We followed this trail around the base of the hills & soon found ourselves going off quite in a contrary direction to the course of the road, & the marsh on our right was entirely uncrossable a perfect quagmire.[26] There was nothing to do but go back some 3 miles or follow the path & see the End of the Adventure & the latter we decided to do, & adventure indeed it nearly proved to us. About 5 1/2 we came upon an Indian fishing nets of willow twigs being set in the creek which wound through the marsh, & their camping place among the rocks close by having fresh ashes in its fire place, everything indicated their recent presence & gave promise of their speedy return.

This then was our "Cutoff" a trail which the Indians carried

[25] A plate of Pyramid Lake appears in Frémont's *Report* opposite page 216. Frémont's party traveled along the shores of this body of water, and followed the course of the Salmon Trout away from it. There is an extensive account of the country and the Indians about the lake and river, pp. 216-19.

[26] There are numerous flat places in this area that catch and hold water. For instance, the ground surrounding the sites of some of the emigrant cabins at the Donner camp contained marshy spots.

their fish to dispose of to trains. It was too late for us to return or better our condition. So we unpacked & picketed our mules in the magnificent & untouched grass which surrounded us & quietly cooked our supper hoping the "Varmints" would not discover us but very apprehensive lest they should. Guns & pistols have received special attention this evening & we have made every preparation to defend ourselves should the Indians resent this inroad upon their domains.

The view from our present camp is grand. In front of us towers a range of the loftiest mountains we have yet seen their tops covered with perpetual snow, & some places covered on their sides. Through these mountains & over some of them will be our journey tomorrow. Behind & around us are the many colored mts we have left & in front between us & the Sierra Nevada is the beautiful valley. There again the sunset this evening was beyond description. The varieties of such & changing hues of the clouds as they reflected back the rays of the setting orb of day, would that I could have daguerrotyped them!

Camped this evening at 5 1/2, distance 16 miles, crossing Ebro River 11 times again & water getting colder every ford! Saw the grave of a young man this P.M. who was drowned a few days since by losing his footing on a rapid & being carried down into a deep pool below. This evening cloudy & some appearance of rain. Wind W.

Thursday Sept 13. Morn cloudy & dark. Wind N.W. Night mild, clouds. Start 7. Unexpectedly our nights rest was undisturbed by Indians except in dreams, & we left our camping place not a little rejoiced at what we considered our lucky escape, & fortunate indeed we were, for these rascals though ever so civil on the road & among the trains of Emigrants are always on the lookout for the scalp of any unlucky white who may chance to stray. We are told this P.M. when we gave an account of our adventures to a train we passed that some of their men riding over the hills in pursuit of antelope discovered an Indian village only about a mile back from our camp. This was much more comforting news to us at the time we heard it than it would have been the evening before!

As the slough spoken of yesterday lay directly across the course we must take to regain our road, we explored various places to find out where we might attempt a passage, & taking my little "Vic" I pioneered the road the rest following in single file at a rapid pace. We had to force our way through reeds & bulrushes 10 or 12 feet high, & ankle deep water all the time. In fact the reeds which we trampled under our feet only prevented us from sinking in the springing and miry ground. A travel of about a mile of this description & I espied the open plain through the weeds before me & soon stood on firm ground & had the satisfaction of seeing the rest of my party one by one safely landed by my side. Our attempt was rather hazardous as far as our animals were concerned. Had one of them broken through & mired we should have been compelled to abandon him immediately as there was not foothold from which we could render any assistance.

Struck the main road again about 9 soon entered the mountains following the River valley. After leaving the valley the Mts. began to have some small pines & cedars scattered over their sides, & we noticed an occasional pine on the bank of the stream among the cottonwoods & crossed the river today 5 times, at one of the crossings a beautiful mt torrent tumbled into the river from above water cold as melted snow could make it.

At noon today we came to a majestic pine standing alone on the bank of the River, measuring nearly six feet through at the base & running up at least 70 feet without a limb. This old fellow I take it is a kind of advance guard to the immense pines of the Sierra. Views of the peaks of which we began to catch one after another as they were in the distance. Camped at 5 distance 15 m. Strong wind from W. Evening clear, day cool.

I observe in looking back on my diary that I have omitted a description of the Humboldt Sink, & as it is a singular place & well worth noting I will insert it here. There is considerable dispute among Emigrants as to what the Sink is, & also a difference in the various guide books on the same subject making distances to the Hot Springs & Salmon Trout vary from 4 to 10 miles according as they date from one or another of the ponds, marshes &c of the region. Before reaching the Sink the road leaves the River on account of the marshes & swampy ground in its vicinity & passes through the plain covered with swamp grass, & saline crustations. Some 10 miles travel brings you to what we called the "first Wells" & here I take it you get a view of the Sink proper being a vast marsh some 4 or 5 miles across with a lake or lakes in the middle—the ground white with sulphurous & saline deposits & the water milky for the same reason. The wells were holes 5 or 6 feet deep some 1/2 mile from the shore of the first pond, & the water filtered into them is strongly sulphurous & unpalatable. The ponds of the Sink were covered with all kinds of wild fowl, geese ducks curlews, snipes cranes &c. Perfectly secure from man or beast, as the ground is a perfect mire in every direction.[27] Continuing around the Sink or marsh in a South East course you come to the "last wells" at the foot of the marsh & ponds being the last place where water can be obtained before crossing the desert to Salmon Trout. The water is exceedingly nauseous, but as none other can be had it must be used & canteens for desert travel have to be filled with it. We camped at these wells from evening till moon rose at 10.[28] The sink was all north of us, & covered a vast extent of plains. It is by our travel some 10 miles long from where the river begins to spread in the marsh, to its foot & about 5 miles across, within this space the great quantity of water brought down by the Humboldt is certainly absorbed, sinks & goes no one knows where. The soil through which it sinks is sandy, light & strongly impregnated with potash & its carbonates.

Leaving the "last wells" the country seems to become more broken with ravines and ridges. Many of these ravines are half full of water forming ponds of half mile in extent. The largest one of these holes is some six miles from the marsh & the Mormon guide & others contend that it is the "Sink" proper. I think though it is only like the holes called wells—a ravine below the level of the marsh & filled with water filtered through the soil from the Sinks above. The marsh of the plain at the "wells" is covered with a stringy growth of reeds & also cane, grasses, bulrushes &c, & makes fine feeding for the water fowl.

While we were encamped at the last wells, after dark, one of our party fired a gun heavily loaded in the direction of the lake which was only some 100 yards from us, & the noise made by the wings of the frightened birds was like thunder, & we could hear it continuing up the plain as flock after flock take the alarm like the rumbling of thunder after the first heavy roll. I should like much to explore the innermost recesses of this remarkable place did time & means permit by a boat from the mouth of the River. This will no doubt be done yet by some fortunate individual & I shall look for the account of his discoveries with interest.

Friday Sept. 14. Morn some cloudy. Wind W. cool. Night quite cool. Start 7 1/2. Travelling along valley shortly after leaving camp I being some miles in advance of my company,

[27] Ware informed emigrants that "After crossing the plain you reach the [Humboldt] 'Sink' (thus called from the river loosing itself in the sand at this place,) it is a low marsh, surrounded with bulrushes and saline incrustations and emits a most disagreeable effluvia; the water cannot be used for man or beast." *Emigrants' Guide,* 32-34.

[28] These wells were shallow pits dug in the ground by earlier emigrants. Water filtering through the soil shed some of its chemical content. The last one was beyond the sink.

came up to a spring carriage standing still in the road surrounded by Indians, some trying to climb the wheels to get into it. As I came opposite I looked in & saw a single man with two horse pistols lying on the seat by his side, & asked him if I should stop till the teams behind came up which he was glad to have me do, I staid with him some 15 or 20 minutes pistol in hand, & we both watching the Indians closely & occasionally as one or two would attempt to climb on the wheels pushing them back. They stood around as if irresolute, looked at the priming of their rifles such as had them, or fixed their arrows in their bows & muttered among themselves we said nothing to them. Kept them from attempting to enter the carriage, & one of them putting his hand on "Vic" I pointed my pistol at his head raising the hammer partially, on which he grinned like a monkey & fell back among the rest. On the wagons coming up the rascals cleared for the Mountains & I travelled on. I afterward learned that the person in the carriage was Mr. Peters of St. Louis, a gentleman of wealth & travelling for his health. The Indians were Shoshones or Pah Utah,[29] the latter word pronounced as if spelt pyute, & was deterred from attack upon us by fear of the consequences should any wagon come up. Several of them have been shot by Emigrants & they are very much afraid of the American Rifle.

About 10 we began the ascent of the first mountains, road very steep & pine timber beginning to be very abundant so as to make us some shade. We ascended some 4 miles & attained the top from whence we had a fair view of the valley we had left & on yesterday's travel & mountains in the distance, looking immense from our elevated point of observation. Travelled about a mile on the side of the mountain & began our descent to the second valley. Mountain sides here were covered with fine growth of pines, cedars arbor vitae of large size, Firs &c, evergreen bushes of all descriptions formed an undergrowth of great beauty. The contrast to our level sandy dreary roads & parched sage bushes of the last few weeks. Made our mornings previous very exhilirating & delightful & we scrambled on over the mountain road shouting & singing like schoolboys set free.

Descending into the second valley about a mile down a hill almost perpendicular we found it like the first, set in the midst of mountains, covered with grass, & very beautiful but smaller, only about half a mile across. Our course continued after leaving this valley through pine woods & rolling country, camped at 5, distance 19 m. Evening clear cold.

[29] The "Pauta Indians" were noted by T. H. Jefferson on his map of 1849 in the Humboldt Valley, south of the Great Trail. T. H. Jefferson, "Map of the Emigrant Road from Independence, Mo., to St Francisco California, 1849." Bruff noted that the "Pautahs" were only a band of Diggers. They and their snug brush arbors were seen along the way. Bruff, *Journals*, II, 1183.

September 15–September 27, 1849

SATURDAY SEPT. 15. Morn very cold, night do, ice made 2 in. thick. Thermometer I should suppose would have stood about 15°. The ground our bed packs &c were perfectly white with frost & our ears & faces looked not exactly white but considerably blue with cold. We were camped in a valley near a Miss. train. Going over to warm myself by their fire I found a soldier who belonged to Gen. Kearneys detachment some years since entered Cal. from Taos. He was on his way home & brought us news from Cal. new & interesting. The prices of provisions & clothing low, Gold was not as abundant as had been represented, miners & Emigrants disappointed & going home in great numbers &c. He intends spending the winter at the Mormon city, & inquired of us how far back he should find Emigration. He seemed to think that many of the teams we had passed so far back would never get through.

Start at 9. The wolves kept up a great howling all night & this morning one lingering about our camp Cross undertook to shoot him for soup & a roast, but did not succeed. Soon after leaving camp we came to the huts built by the unfortunate Donner Party in the winter of '45 ['46], & where so many of them perished.[1] The cabins stand to the left of the road down in a dense grove of fir trees of which they are built. The roofs have fallen in & nothing stands but the square enclosures with patches of newspaper hanging to the logs on the inside with which they build their huts & to keep out cold around were scattered shreds of female dresses some scraps of jeans cloth, bones of all descriptions, human & other animals they had eaten. Pieces of iron &c &c. The stumps standing in the vicinity showed well enough the depth of the snow that stopped them. Many were cut off 12 feet from the ground. Others probably cut after first snow was six feet high. Twas a most melancholy & gloomy spot, & the imagination could find full scope in the indications of human suffering scattered around. I gathered some relics as curiosities & left, thankful that late as my journey had been prolonged, I was still safe from any such catastrophe as befell those unfortunates.[2]

The road from the Donner huts has been changed, instead of going around Truckee's Lake as formerly it begins to ascend the mountain immediately, being a saving of some 4 or 5 miles. The Lake is some two miles to the right & was described to me by those who visited it as a beautiful sheet of clear water about 3 miles long by about 1 broad. The ascent to the pass from Donner cabins is about 5 miles over rocks & steep bluff & through majestic forests of fine cedar. Fir, arbor vitae &c, &

1 The Jayhawkers reached the cabins below Donner Lake on August 28, and Joseph Wood visited this site, as Douglass Perkins did eighteen days later, September 15. "We soon came to a swamp on our left where stood the walls of a double log cabin in which a division of the unfortunate Reed & Donner party attempted to winter 3 or 4 years ago. The ground around was strewn with bones of the oxen. They were smashed up fine for the purpose of getting the last particle of marrow or anything which would sustain life. The stumps were 8 or 10 feet high indicating that the snow was deep. It was very hot when we passed & it did not seem possible that we were in the vicinity of so much suffering. We spent a short time in looking at the ruin." Wood Diary, August 28, 1849.

2 Perkins had probably read the account of the Donner tragedy in Edwin Bryant's *What I Saw in California*. Almost every emigrant who started over the California Trail in 1849 had this tragedy in mind, and drove hard to get across the Sierras before the first snowfall. This tragedy has been treated in several modern books, among them George R. Stewart, *Ordeal by Hunger, the Story of the Donner Party*; C. F. Mc-Glashan, *History of the Donner Party, A Tragedy of the Sierra*; Hoffman Birney, *Grim Journey*; and Bernard DeVoto, *Year of Decision*.

a rich luxuriant undergrowth of laurel & various other ever-greens. The journey is wild & magnificent beyond description. I was perfectly in raptures during the whole of the toilsome ascent, & wished often that some of my enthusiastic friends at home, who go into ecstacies over our hills could by some air balloons or other labor saving machine be placed by my side. The trees exceeded anything I had ever seen & fully realized my expectations of a Cal. forest. Hundreds of them were six feet in diameter & standing so densely together that I could hardly get myself & mule through them.[3] The road in finding a passage through the trees & among the rocks lengthened the distance to the foot of the pass at least one half.

Up, up, we toiled wondering every five minutes how "the dickens" ox teams & wagons can get over here, & it is a wonder indeed, until at 3 P.M. we arrived at the foot of the terrible "Passage on the backbone." For half an hour before arriving we could hear the shouts of teamsters urging their cattle up the steep & when we were near enough to see through the forest we could look up nearly over our heads & see wagons & cattle looking like pigmies, & as if almost suspended in the air. The "Pass" is through a *slight* depression in the mountains being some 1500 or 2000 feet lower than the tops in its immediate vicinity. As we came up to it the appearance was exactly like marching up to some immense wall built directly across our path so perpendicular is this dividing ridge & the road going up to its very base turns short to the right & ascends by a track cut in the side of the mountains till two thirds up when it turns left again & goes directly over the summit.

The distance to the top of the pass I should judge to be about 1/2 mile, & in this short space the elevation attained is somewhere near 2000 feet! The mountain is mostly rock. Where the road is cut tho' it is red clay & stone, which by travel & sliding of animals feet has been much cut & powdered up making a deep dust on the first half of the steep. At the foot of the ascent we found the Missionary Train from Indiana, preparing for the Enterprise. One wagon had already started with 13 yoke of cattle attached, the load in the wagon not exceeding 600 lbs, & they could get but a few yards at a time stopping to rest their team. They were about half way up when in an inclining place the wagon began to slide over the precipice! The men seizing hold at all points stopped its progress to destruction, & by some management it was placed upon the road again. Had it got a fair start over the hillside it must have dragged all the cattle with it down upon the

3 John Edwin Banks shared Perkins' admiration of the large trees in this area. "Some trees we see," he wrote, "are truly astonishing, measuring in some instances from five to nine feet in diameter. They are all pines and cedar, their height one hundred fifty to two hundred feet." Scamehorn (ed.), *Buckeye Rovers*, 88.

rocks below. We leading each his mule, began to scramble up sometimes upon "all fours" like our animals, & glad enough were we to stop "to blow" several times before reaching the top.[4]

At last the summit was gained & we attempted 3 cheers for our success which unfortunately failed for want of breath, but sitting down for 1/2 hour we enjoyed the magnificent prospect on either side of us. Our route back could be traced for miles, & the mountains among which we had been winding our way. Far below us was snow in vast quantities which never melts & on either side were peaks some thousands of feet higher than our position. Before us we could see the mountains of Bear River & Yuba Valleys descending in size towards the coast, & the Yuba Valley some 5 or 6 miles distant with its green grass & camps, lay almost under us. I could have spent hours on this spot so many thousand feet higher than I ever was before or ever expect to be again, but the coldness of the air tho it was Sept, & a bright sun shining compelling us to "button up" to the chin, & the growing lateness warned us to be moving.

Descending the west side of the Pass was, tho very steep nothing compared to the Eastern ascent, & we encamped some 2 or 3000 feet below our recent elevation, near a large train which crossed the day previous, at 6, distance 15 miles. The time usually occupied by an ox team to get to the top of the pass is from 1 1/2 to 2 hours. About 8 in the evening some of the Missionary wagons which had made the ascent came down by torch light it being rather uncomfortable lodgings on top, & the Effect of the Blazing pine knots in the dense forest above us, the shouts of the men & rumbling wagons &c was very picturesque. It reminded me much of Maelzels famous exhibition of the Burning of Moscow, which I saw years ago, where the French baggage trains evacuate the city by torch-light.[5]

We had quite a treat this evening—a man belonging to the train near us struck up some lively tunes on the valve trumpet, the notes of which rang & echoed among the hills & trees most gloriously seeming much like "sounds from home." Evening quite cool. W. W.

Sunday Sept. 16. Morn cloudy—cold—appearance of snow—night cloudy. Started at 7 1/2, continued our descent down ragged & steep hills to the Yuba valley where we found some 40 wagons encamped. Passed through the valley & up into another range of mts. with more magnificent & grand scenery.

[4] Ware did not minimize the difficulties of crossing the pass. "We assure you that you will be tried to the utmost, in view of the appalling obstacles to be surmounted, but never despair, [if] others have [gone] over triumphantly, you can! Commence and unload, at once pack everything over the summit, on whatever mule you have in your party, then haul your wagons up the precipieces [sic] with ropes. By adopting this course you will certainly save time, and perhaps hundreds of dollars, from breakage of wagon, if not total loss of some of your teams." Ware, *Emigrants' Guide*, 39. Because of the Donner tragedy, all the emigrants had a great respect for the early and furious winter storms that could be their undoing at this place. Bryant described graphically the arrival of his party at the blue lake now known as Donner Lake. He was especially impressed by the forest. "We reached the upper end of the lake at four o'clock, and encamped on the left of the trail, in a small grassy opening surrounded by tall and dense timber. The forest in the narrow but fertile bottom of the lake, and on the sides of the mountains, where there is any soil for its sustenance, is dense, and the trees are of immense size. A brilliantly green and highly ornamental moss covers the limbs of many of the trees." Bryant, *What I Saw in California*, 229. "Today at 8 o'clock A. M. we reached the summit of the long looked for Sierra Nevada. It was high and difficult of ascent. The snow N. & South of us lay in patches & below the snow the grass was green. The top rocks were trap rocks. We ascended the peak in the night the wind blew strong & cold from the east & we were glad to descend on the western side." Wood Diary, August 29, 1849. Joseph C. Buffum gave a realistic description of the labor involved in crossing the Sierra through Donner Pass. "We let the wagons down by ropes and came into the valley about 10 oclock and took breakfast and a nap for I was never so weary in my life having eat but little for 20 hours. P. M. Crossed the valley and camped in edge of timber as there is no prospect of grass ahead. the mountains are less abrupt here and are covered with noble timber." Buffum Diary, August 14, 1849. A modern highway marker states the

At 11 came upon a most lovely lake embossomed in the Mts & surrounded by forests & evergreen shrubs. At its foot was a lofty & solitary rocky peak, rising from the woods, & looking as if the guardian genius of the spot. The famous "lochs" of Scottish scenery cannot be more beautiful or romantic than was this & I doubt much whether the differences between the two could be pointed out.

Camp down an almost perpendicular descent where wagons are let down with ropes & the trees at the top cut into by their frictionlike "checkposts," & encamped in Yuba valley again—in the midst of enormous masses of rock through which the Yuba tumbles & foams, a good sized creek of pure cold water, at 5 1/2 distance 15 miles. Evening clear wind west. No grass.

Monday Sept 17. Morn clear cold, night cold Wind West. Could find nothing of our mules this morning we turned them loose last evening to find something to eat, the prospects of which were rather slender, & they must either have wandered far in search, or were stolen by the Indians & it was not without anxiety that we started in different directions to find their trail. I went up the valley over rocks & bushes, finding some spots of good grass but nothing of our mules & after a walk of about 5 miles returned to camp to find Sam & Gilbert as unsuccessful as myself. We sat down rather non compas, to deliberate on our course should the others be likewise empty-handed. About 12 however we were cheered by the sight of John & Coleman coming down the mountain with all the animals having found them 5 miles on our yesterday's road, on some good grass which they probably noticed & remembered.

Packed up & started a little past 12. The road this P.M. exceeded anything for roughness that I ever conceived of. Sometimes passing directly over huge masses of rocks where wagons had to be unloaded & lowered down 30 feet or more, & cattle driven round, at others a descent over a rocky hill like stairs jumping down two or 3 feet at a step & the wagon lifted down each jumping place. At others masses of stone & pebbles boulders &c for a mile with nothing of dirt to be seen. Any man who gets his wagon safely over these roads to California deserves a good price for it to pay him for his trouble. How triumphantly we pass all wagons & slide down the hills & over the rocks, leaving them to make their 5 or 6 miles per diem.

Passed a number of beautiful little lakes today, occasionally an open space in the forest covered with fine grass, called

highest point in the saddle of Donner Pass to be 7,089 feet. A United States Geological map of California represents the Donner Pass as being 7,735 feet. Perkins reached the divide about 1 1/2 miles southeast of Donner Pass proper.

5 The towns along the Ohio were visited frequently by artists who displayed panoramas of famous geographical features of the country, of legendary scenes, and of scenes in biblical literature and history. Perkins refers to one of these wandering panoramists who visited Marietta with his rolls of canvas.

"glades" but generally without water. Camped at 5 172 12 miles. Some grass & cold water. No stream.

Tuesday Sept 18. Morn clear cold, breeze from East. Night cold, considerable frost. Start 8. Came upon a train camped which had just killed 3 Blk tailed deer, bought a hind quarter for 1.00$. Saw also some of the pretty spotted squirrels of the mountains, about size of our grey squirrels, grey also, but spotted like a fawn. Noticed numbers of Blue Jays, not as handsome as our bird however. Had another steep descent where the trees were worn by the friction of ropes & we with our animals had nothing to do but slide down half a mile or thereabouts.

Found some of the nuts of the nut bearing pine, resembling a bean in shape & size, & very sweet & pleasant. Camped at 7 in a barren place without the least appearance of grass or oak bushes which latter have begun to show themselves, & on the leaves of which we shall have to feed our animals here after. Tied our poor mules to trees & let them stand to ruminate on their hard fate. We were not much better off ourselves. Our provisions were giving out fast, only a few crackers left. No salt meat & some tea, on this we made our supper & lay down in the dust & slept soundly.[8] Distance 18 m.

Wednesday Sept. 19. Morn clear mild. Start 7, two miles travel brought us to the first "diggings" on Bear River at "Steep Hollow." A tremendous descent & we were on the River among the gold washers & here we saw the first of our future life. Found numbers of teams camped in the valley, among them the Zanesville Company. With them I found Doctor Cormyn one of Charley's students. He was doing very well— had considerable practice for which he charged   per visit. The Doct went with me two miles down the stream to enable me to see the operations of the mines, showed me how to wash off the black sand &c, & was very attentive.

Leaving the "Hollow" we went over the hills to the River again at a store established for the convenience of miners. Here being entirely out of provisions, having had two good messes of our venison, we expected to be able with our little remaining funds to buy enough to take us to the City, but such prices! Flour 60 cents per lb, bacon 100. Ham 1.50, fresh beef 1.00. Crackers 80. Beans 50 per lb, &c.[7] On comparing notes twas found that we could not buy enough to last us half way. What was to be done? Nothing but good work in the

6 As though in common refrain emigrants rejoiced at getting past the Sierra Nevada summit, but they were then confronted by a new obstacle equally as threatening, shortage of food. Perkins and Huntington, down to their last morsel, were not alone. Joseph Warren Wood lamented, "Our provisions were now nearly exhausted & we boiled the last of our rice in our tin cups & eat it up. We had reason to believe our teams were close at hand and they proved to be so. We met a man in the road who had not eaten in 48 hours, he said he had tried to buy Bread but could not obtain it. I gave him some which shortened our stock materially." Wood *Diary,* September 1, 1849.

7 The high prices that prevailed on the gold fields apparently took many of the eager emigrants completely by surprise. There is practically no indication that any of them considered this possibility before leaving home. On August 13, 1849, James A. Pritchard gave an accounting of prices he had received for provisions he had sold in the vicinity of Coloma: flour 60¢ a pound; sugar 65¢; pork 75¢; hard bread 75¢; molasses $4 a gallon; whiskey $4 a gallon; brandy, wine, and syrup $45–$65 a case; coffee 55¢ a pound; tea $4–$6 a pound. Pritchard, *Diary,* 138.

Joseph S. Wallis, of the Naumkeag Company of Massachusetts wrote: "The Elephant House on the levee is the most famous hotel in Sacramento. It is one story in height, 150 feet long and 40 broad. There are three tiers of bunks on each side. They charge two dollars a day without meals. We pay fifty cents for a cigar or drink. At Hangtown [Placerville] we stopped at the Eldorado Hotel, M. Eastner, Proprietor. Here is the bill of fare:

Soup
Bean	$1.00
Ox Tail (short)	1.50

Roast
Beef, Mexican (prime cut)	1.50
Beef, Mexican (up along)	1.00
Beef, plain with one potatoe fair size	1.25
Same from the States	

Vegetables
Baked beans, plain	.75
Baked beans, greased	1.00
Two baked potatoes, medium size	.50

Entrees
Saur kraut	1.00
Bacon fried	1.00
Bacon stuffed	1.00
Hash, low grade	.75
Hash, 18 carats	1.00

mines for a week, hire ourselves out & make whatever we could. This was decided to do & also determined to go on to Yuba about 3 p.m. Distant 12 miles, while I went on another road to "Finley's" a store two miles off, where there was said to be a PO to deposit a letter for Doc, telling of my whereabouts.

Arrived at Finley's & found no office or means of sending word down. So camped near the store, feeling blue enough. Distance 10 m. Day hot, Evening cloudy.

Thursday Sept. 20. Morn clear warm, night mild. Wind E. Start at 9 for Yuba. Our road is improving rapidly, hills becoming less lofty & steep, country more rolling. Pines & their kind are disappearing & in their place we have scrub oaks. Saw in my mornings march several large vultures, such as I never before met with, perched on a dead tree. Their heads & necks entirely bare of feathers & red—body black, & immense claws & beak. A buzzard by the side of one looked like a black bird, & as one sailed close over my head I judged his spread of wing to be seven foot. I should take them to be a species of condor from the recollections I have of the bird.

Overtook the rest of our party at noon encamped in a little valley & blue as can be imagined. Sam C's mule had fallen down from weakness for want of grass, & could hardly carry himself along after his pack was taken off. After I had rested we packed up & travelled the remaining three miles to Deer Creek determined to work there if work could be had. Camped on Deer Creek at 6. Distance 12 miles. No grass, were forced to feed our mules on oak leaves, which they don't fancy much, but are compelled to eat to keep from starvation. When we gathered around our half cracker each & cup of tea, a more "blue" looking set of poor devils could not be picked up. S.C. in particular was just ready to cut his throat, & had given up in perfect despair.[8]

Friday Sept 21. Morn clear & mild. W. E. Saw numbers of the miners this morning as we were eating our apology for breakfast, going down the creek to their respective places of washing. They all had a sallow sickly appearance & we were told that there had been a great amount of sickness during the summer, the prevailing complaints being of a bilious character, rather tedious & expensive though seldom fatal.

After breakfast dispersed in various directions to seek Employment of any kind but were all unsuccessful. None making enough to warrant his employing hands. Our next resource

Game	
Codfish balls, per pair	.75
Grizzly roast	1.00
Jackass Rabbit (whole)	1.00
Pastry	
Rice pudding, plain	.75
Rice pudding with molasses	1.00
Rice pudding with brandy peaches	2.00
Square meal	3.00

Payable in advance. Gold scales at the end of the bar. Howe, *Argonauts of '49,* 156-57. William North Steuben was impressed also with the inflated prices. He gave the following tables:

Pork	$1.00
Dried beef	1.00
do fresh	35
Flour	50
Dried Appels	1.50
Rum, brandy pr bottle	5.00
Ale per bottle	50
Suger per lb	50
Molasses per gallon	5.00

And all things in proportion Steuben Diary. This accounting was scribbled on a back cover. "Some providings bot. in Sac City at the following prices: Flour 10cts—rice 12 1/2—bread 15—cheese 75¢—Pork 30 c—sugar 15—2c/ Vinegar 1.25. Molasses 1.25, paid 25¢ for washer pan 5$ pick Crowbar 4.50. Sail needle 50 cts ea. bread 50¢ per loaf." Chamberlain Diary, August 20, 1849.

8 Joseph Wood and his companions did not give up in "despair" when they reached the upper diggings on August 31, 1849. They did, however, suffer a rude shock when they saw how little gold five men washed out in a day and how much they had to pay for food. Wood took some satisfaction, as did Perkins, that it was said, "Provisions are said to be far cheaper lower down, or at Sacramento City 75 miles from here." Wood Diary, August 31, 1849.

Joseph Sedgely had even more reason to despair. "We remained in Sacramento until the 9th inst. We then left for the dry diggings, on the North Fork, where we arrived on the 11th. I was taken sick, that evening, with fever and ague, and Hawkes was taken sick at the same time, and there we lay on the ground, until the 21st. We then hired an ox-team to take us back to Sacramento, and were dragged two days under the burning sun, and reached the city on the 23rd. Finding that we did not improve, we took passage in an open boat for San Francisco, where we arrived in an exhausted condition on the 28th, and took board at the Pacific House at $12 per week. We are sick and feeble." *Overland to California,* 65.

was "prospecting," or endeavoring to find a "hole" to dig for ourselves, Coleman & myself, taking our only tin pan & pick, started up the creek & tried various places, he digging & I washing. The produce of our mornings labor was about 1.00$ This would not do of course, not half paying expenses.[9]

On returning to camp we found Cross & Co still more unsuccessful, not having made a cent, we were pretty nearly in despair, & I determined for myself to start for Sutters at all hazards & trust to chances to get something to eat on the road. Here Gilbert who was very anxious to go down said he had a few dollars left, & if I would carry his baggage he would buy something to last us down. To this I agreed & we went up to "the store" to see what could be had *cheap*. The prices of everything were very discouraging till the storekeeper mentioned that he had some flour that had been wet & heated & some sour & lumpy, but we could have it if would answer at 15c per lb. We immediately invested all our capital in this article, purchasing 15 lb. On hearing of our success Cross was seized with a desire to travel, & took his tin pan up to the store & succeeded in trading it for 3.00 worth of the same flour. So once more we were provisioned. These tin pans by the way are as indispensable to the miner as his pick or gold washer, as only by their aid can he separate the grains of metal from the black sand in which it is found.

Packed up once more & left at 1, happy to turn our backs on the scene of our mishaps & adversity. Leaving Deer Creek valley road ascended a moderate hill & country we found much more level than any we had yet passed over this side the Sierra. The majestic pines had entirely disappeared & we saw only scrub oaks, & some undergrowth of various bushes. Camped at 5 on a little dirty puddle of water. Some coarse grass. Distance 8 m. Day very hot.

Saturday Sept. 22. Morn clear warm W. E. Night mild, start 7 1/2, road was rolling country becoming more level as we advanced. Started a flock of quails this morning, the first I have seen. They resemble our bird much tho rather smaller & darker colored. Their note is quite unlike, no clear whistle, but a crow more like the pheasant or prairie hen.[10] Saw also some California pigeons, a fine bird considerably heavier than the eastern pigeon, & without the red breast of the males. Camped at 5. Distance 10 m. Day hot.

Sunday Sept 23. Morn clear warm, night warm. Wind East. Start 7 1/2. Was awaked this morning by "Watch" running

[9] This entry of Joseph Warren Wood in his diary on March 3, 1850, might well have epitomized the disappointment of thousands of miners. "Today I worked on the Quicksilver machine & made 3 dols. poor, poor, poor." "Today we done as yesterday, in the morning it rained enough to wet a man through his clothes, the first rain I have saw fall for some time. Wednesday 10, Still working at the same place. Robbins found a lump worth 12 1/2 dollars in travelling from our camp to the diggins we saw a pismire flitting there were millions of them going from low ground to high." He was on the middle fork of the Cosumnes. Others had similar experiences. Sunday, October 7, "Today I was wandering around in the dry diggings and I succeeded in picking out a lump worth from three to four dollars. I then gathered up a gallon of dirt and washed it, and found about three dollars more in it." Markle Diary, October 7, 1849.

[10] It is probable that Perkins saw a covey of northern valley crested quail. Dawson, *The Birds of California*, IV, 1575-86, indicates that this, rather than the desert quail, was in the range crossed by the emigrants.

furiously after something near my head, & raising up saw a wolf making off over a little ridge, the dog close behind him. At the top of the hill the wolf stopped & Watch trotted up to him & thus they exchanged very cordial salutations a la mode Dog! after which our animal trotted back towards us, & the wolf apparently accepting the invitation to visit Watch in camp followed after! & came up to within 20 feet of us as I leaned on my elbow before he made the discovery that Watch was betraying him.[11] A motion I made to seize my pistol alarmed him & away he went, no doubt wishing Watch all manner of evil. These rascals are very numerous & very saucy also in the night. They make all manners of demonstrations round us every night but hardly venture an open attack.

Had a distant view from the top of a hill of the valley of the Sacramento this P.M. could distinguish nothing however but an indistinct looking plain with some trees scattered over. Camped, having "laid by" most all day. Distance only 6 miles Day hot, Evening clear. Wind East.

Monday Sept. 24. Morn cloudy, wind S.E., warm. Start 7 1/2. Soon after leaving our last night camp we emerged from the hills upon the barren ridges bordering the Sacramento Valley. A few small oaks were scattered here & there & a scanty grass found some nourishment in the red clayey soil. This region looked rather desolate & dreary far different from anything we were prepared to see in this "beautiful country." At 9 we descended from these "barrens" and were at last in the long hoped for "Valley of the Sacramento" & at 11 arrived at Johnson's famous "ranch."[12]

A mile above Johnson's we met a party of native Californians in their fancy colored dresses & ponchos with huge spurs & fine horses & all the accompaniments of tinkling bells, dangling tassels, lasso &c. They were driving a herd of cattle, & shouting & singing most merrily. Here I first saw the use of the lassoe, & beautifully was it done. A Mexican was riding along singing cheerily & occasionally twirling his noose around his head most gracefully & easily, till the hoop spread out to 4 or 5 feet in diameter, when off it would shoot wavering through the air like a soap bubble till it dropped directly over the object aimed at.

"Johnsons" has been an extensively cultivated "ranch," but is now neglected, & occupied by some Americans with goods & groceries. Acres of land formerly under cultivation are fenced in by a deep trench 6 to 8 feet deep & 4 or 5 across, the dirt thrown out being piled upon the inside making a kind of wall

[11] "The wolves howl around us in a manner which I have never heard equalled. We found in the morning that we had camped near a dead horse, which they had undoubtedly gathered around for the purpose of feasting upon its carcass." Wood Diary, November 19, 1849.

[12] Johnson's Ranch, the site of present Wheatland, California, contained 22,197.31 acres, or five square leagues. It was carved from the Rancho de Yuba on the Yuba River, or the Foldos de los Sierra Nevada, and on the Arros de los Osos, or Bear River. This information was taken from the abstract book of the Rancho de Yuba now in the possession of Lowell Swetzer of Wheatland. Some kind of legal transfer was made of this property at Hock Farm in 1851, after it had been sold by Johnson and Keyser in 1849.

& fasse. This expensive & laborious kind of fencing I am told is the only one used by the "rancheroes." However as they can employ hundreds of Indians for a mere song in clothing & beads, perhaps these trenches may be better than a wood fence, especially as timber is scarce in the valley.

We were compelled to encamp at Johnsons with Cross, who was sick from eating our unhealthy fare. When we bought our breadstuff we also ventured to ask the price of saluatus. 5.00$ per lb! & so we could not afford to indulge in such a luxury, & mixed up our sour flour with a little water & salt & baked it before the fire, & I only wonder it has not made us all sick, a heavy unhealthy mess. Camped at 11. Distance 11 miles. Day hot.

Tuesday Sept 25. Morn clear hot, wind S.W. Start 8. Johnsons is situated on Bear River. Our road follows the stream down to Vernon where it empties into Feather River, & where Feather River empties into the Sacramento. We chose the River road via Vernon instead of going across the plains without water or grass, tho some 15 miles nearer.

Bear River here is a small but pretty stream, would be called a mill creek in New England. We encamped on its banks at noon today on most luxurious green grass, which we found in a thicket of briar bushes among the trees. At evening made Nichols ranch at the mouth of Bear River, now owned by a couple of enterprising Dutchmen from the "Shtates" who are putting up quite a respectable adobe house. Camped at 7, on Feather River. Distance 18 m. Day very hot. Wind S.W.

Wednesday Sept 26. Morn clear warm. Wind E. Night warm. Start 8. Met a mile from camp the first of the Government relief parties, & learned that some hundred thousand had been appropriated to be dispensed for the relief of those Emigrants whose late start or loss of cattle renders it doubtful whether they will pass the Nevada before winter. Hundreds of such there are as I know having passed multitudes who must now be 600 miles behind, & other parties will be joyfully hailed on their errand of mercy. Pity but the Government treasures could always be appropriated to a good purpose, in relieving misery instead of causing it.

Arrived at Vernon a small town at the junction of Feather & Sacramento Rivers,[13] at 9 A.M. & first beheld the stream whose name & fame are known all over the world & whose golden sands seduced me from all I hold dear into the wilderness.

13 "About noon we came to Vernon situated at the junction of Feather and Sacramento rivers, a few small boats lay along the shore. A few tents & a half finished 2 story pumpkin shell composed the town. Business was brisk. Pork was $80.00 per Bbl. & flour $24. per hundred. A small loaf of bread was offered us for .75¢. Whiskey was $1.50 per pint." Wood Diary, October 17, 1849.

At this place the Sacramento is a clear still beautiful River about 1/4 or 1/3 of a mile across, the banks lined with oaks & various vines & bushes, willows &c. Its valley of the beauty of which we have heard & read so much, is a vast waste plain covered with a scanty wiry grass, with occasional marshes in patches of trees.

Never was there such misrepresentation as about this country, both as to the futility, fertility or capability of cultivation, & richness of the mines, & all that a few men might make fortunes. Among the Emigrants you will hear Bryant, Frémont, Robinson & others whose published accounts were the chief inducement to many to leave their comfortable homes, cussed up & down, & loaded with all kinds of opprobrious names. They have all amassed fortunes off of the Emigration they have induced.[14] This valley presents few attractions to any one who has lived in the states. No beautiful forests, or rich meadows but very few singing birds, *Except Owls,* & these abound. There are some Elk in the plains & any quantity of wolves, also in the sloughs great numbers of cranes, geese, ducks &c, but every one without exception is disappointed both in the appearance of the country & the richness of vegetable or mineral productions.

We crossed the River at Vernon without any ferriage to pay owing to the competition between Vernon & a little town opposite called Frémont.[15] I saw just below Frémont a sight cheering to my eyes, in the shape of vessels, schooners &c, which navigate this far up, with various cargoes at enormous prices. I felt as if we were once more within the pale of civilization & our desert journey was indeed at an end. Traveled down the Sacramento & encamped at 4. Distance 16 miles. Day hot, wind S.W.

Thursday Sept 27. Morn clear cool, wind E. Night cloudy windy. Start 8. Noticed a curious bulbous root growing directly in the road & paths, and nowhere else. A little tuft of hairlike horse hair is all that appears above ground. On pulling it up resembled a cocoanut with the husk on only some smaller, & on peeling the bulb resembled that of tulip. There appeared to be no sign of any stalk or stem nor could I determine their nature or use.

At about 10 in attempting to water our animals "Old Jack" Sam's mule sank in the mud & was mired fast enough besides getting all his packs wet. We had to work to get the mule out of the river & when we at last got him to the top he was sadly strained.

[14] Perkins may have been correct in saying that Bryant, Frémont, and Robinson were cussed up and down the trail for their humbuggery. He was also correct in saying that Bryant and Frémont's works had sold well. The fault lay with the emigrants, who read into these guides—if such they might be called—what they wanted to find. Bryant's *What I Saw in California* could hardly have been said to soft-pedal the hardships that travelers along the trails to California might expect to undergo. Bryant's instructions, which were published in the Louisville *Daily Journal,* February 26, 1849, and in the Louisville *Courier,* December 9, 1848, were commonsense advice, as he would prove on the trail itself. It is indeed doubtful that Perkins saw the abbreviated English edition of *What I Saw in California* (G. Routledge and Co., London, 1849), in which an inserted chapter, not of Bryant's authorship, made extravagant and unreliable promises. Frémont's description of his second expedition was not a cheerful invitation for greenhorn travelers to take to the road, especially in the Sierra Nevada in the winter months. Perkins had read the section in John C. Frémont, *The Exploring Expedition to the Rocky Mountains, Oregon and California,* in which was appended to the 1845 narrative, pp. 234-55, a description of the territory where subsequently would be located the gold fields; Colonel Mason's famous report was also included here. This report did give a warning about prices. Flour, he wrote, was worth $80.00 a barrel, and clothing was high. There was also advice to emigrants making the journey to California, pp. 454-55. Alfred Robinson's *Life in California During a Residence of several years in That Territory comprising a Description of the Country and the Missionary Establishments with Incidents, Observations, etc etc,* was published in New York in 1846. It could hardly be called a guidebook for overland travelers.

[15] "Frémont, a town directly opposite [Vernon] appears to be about equal in size & Business to Vernon." Wood Diary, October 17, 1849.

We continued our last days journey down River. Timber pretty plenty on its banks with wild grapes in abundance running over the willows & other bushes & a tolerably fair small grape just ripe, of which we eat plentifully, occasional spots of good grass & the broad plain stretching way to my right to the coast Mountains, the faint outline of which we could just perceive. At 4 we turned to the left through the timber & came suddenly upon the banks of the River at the ferry & directly opposite the far-famed & busy Sacramento City, with its vessels & cloth houses & all the scenes & sounds of civilized life, Oh, how rejoiced we were at the prospect! & what a comfort there was in the anticipation of letters to be received & friends met, *& many made!*

Crossed the ferry for which we paid the last 50 cents that could be scraped together, & took our way up the river through the streets, feeling a little awkward in our dirty & ragged prairie costumes, & encamped just in the edge of the city on the American River, under some large sycamores & overhanging vines, & here we are at last, at the end of our long & tedious journey, & having for the last time thrown off our packs & turned our mules out to graze.[16] We laid ourselves down in the shade to luxuriate in the glowing thought. No more deserts to cross & no more thirst on those deserts, no more getting up at daylight to swallow a cracker & pack up for a tedious days tramp through the heavy sand & in the hot sun. No more fear for the safety of the mules & ourselves & doubts as to our getting through, &c &c. Here we are at last through without a cent in our pockets, but here money can be made & we are in no danger of starvation, however I must see what I can hear of "Doc" & whether letters are to be had & try to dispose of "Dav" or "make a raise" some way, as our last flour goes into tonights baking—14 miles—

On a review of our journey & its incidents now that it is all over & our sufferings & privations at an end, I would not have it differ in any respect from what it was, we saw everything of frontier travelling that could be seen & struck the life in all its varieties, with wagon, packs & on foot, & the harder the times we had the pleasanter the restrospect, by contrast. Dearly have I paid for my experience to be sure, both pecuniarily & physically, but I should know now exactly how to go back by the same route both pleasantly & speedily, & at much less expense, & consider myself pretty well qualified to give advice to any of my friends who wish to try the same journey.

[16] Sacramento for most gold seekers was a mecca. When the trail-worn emigrant reached this point, he felt that once again he was on at least the fringes of civilization. On Friday, October 19, 1849, Joseph Warren Wood entered his impressions in his voluminous diary. "Today I have visited Sacramento City. It is situated on the East side of the river, just below the junction of the American Fork. It is a strange place of great promises. It must contain 4 or 5,000 inhabitants, mostly men. The houses are in a great measure built of cloth. 2 or 3 are of sheet iron. Some are of boards. It is useless for me to enter into a minute description of all I saw. Men were engaged in buying and selling. There were hotels with great names. Stores & Doctors & Lawyers shops. They were universally patronized & their owners undoubtedly cleared their ounce per day. Things that were wanted were high. While things not needed were sold or thrown away." *Ibid.,* October 19, 1849.

Margaret Catherine Haun, the Iowa housewife, was somewhat more charitable in her retrospective description of Sacramento than were Charles G. Gray and Dr. George O. Hildreth. She found the trade center of the gold-rush country a "goodsized, hustling, thriving town of about an equal number of houses and tents. It was the principal distributing point of merchandise for the mines. Although very tired of tent life many of us spent Thanksgiving and Christmas in our canvas houses. I do not remember having had a happier holiday times. For Christmas dinner we had a grizzly bear steak for which we paid $2.50, one cabbage for $1.00 and—oι horrors—some *more* dried apples. And for a New Year's present the Sacramento River rose very high and flooded the whole town." Haun Diary, 41. "Sacramento City to use a *Willisism* (N. Parker Willis) should be called Ragdom as almost all the dwellings are built of slender joists of wood, cover'd with canvass or cloth, many tents were also in use & they gave the place the looks & air of *an immense camp,* & about a mile below the American River, the streets are laid out regularly, those running *east & west* being named alphabetically, & those *north and south* numerically." Gray Diary, November 9, 1849.

From my own observation I am satisfied, that with good oxen & *light teams* a man can travel with more personal comfort than in any other way, & by starting early enough is safe for getting thro. If I were going with a wagon I should take cattle in preference to mules & they travel on a long journey nearly as fast & are much better for the steep hills & rough roads, pulling steadily & patiently, with an ox team many comforts & luxuries, cooking apparatus &c can be carried taking much from the hardships of frontier life.

But the only way to go in my estimation is with good pack mules & Canadian ponies for riding animals, trailing speedily & without any apprehensiveness as to the roads or hills or streams & no fear of breaking tongues or axles. The anxiety of mind consequent upon having to carry a wagon over such roads, more than counterbalances all the comforts & luxuries attending it, & for my part I prefer the simple fare of soft crackers & ham with cold water to the nicer bread, beans &c of the trains, with the cooking, washing dishes &c necessarily attendant & prefer to rest after a long day's journey on simple fare, to spending time cooking richer. All necessary to make a trip very comfortable, is some good *soft* crackers which keep quite as well as hard ones & are much more palatable beside soaking readily in water, good hams, the bones taken out & the meat sewed up in cloth, some tea or choclate, plenty of sugar, which is a really necessary article, the lunch at noon being sugar & water with crackers crumbled in it, & when a man is hungry forms no unpalatable dish, a little pepper & salt in case any game is killed along the road or fresh meat bought, some soap, matches, &c, & the stock is complete.

For clothing a couple of pair of *woolen* pants some half worn cloth ones, which almost every man has, 3 or 4 hiking shirts & 12 woolen ones, some prefer woolen altogether, a stout woolen roundabout, with large pockets, strong boots & good "Palo Alto" hat. Buckskin gloves, woolen socks a couple of towels &c, very little need be spent on personal outfit, most every one has old clothes Enough to supply all his wants. For bedding some prefer a Buffalo & blankets, this is my own choice—& others three or 4 blankets. The pad of the pack saddles makes a fine pillow.

Two good active medium sized mules are sufficient for one man & a Canadian pony accustomed to fast walking. The mules should not be packed over 100 lbs each. They break down if overloaded. Pack saddles should be large & fit well & well padded with hair or moss, & packs should be made of leather, to hang to the horn of the saddle one on each side large roomy & with a top to buckle like a Doctors saddle bags. Thus completely fitted out a man can "roll ahead" without any regard to wind, weather or country, & feel perfectly independent.

September 28, 1849–
February 28, 1850

SATURDAY OCTOBER 6. Here I am still in Sacramento City—feeling quite well & enjoying myself finely. The day after we arrived went down town to try & pawn my watch & not be obliged to force my mule into market. In course of my travels I stepped into a store & saw a young man behind the counter whom I recognized as an acquaintance I made in New Orleans, a fine clean fellow, a nephew of Col. Bailie Peyton & named Peyton himself.[1] He also recognized me & after mutual greetings & congratulations on finding each other in the *famed!* land, I told him my "fix" & offered my watch till I could redeem it. Peyton only laughed at me, "Why," says he, "do you think I would take your watch for security for anything you want? Just make a bill of provisions, clothing, anything you need. Tell me how much money you want & my store & purse are at your service." Here was a "hit" indeed, & I hesitated at first about accepting his generous offer but he insisted & so I bought a lot of flour, crackers, pork, fish & little sundries & took some money for my present necessities, feeling very happy & gratified at finding such a friend when I had expected only strangers.

Two days after I sold "Old Dave" for 120$ & was able to pay Peyton & thank him for his generosity. He is a noble fellow & I shall always remember him. Met in the street the same day Doct Jackson, Harlan & others of our traveling companions who were about leaving for the mines. They arrived a few days before us. From Harlan I learned that Zeb & Stephens arrived three weeks since, & had gone to Redding diggings on the upper Sacramento. Zeb having been sick for some time previous to leaving. No letters in the P.O., or news from Doc though saw some of the passengers who came round in the same ship & who said he was coming up soon so stuck a notice on the P.O. door telling where I could be found, & two mornings after who should walk into camp but the old fellow himself looking fine & hearty. Of course we were rejoiced to see each other & Doc & Hollister who came up with him moved their baggage to our camp & have been with us ever since.[2] We are waiting now for the remainder of Docs freight to come up from San Francisco consisting of provisions &c, & then shall leave for the mines. Met in the street yesterday our old companion Doct Riggs who was just in. He looked rather thin & said they had had a hard time. Received by Doct a letter from Hat, the only word from home since I left. No mail has been brought from Panama for three months & my letters must be waiting there in quantity.

The city is a remarkable place indeed, built almost entirely

[1] B. K. Peyton's business in Sacramento was located between Third and Fourth streets in 1850. J. Horace Culver, *The Sacramento City Directory*, January, 1851. Perkins had become acquainted with Peyton on his visit to New Orleans during the fall and winter of 1848.

[2] Dr. George O. Hildreth, who had arrived in San Francisco by way of the Isthmus of Panama and W. D. Hollister of Marietta. Hollister wrote his family at Marietta on December 9, 1849, and Dr. Hildreth wrote his father on December 24. Their letters reveal much about conditions as they viewed them in California (See Appendix).

of cloth spread upon light frames. In this way are built stores doing large business. Hotels, gambling houses in abundance. Theaters, &c, & almost everything can be had & seen in this city of 6 months growth as in an eastern city. Lynch law is the only law known & yet though there are in the population of 7 or 8000, 20 or 30 large gambling houses & liquor bars, there has been no case of riot & bloodshed.[3] Half the stores have no front, & no way of closing up at night, & provisions, clothing, fancy articles &c are left thus exposed but nothing ever stolen or disturbed.

Rents & prices of land & horses are very high.[4] The "City Hotel"[5] a new building of wood cost 100,000$ & rents for 50,000$. Board at it 25.00 per week. Common store rooms rent from 300 to 1000$ per month, &c.

The city is built at the junction of the American and Sacramento rivers, is laid off in squares & streets lettered & numbered, & is growing fast, but I think it will as suddenly go down as it has risen, before two years, there being nothing to support it when the Emigration is over & returning as hundreds are now doing. Suppose a tent in the windward side of the city should take fire, as no doubt some will do as soon as cool weather sets in & stoves are introduced, in ten minutes where would the City be? The gold miners are now going out rapidly & the richest diggings have disappeared & even if the mines should yield for some time to come, it will not be in such quantities & with such ease as to induce emigration or keep that now in here, the mining gold decreases in quantity must be done by a few companies with capital & heavy machinery. I find on conversing with miners, business men, & Emigrants that everyone is disappointed & wants to go home. Nothing like the amount of money can be made or mined here that was represented or supposed & every man who was doing anything at home regrets leaving it to come here. Some few first comers are doing well, the majority can hardly save enough to take them home.

Was awakened this morning by Watch who seemed to be at swords point with some "varmint" & got up to reconoitre. Though hardly daylight I could see some objects up a small tree over our fireplace which I made out to be "coons" & with my revolver I succeeded in bringing them all down, three in number. Sold two to the City Hotel for 2$ & eat the other which we found delicious, tender juicy & fat. They are very plenty along the River bank & we have seen their tracks in our camp frequently.

[3] Gambling was one of the most exciting forms of relaxation. Men away from home, freed of the inhibitions imposed upon them by a puritanical society, threw off most restraints. For some, wealth came easily and went as easily. Everywhere in the San Francisco–Sacramento area gambling and carousing were common. Coy, *Gold Days*, 156-272, and Paul, *California Gold*, 81.

Charles G. Gray agreed with Perkins that gambling and drinking were absorbing pastimes in Sacramento. "Had a pretty good look at the city of Ragdom, with its City Hotel & General Jackson's House & Frémonts House, & the Elephant who appears to have taken up his quarters here, besides its theatres & markets & hells & drinking saloons without number, rejoicing in the names of the Empire, the Golconda, the Shades, the Plains & a dozen others, & where gold & silver coin & lumps of native gold from the mines also, were piled up in masses on the tables.

"Gambling here is a perfectly regular business & carried to a great excess, by all conceivable games." Gray Diary, November 10, 1849.

[4] "I heard in Messrs. Symmons and Hutchinson's office that it would come up to $2,000,000. Don't stare! This country raises gold and they have not begun to dig it up yet. If you had seen the heavy valises that came down the river when I came you would think there was some here, if not more. I know of three men who are going 'home' with $150,000. There are some rich men here. Samuel Brannan, one of the proprietors of Sacramento City and who owns the City Hotel there, has an income from rents alone of $160,000 a year, besides a store here and at the place." White (comp.), *A Yankee Trader in the Gold Rush*, 54.

[5] By 1849 the City Hotel was a prominent landmark in Sacramento. It was located between I and J streets on Front Street. *Sacramento Guide Book*, 41. A lithograph by G. V. Cooper, made in 1849, shows it as a two-story, oblong wooden structure with a peaked roof. Copy of lithograph in California State Library, and in the Honeyman Collection, Bancroft Library, Berkeley.

Yesterday evening a negro killed an enormous animal of the ocelot species opposite us, & about 100 yards off twas a beautiful creature spotted like a leopard, but would be a rather savage customer to handle if wounded or provoked. Some of my old friends the wolves are still extant in this vicinity & for two nights past have given us their familiar serenade, almost transporting us back to our camps on the plains again.

October 12. Still in Sacramento City waiting for the goods & provisions, which is very provoking as if the mines are half as good as they ought to be we could have made more than enough to buy them here, in fact it was a great mistake bringing such articles round at least so it has proved, as all kinds of provisions can be procured here at a small advance upon expenses, & without any delay.[6]

I made a new acquaintance, or rather met an old one, a few days since, whom I was most agreeably surprised to meet in this country, nobody but Miss G[eorgiana] C[utler] My fair friend from Quincy Ill. who visited Miss H. B——— in Marietta, some time since, & whom I found in a fine large tent spread with Buffaloe robes &c quite pretty bower of a place looking as blooming and sprightly as of yore. She was with her step-father Mr. Dan Whitney, & came with him across the plains, travelling most of the distance on horseback. She must be an enterprising character indeed to undertake such a journey & enjoy it too as she professes to have done, I sat down & had a delightful talk with her of old times & new & enjoyed her society the more, that I have not spoken to a lady before for 5 months, fond as I am of them. "Friend G———" as I call her, dresses quite stylishly, as at home, wears her silks, laces &c, & has already attracted a host of admirers, & if so disposed can no doubt marry a California fortune whenever she chooses. Since my first interview I have called on her several times, & pass my most agreeable hours at her bower.[7] Though my costume is rather "outre" I have nothing but my frontier uniform which, however has *now* the merit of being *clean* if not fashionable—

Stepped into a celebrated gambling house a few evenings since, as I often do to hear the fine music they employ, & saw pretty heavy betting at a Monte Bank. One man bet 25 & 50 dollars for several bets & lost all he had, then borrowed of a companion & lost all again & borrowed a second time 50 dollars, & lost all but 10. With this his luck turned & he began to win, & doubling his bets as fast as the cards were laid out, he soon had an enormous pile of specie before him, which he raked into his

6 Many an emigrant had reason to agree with Perkins on this point. Charles G. Gray shipped his goods aboard the sailing ship *Orb*, but when he arrived in Sacramento, he found that there were three vessels named *Orb*—a barkentine and two sailing ships. Which one had brought his goods was a question. Too, if the ships were unloaded in San Francisco there was grave danger that their cargoes would be stolen outright or pilfered. Gray Diary, November [undated], 1849,

7 Miss Georgianna seems to have made no other mark in California or Illinois history under her maiden name. Her presence in Sacramento was without a doubt an honorable one. Perkins knew her as a member of the famous Manasseh Cutler family, which had helped to settle Marietta. There seems to be evidence that Georgianna was the daughter of Charles Cutler who died in 1849 on the trail to California. He had married Maria Walker in 1819 and was the father of six children. Information supplied by Carol R. Prisland, Chairman, Research Committee, Evanston (Illinois) Historical Society, October 15, 1964, from the privately printed *Gates Ancestral Lines*, I (1843).

8 "There is, however, a large floating community of overland emigrants, miners and sporting characters, who prolong the wakefulness of the streets far into the night. The door of many a gambling-hell on the levee, and J and K streets stand invitingly open; the wail of torture from innumerable musical instruments peals from all quarters through the fog and darkness." "In the gambling-hells, under the excitement of liquor and play, a fight was no unusual occurrence. More than once, while walking in the streets at a late hour, I heard the report of a pistol; once, indeed, I came near witnessing a horrid affray, in which one of the parties was so much injured that he lay for many days blind, and at the point of death." Taylor, *Eldorado, or, Adventures in the Path of Empire*, II, 28, 33.

9 A movement was begun in July, 1849, to organize the city government of Sacramento.

boot which he took off for the purpose, nearly filling it. The play still went on, our man betting 150 & 200$ generally winning till at last the banker dared him to "take the Bank" that is bet all he had against the Bank. He agreed & the cards were thrown down—a deuce & quartre, taking up the "deuce" the lucky fellow turned it bottom side up upon the Bank & the cards were drawn & the third one came up "deuce," & the bank was broken.[8] Taking off his other boot, our man raked the whole pile gold silver & all into it & politely wishing the banker a "very good evening" he left with a couple of friends —his winnings amounted to about 2500$. Thus twas I looked at him with envious eyes, thought of my dear wife & how happy I should be to lay the whole in her lap, & how much more good this sum would do me, than the man who won it, & who would probably lose it all the next night.

The "charter Elections" came off last Monday, with a good deal of excitement, the principal issue being, between the gamblers & those disposed to tax them for licenses. The "Chartists" carried the day, & the gamblers are much chagrined of course.[9] A duel was near coming off between Lundy, a prominent gambler & notorious duelist & Col. Winn, President of the City Council, growing out of a dispute at the polls, which was prevented through the intervention of friends after the challenged had passed.[10] Had it taken place both would probably have fallen as both are good shots & both have killed their man.

November 1. Well here we are in the gold mines of California, & mining has been tried "& found wanting!" We left Sac. City October 18, with our provisions &c in Chapins wagons en route for the Cosumne River[11] distant some 28 or 30 miles & arrived here the 21. We are about S.E. from the city, in a rolling country & on a small rapid stream tumbling over a rocky bed. The appearance of the country through which we passed was somewhat better than that down the Sac. River, as we saw it, but yet I have not been in any part of the "beautiful valley" of which we used to hear. On locating here we immediately went to work making "washers." Doc & John being somewhat "under the weather" I did nearly all the work on the mine alone. I cut down a pine tree, cut it off the proper length peeled & cut down one side & with axe & adz hollowed it out till I reduced it to about 1/2 inch in thickness by nearly two days of hard labor & blistering of hands &c, & another day put in the "ripples" dash screen &c & took it down to the rocky

The population was expanding rapidly, and lawlessness was threatening. An election was held in the St. Louis Exchange on Second Street, and nine town councilmen were elected, among them Albert Maver Winn. On August 1, Winn and six councilmen issued a proclamation to the people of Sacramento saying that on July 13 a charter had been submitted for the people's consideration, and they had rejected it by 146 votes. They requested a second meeting in the St. Louis Exchange to reconsider the charter, this meeting to occur on October 10, 1849. A "Law and Order Party" was organized to contend with the rowdy element, especially the gamblers. In the next election the Charter or Law and Order Party carried the day by 296 votes, and the Peoples Charter was adopted. Morse, *The First History of Sacramento*, 41-43. An extensive discussion of the charter issue appears in the Sacramento *Transcript*, April 3, 1850.

10 Brigadier General Winn was a prominent businessman in early Sacramento. Here Perkins refers to the riot among squatters, who were protesting because one of their number had been dispossessed from a plot of ground. General Winn issued a proclamation declaring Sacramento under martial law. He enlisted the peace-loving citizens to help him. Quiet was restored quickly. Morse, *First History of Sacramento*, 63, 78-79, 106-107.

11 The Cosumnes River, an eastern branch of the San Joaquin River, branching off from the Mokelumne River south of Sacramento.

12 Many men became so disillusioned at the beginning of their search for gold that they made immediate plans to return home. Joseph Sedgely, in discussing his plight from October 28 to November 28, 1849, gave a splendid account of the kind of despair men suffered. *Overland to California*, 64-65.

13 "How vain are the hopes of man. This morning we are drenched with rain, & the prospect is that it will continue through the day. My socks which I hung by the fire to dry are completely wet. It is really discouraging. Mr. Pollard whose anticipations of a fortune have always been moderate, has fallen this morning, before breakfast $3000.00 he thinks that if he had 2000 he would go home contented. I should hate to have a

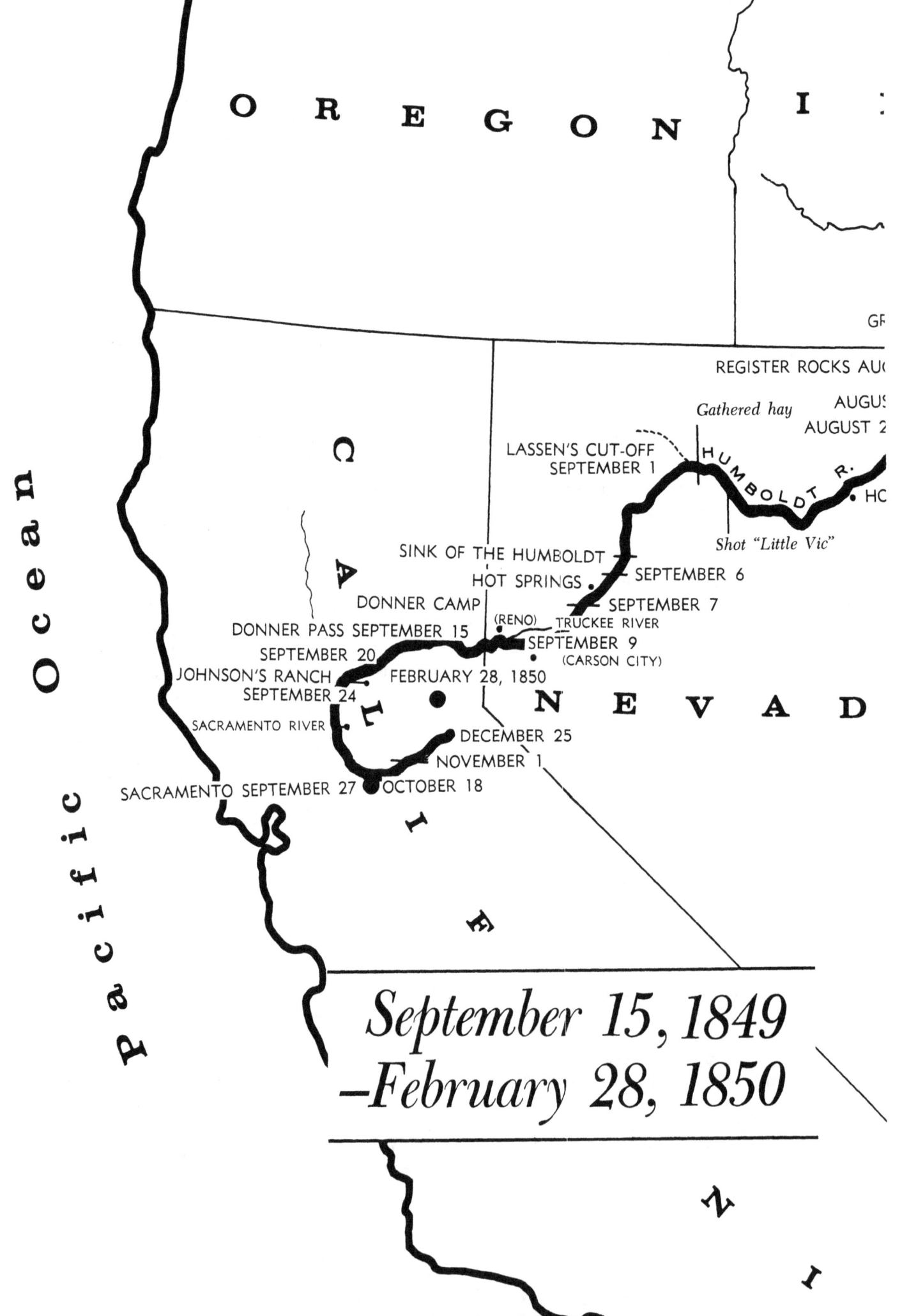

OREGON

I

Pacific Ocean

GR

REGISTER ROCKS AUG

Gathered hay   AUGUS

AUGUST 2

LASSEN'S CUT-OFF   HUMBOLDT R.
SEPTEMBER 1

HO

Shot "Little Vic"

SINK OF THE HUMBOLDT

C

HOT SPRINGS   SEPTEMBER 6

A   DONNER CAMP   SEPTEMBER 7

L   DONNER PASS SEPTEMBER 15   (RENO) TRUCKEE RIVER
SEPTEMBER 20   SEPTEMBER 9
I   JOHNSON'S RANCH   (CARSON CITY)
SEPTEMBER 24   FEBRUARY 28, 1850

F   SACRAMENTO RIVER   N E V A D

O   DECEMBER 25

R   NOVEMBER 1

N   SACRAMENTO SEPTEMBER 27 OCTOBER 18

I

A

*September 15, 1849
–February 28, 1850*

N

I

bar where we are to commence our fortunes! & since have worked away like a trooper, rain or shine, with but indifferent success.[12]

The first three days Doc took hold with pick & shovel we excavated a hole about 4 feet deep & made in that time about 2.00! Here Doc broke down & was taken sick & John being about recruited he commenced with me in another place & we have made something everyday, the highest 11.00 lowest 3 each. This wont do & we shall probably leave soon for richer diggings next week. Tis terrible hard work, & such a backache as we have every night! We are below the bed of the creek & have to bale out water from our "hole" every hour, & work in the mud & wet at its bottom.

I am pretty well satisfied that fortunes in Cal. as anywhere else, with some exceptions, take *time* & hard work to get, & I must go home without mine I'm afraid, as I would hardly lead this kind of life 5 years for any fortune, & I certainly would not be separated from H. that length of time.[13] Since we have been here every man of our company has been sick but myself, Dwight, Sam & Doc are all laid up at the present time & John is but just up, bilious fevers are the prevailing disorders. Dwight & John have been quite sick.[14] John was partially delirious several days. I have been perfectly hardy & able to do my days work all the time, & hope I may continue so.

Game abounds around us, quails rabbits & deer. The Harmar Co. has Killed 4 deer, a company near them Killed two this morning, & you can hardly walk a mile in any direction without stumbling upon them. Two ran almost into our camp a few evenings since & stopped about 20 yards off to look at us. I tried to get a shot at them with my revolver but could not see distinctly enough & they ran off. I am without any gun having lost mine on the plains as stated in my journal, nor is there a gun in our camp. I must take time before long & kill some venison, borrowing a gun from the Harmar Co. who are camped about 300 yards off. I hate to lose a day though as every dollar I make I look upon as so much towards getting home.

The weather since we have been here has varied. Much at first the sun was very hot during the day, afterward it became somewhat cooler & the last three days have been rainy, considerable water falling. I have however worked every day so far, & dont think I shall lose many rainy days in Cal. I cant bear to stay here & think I am making nothing. Nights are generally quite cool, but no dew. The country in this vicinity

man offer me 2000, this morning to go home for I want to go with a larger pile." Wood Diary, November 20, 1849.

[14] Bad weather, poor health, and poor returns from mining operations under impossible conditions frustrated many a miner. William E. Chamberlain, who had struggled so manfully over the trail in 1849, wondered in February, 1850, if his effort was worth it. Cryptically he wrote in his journal, "Diorhea (50 buckets) $10.50." Chamberlain Diary, February 13, 1850. "Last night we had a very severe storm—heavy rains and high winds—Capt. V's tent was nearly blown from its moorings and occupants were kept up nearly half the night to prevent its being carried entirely away. The rainy season has now commenced with considerable severity." *Ibid.*, November, 1849. The cold rains and exposure of the winter of 1849-1850 were serious threats to the miners, as was the poor food they ate. Many died from dysentery and diarrhea. "Still on the sick list, and J. Teats little better. The fever has been checked, but camp disentary still hangs to both of us, and is attended with great pain. Evans, Notes of Defiance Gold Hunters, January 21, 1850. Death came to many an argonaut, and his friends often mourned his passing by noting that he had left a wife and children at home. Frank F. Kent described almost precisely the fate of Douglass Perkins. "Almost all of us had severe attacks of the diarrhea or dysentery before or after our arrival at this place [Sacramento]. About the middle of October the first of our number died. Andrew Watkins of Boston, a member of the L & C. Co'y. He was attacked with the dysentery coming up the river, and died on board the brig some ten days after we reached Sacramento. He was formerly in the employ of Varnum Warren, drygoods dealer in Hanover St. Boston —was 37 years old, and has left a wife & I believe one or two young children to mourn his loss. This will be sad news indeed to his friends at home, and it threw a gloom over our small band of emigrants. We dug his grave in the burying ground near the city hospital & about a half a mile above Sutter's Fort. I little thought that my first digging in California would be digging a grave." Kent Journal, October, 1849.

is rolling, some hills quite high, soil light & rather gravelly except in the ravines where a reddish clay abounds, & covered with a scanty growth of scrub oaks, presenting no other attractive features.[15]

Thursday Nov. Still in our camp on the Cosumne & poor as the "diggings" have proved are likely to remain here all winter. The rains have set in in earnest & teams cannot travel. Chapins teams went to town for our provisions & sundries & are stuck fast about 5 miles from here, not able to move a step. I fear we may have difficulty in getting anything to eat. Doc has finally left us for good, & will probably go home in the next steamer, couldn't stand the hard work & went to town last week. His leaving puts the finishing stroke to my list of disappointments "& now I'm all alone."[16] Shall have to give up my expectations of accumulating sufficient to carry me home in the spring & be thankful if I get enough to pay my expenses this winter. Well I'm here & must take the country as I find it.

Witnessed the ceremony yesterday of an Indian burial, one of the wives of the chief died of consumption. The corpse was dressed in beads, fine blankets shawls &c, a hole dug in the ground & wood piled on it on which the body was laid & then *burned* to ashes, till it all fell into the hole which was finally filled up. The mourners sat around keeping up a most lugubrious howling & singing while a woman nearly naked who acted as priestess, fire stirrer &c. kept dancing around the hole chanting a sort of a song. Various presents of clothing ornaments food &c were made by friends, which were all burnt with the body that they might serve its wants in the other world. I should suppose the articles destroyed must have cost the Indians over 200$. The poor fools had much better have kept them for their own necessities this winter. The chief himself was off after another wife when this one died & came back just in time to do his share of the howling &c.[17]

The last two weeks with the exception of two or three days have been almost constant rains, & our situation begins to be uncomfortable. Nights sometimes quite cold & lately there has been heavy fogs in the morning. From the top of the hills immediately above us can be seen the Sierra Nevada completely covered with snow.

Christmas Day, 1849. Oh how I wish I could spend this day at home, what a "merry Christmas" I would have of it, & what happy faces I should see instead of these of the disappointed

[15] The Reverend Walter Colton gave a good description of the gold-bearing region in his book, *Three Years in California.* "The gold region, which contains deposits of sufficient richness to reward the labor or working them, is strongly defined by nature. It lies along the foot hills of the Sierra Nevada —a mountain range running nearly parallel with the coast, and extends on these hills about five hundred miles north and south, by thirty or forty east and west. From the slopes of the Sierra, a large number of streams issue, which cut their channels through these hills, and roll with greater or less volume to the Sacramento and San Joaquin rivers. The Sacramento rises in the north, and flowing south two hundred and fifty miles, empties itself into the Suisun, or upper bay of San Francisco. The San Joaquin rises in the south, and flowing two hundred miles discharges itself into the same bay. . . .

"The streams which break into these rivers from Sierra Nevada, are from ten to thirty miles distance from each other. They commence with the Feather River on the north, and end with the river Reys on the South. . . . The alluvial deposits of gold are confined mainly to the banks and bars of these mountain streams, and the channels and gorges, which intersect them, and through which the streams are forced when swollen by the winter rains." Pp. 365-66.

[16] Hildreth returned to Sacramento but did not go back to Marietta, however, until 1853. He was present when Perkins died aboard the wharf boat *Orb*.

[17] The practices of California Indians in connection with death were described by Father Boscana in his pamphlet on the Indians about his Mission in San Francisco. "They did not put into immediate execution," wrote the priest, "the solemn duties and funeral performances, but suffered several hours to elapse, that they might be assured of his death. In the meantime the pile was prepared, and the person summoned, who officiated on such occasions in the applying of the torch; for it was usual, in this neighborhood, to employ certain characters, who make their livelihood by it, and who, generally, were confined to particular families. As soon as everything was prepared, and the time had arrived for the ceremony,

set now around me. The day here ushered in by firing of guns pistols &c & some blasting of heavy logs in lieu of cannon & this is about the amount of our celebrating. Some have passed the time in drinking & gambling, which latter vice is very prevalent everywhere in this country, a relic of old Spanish rule & customs.

We are now situated about 50 miles from the Sierra Nevada, about 5 from the McCosme & near a famous diggings called "Matheney's Creek" or "Dead Man's Hollow,"[18] from a couple of men having been murdered here for their gold this past summer. The rain ceasing somewhat & the weather becoming pleasant after the date of my last entry in this journal, we determined to make an effort to get away from our last poor place & hearing much of our present situation we started for it & with some difficulty from muddy roads & mire we succeeded in landing ourselves with our goods & chattels where we now are about the first of Dec.

Shortly after our arrival we were greeted in our new home in these higher altitudes by a heavy fall of snow & as we found our light clothes rather uncomfortable, John & myself went to work immediately building a cabin & after a week or ten days hard labor succeeded in erecting a very snug neat little house about 7 X 9 inside, & having most everything in are now comfortably installed—a good fire blazing in our own stone fireplace, & are sitting by it congratulating ourselves that we are not exposed to the cold rainy weather which is now prevailing.[19] As yet we have had no opportunity to try the diggings here, between the unpleasant weather & our building, our time has all been taken up. We are encamped on a small creek running north. "Matheney's" famous hollow is just over a ridge east of us but has too much water to be worked this winter. We shall probably try it in the spring. Our life is monotonous enough. We have to amuse ourselves the best way we can by reading over & over the few books we have, playing cards, smoking &c.

The members of the Harmar Co. are near us in two cabins & we visit each other occasionally. As yet I have heard nothing from home & know not whether to think all is well with those I love most. Hope to hear however before long though from the miserable character of the Post Office at San F. think it doubtful whether many of my letters will even reach me.[20]

January 8, 1850, Thursday P.M. Have just returned from assisting in the performance of the last sad duties for one of our little party which left Marietta last spring so full of health

they bore the corpse to the place of sacrifice, where it was laid upon the faggots. Then the friends of the deceased retired, and the burner (socalled) set fire to the pile, and remained near the spot until all was consumed to ashes. The ceremony being concluded on his part, he was paid for his services, and withdrew. Everything of use, belonging to the deceased, such as his bow and arrows, feathers, beads, skins &c. were consumed with him, whilst his relatives and friends added, also, other articles of value to the sacrifice, but during the scene of burning they did not observe any particular ceremony." Father Geronimo Boscana, *Chinigchinich: A Historical Account of the Origins, Customs, and Traditions of the Indians at the Missionary Establishment of St. Juan Capistrano, Alta California.* Translated and published in Robinson, *Life in California,* 314. "We saw Indians perform their funeral rites. There are a great number of Indians here. Many of them wear no clothing at all." Sedgely, *Overland to California,* 61.

[18] A long story appeared in the Sacramento *Transcript,* April 15, 1850, describing the dispute over Oregon Bar, in which a rich vein of gold was discovered by four men from Oregon. Located on the forks of the American River were Spanish, Oregon, Kelly's North Fork, and Dead Man's bars. The Oregon men claimed two-thirds of the richest bar and succeeded in holding it by bluff until June. Benjamin Brown of Cleveland, Ohio, discovered that the Oregon claims were perhaps invalid. The Oregon and the Ohio men met in a face-to-face gun fight in which the captain of the Oregonians was shot in the chest. The captain was nursed by both Buckeyes and Oregonians until he died. This gave the bar the name "Dead Man's," largely from the pile of stones over the murdered captain's grave. Rumors of strikes or of color frequently sent farmer-miners rushing from one diggings to another. One wrote, "I, at length, followed the crowd to Weaverville. Here, about a mile from town, between Ten Cent Gulch and East Weaver Creek, among the pines, I put up for myself an eight-by-twelve log cabin, with shake roof, and generous stick chimney. The mountain lions were very numerous in the vicinity, as their nightly serenades kept me

constantly reminded; but with a strong door securely pinned, I felt amply assured against any undue intrusions on their part." Leeper, *Argonauts of '49*, 115. J. Goldsborough Bruff reported having heard of an emigrant train of six who had made a "beeline" from Emigrant Pass toward Sacramento, and assertedly came on a lake, "deeply basined with high mountains, with plenty of golden pebbles, and hostile indians, near the headwaters of the Yuva [*sic*]." One of the six "gathered about *$30,000 worth of ore;* his comrades were murdered by the indians, and he escaped.—He travelled at night, and hid himself by day. Cached $20,000 worth of dust, followed the Yuva river down till he reached Yubaville, taking *$10,000 worth of dust in there.*" The man who told Bruff the story said that he was "personally acquainted with the man, and can vouch for his veracity. This story clinched the matter; it fully corroberated Gibbs' statement!" Bruff, *Journals*, II, 844.

19 While Perkins sat thinking about his problems and wishing that he was back with his loved ones in Marietta, at nearby Weaverville Diggings his fellow Ohioan, David DeWolfe, was writing his wife, "Oh! Matilda oft is the night when laying alone on the hard ground with a blanket under me & one over me that my thoughts go back to Ohio I think of you & little Sis & wish myself with you but I am willing to stand it all to make enough to get us a home & so I can be independent of some of the darned sonabitches that felt themselves above me because I was poor cuss them I say & I understand they prophisy that I will never come back Darn their stinking hides if God spares my life I will show them to be false prophits for as sure as I live we will shake hands and give a warm embrace by spring anyhow & before if you say so." DeWolfe Diary, July 30, 1850. Homesickness was as much a miner's disease as diarrhea and dysentery. George F. Kent laid a tiny slip of paper in his journal on which he had written, "I wish —God only knows how much, to see you again, but I would have you as you were— the children still raising their happy laughing faces—the young still eager for the fulfillment of the future's bright promises, & the old folks finding in the calm performance

& high hopes. S. E. Cross was taken sick about two weeks since with a bilious fever. He had been complaining of the premonitory symptoms for some time previous but supposing it would wear off would take no advice, or do anything for himself until the disease was brought to a crisis by a little exposure, & after a few days assumed a malignant type, & although everything was done for him which our means permitted, proved fatal yesterday January 7, at about 11 o'clock P.M. & we buried him today upon the hill side overlooking our little settlement.21 The death of one of our number here so far away from home & all that makes life dear, casts a gloom on all the survivors & we cannot but think of the possibility of a like fate being reserved for ourselves, & oh how terrible the thought. If I must die, let me but get home & die in the arms of my friends, & I'll not complain, but here with no one to care for me, or shed a tear of affection as my spirit takes its flight, tis horrible. The grim monster appears now doubly grim & the grave has more terrors than I ever before felt.22 I was sad enough as I followed the corpse of my late companion to its last resting place, & thought of the grief & agony of his mother & friends when the dreadful news is first communicated, & could well imagine how my own would feel under like circumstances. California has been, & will be the cause of many broken hearts & much grief, & I look forward to my release from it with great anxiety.

[*Tuesday January 20.*] Since my last date, our life has dragged along with nothing to disturb its monotony but the event recorded above. We during some few pleasant days in the first part of this month worked some in the creek, but found it like all else in Cal. nothing but disappointment.23 We made some 4 or 6 dollars each per day, but little more than our expenses. We hope however when spring opens to be able to commence a more successful life & shall get along as well as our impatience will let us till that time.

New Years day passed off pretty much as did Christmas with some burning of powder, explosions of huge logs, gambling, drinking &c. The amount of gambling done is astonishing. The practice is universal, & the staid old farmers from the states can be seen in crowds staking their hard earned gold upon the tables of the Monte & Faro banks which are established in every good diggings, two thirds at least of all the gold passes from the hands of the diggers into those of the bankers, or through games among themselves, & many a poor fool has gone to the city with a few hundred or thousands on his way home stopped in one of these "Hells" to "do a little betting" & came out completely broken, without a cent. Hundreds from the sober portion of the states who hardly ever saw a card before leaving, now work hard for nothing else but to try their hand at some game of chance. Occasionally one is successful & breaks a bank but of course in the vast majority of instances their money slides through their fingers almost before they know it. Nor is this to be confined to the miners, merchants lawyers & even priests in the cities dip into it as deeply as their means will permit.[24]

I once saw a very wealthy man in Sacramento betting on cards at Monte—10, 15, 20 & finally even $50,000 at a single bet! sometimes losing sometimes winning & putting down his thousand dollar check as did others their dollars & coins. The dollar here is only what the dime is at home & no one considers it as worth saving except as change. Nothing is less than a dollar, & the effect of such extravagant habits as are here contracted must have a bad effect on those returning home with any means.

Saturday Feb 28. Have just returned from an expedition round the richest of the mines of which we have heard to see about our summers digging, & find however that all the rich accounts dwindle down as you approach their location into the success of some few lucky ones & am satisfied by my trip that the gold is pretty equally distributed through the country, & a man had better settle down somewhere & work steadily for what he can get & trust to chance for striking a rich hole.

I went through Weaverton,[25] Hangtown,[26] Georgetown,[27] where the Oregon men did so well & came back by way of Colloma & Sutter's Mill & stopped to see the famous place where the first discovery was made—the news of which had set the American world crazy. The mill stands on the American river, is surrounded by immense & almost perpendicular hills, & the discovery of gold was made in the mill race some hundred or more yards below the mill. The race was dug by Indians employed by Sutter with white superintendants over them. One of the Indians in excavating picked up a lump of some-thing yellow—metallic & heavy which was examined by the superintendant & finally taken to Mrs. Weimar wife of one of the white men. This lady boiled the lump in strong lye for two hours without tarnishing it at all.[28] It was then sent to San Francisco analyzed & examined & pronounced to be fine gold & hence spread the news like wild fire which has filled the wilderness of California with enterprising Yankees.[29]

of daily duties & in other's happiness & a solace for the departed glory of youth, and yet a greater blessing could I have prayed for—that the circle of dear friends whom I left behind might have remained unbroken, but alas! This was to be the fairest, the gentlest, the most gifted is gone, and O!. how much of the beauty of this earth went with her." Note laid in Kent journal, ap-parently referring to the death of the author's wife.

20 Perkins had recorded, however, in the entry of October 6 that he had received a letter "from Hat." Records in Marietta show that his letters did reach home.

21 See also, Appendix, letter of A. G. Hovey, Jan. 6 and 9, 1850.

22 "My Very Dear and beloved wife Eliza-beth name ever precious to me I now take up my pen to wright you the Sorryfull knews of the Death of our mutch beloved Brother Wm. B Price news that will fal heavy on the eares of some of those that is near to us in the States perticular Sister Sarah She was more attached to him than is com-mon they lived together a long time with there poor afflicted Mother and She loved him oncomon for a Brother the circumstance of his affliction and death was as follows he had bot 2 mules and a wagon and had commenced hawling and was at Sacramento Cty at the time of his attack which was Diarrhoae he was taken on monday about the 6th of Oct he had a very severe attacked. . . . My self and uncle A MClintic his nurses we wated on him with all the care and attention the circumstances would admit he died on Saturday knight the 19 Oct at 11 oclock on Sabath morning one hour by sun the nessisary preperations being made he was lade in his cofin then the litle procesion consisting of a fiew relations and friends moved slowly up a steep ascent to the top or brow of one of those lonesome hills of Callifornia wheare he was entered with the usial cerimonies." Marshall (ed.), "The Road to California," 256.

23 That moment when the emigrant who had traveled so far under such great difficul-ties first went into the field to wash gold was one of great frustration. Perkins and his companions were to feel it intensely. All the glitter and excitement had gone; now they

faced grim reality, and found little or no gold. One example out of dozens that could be cited is Joseph Warren Wood's notation in his diary: "We commenced washing gold, 4 of us made 1$ by a very laborious day's work. It was nearly enough to pay for the extra bacon we eat, on account of working hard. We can get gold anywhere here by working for it. The trouble is to find it in quantities sufficient to pay us working it." Wood Diary, September 8, 1849.

24 In the diggings themselves miners were lonely, blue, and often rowdy. Those who succeeded in finding gold often treated it as "easy come and easy go." Men who had struggled so hard together to reach California turned into mortal enemies when drinking, or they risked all they had gained by a turn of the dice or a card. Joseph W. Wood corroborates Perkins' observations. "This has been a pleasant day, & has had quite a reviving effect upon the city. There is nearly as much business done here today as on week days. The gambling tables are surrounded by crowds, & the —— allies are filled to overflowing. (I was there myself & saw them) every grocery Hotel & Boarding house is a gambling shop & is supplied with all the modern implements of gambling Rolettes, Mountebanks dice tables & ordinary cards are played & some of the tables exhibit pretty large piles of money. Each prominent house is supplied with something to attract a crowd. One has pictures representing scenery views while crossing the plains. One has the Minstrels, another some instrumental music &c &c." Wood Diary, November 4, 1849. "As we have lived in a tent and been on the move for nine months, traveling 2400 miles, we were glad to settle down and go to housekeeping in a shed that was built in a day of lumber purchased with that first fee. The ground was given us by some gamblers who lived next door and upon the other side, for neighbors, we had a real live saloon." Haun Diary, 42. This was a realistic beginning of life in California for the wife of a future United States Senator. She, however, was equal to the challenge after the grueling overland journey from Iowa. Drinking seems to have irritated at least one of the boys from Marietta. Z. J. Chesebro, in a letter to the members of the Division of the Sons of

The spring seems to have at last set in with beautiful weather & we are beginning to think of moving & shall probably move over into Matheney's Creek next week. We have had two heavy falls of snow since December & the tops of the high hills are still covered. One snow fell over two feet deep.

From the top of a peak 1/2 of a mile back of our cabin can be seen the Sierra Nevada range, white as snow itself, completely covered, shrubbery & all, distant from 30 to 50 miles East. While westward can be seen the valley of the Sacramento, with its boundary ranges of Cascade Mountains on the Pacific & the bay & entrance to San Francisco distant about 140 miles & the eye can follow the valley N.W. up some 200 miles. The view is Extensive & magnificent. During the late pleasant weather John & myself have been working in some of the little ravines & have made something over 100 dollars—better than nothing, though small wages.[30] —End—

General Statistics of Journey
Country & Winds

General course of Winds

The most frequent winds from any one point run East, next most frequent West up to Fort Laramie the prevaling courses are West-South & South West. From thence to the head of the Humboldt in the high regions & mountains course was North-West- & N.W. While from head of the Humboldt to valley of Sacramento their course was usually East, South & South East.

General Course of Road. To South Pass N.W., thence to head of Humboldt nearly west—thence to Sacramento about S.W. Whole distance from St. Josephs to Sacramento City is 1728 miles.

Weather. Up to Fort Laramie in month of June rains very frequent & sometimes very heavy accompanied by vivid lightning, heavy thunder without wind &c. The rest of the distance with few exceptions was perfectly dry ground parched for want of water—

Temperature. Up to Fort Laramie, days & nights quite warm. Thence to head of Humboldt river both somewhat cooler. Sometimes heavy frost, & days scorching hot & among the mountains of the Sierra Nevada range, ice one or two inches in thickness frequently made in a night & days uncomfortably cool except when exercising.

Characteristics of Country. To Fort Laramie rolling prairie & Platte to Mount. bounded by barren sand hills. Beyond Fort Laramie more hilly & somewhat mountainous in vicinity of South Pass & between that & Humboldt. Humboldt Valley not very broad & bounded by barren mountains & hills of sand or gravel. From foot of Humboldt east desert plains to foot of commencement of Sierra Nevada range. Then come mountains which are mountains, rocky wild & covered with magnificent forest trees. From St. Joseph to the Sierra Nevada the region is entirely destitute of trees except on the Platte in places, or on some Creek bottom, or mountainside. At Fort Laramie the wild sage is first seen & forms your constant companion the rest of the journey, being about the only plant which can derive sustenance from the barren soil.

From the Upper Platte ferry to the Sink, volcanic remains are abundant and alkaline deposits in places under the ground white as snow & water completely saturated in places with alkaline & saline carbonates & sulphates. It is on the whole a dreary—cheerless journey & presents little to attract travellers for pleasure.

Temperance in Marietta said, "I have not a single piece of good news for you. The whole country is filled with drunkards. Almost every house is a drinking, gambling, house. I have not seen a single son of temperance since I have been here. I remain a faithful brother, in a land of liquor." Marietta *Intelligencer*, December 9, 1849.

25 The "Weaverton" mentioned by Perkins doubtless was "Weaverville," located in El Dorado County.

26 This place was first known as Old Dry Diggings, then Hangtown, and later Placerville. Coy, *Gold Days*, 76.

27 Georgetown, California, is located in El Dorado County.

28 This story is verified in part by Owen Cochrane Coy. He says that the woman's name was Mrs. Wimmer, who was then engaged in making soap. She also soaked the ore in vinegar. *Gold Days*, 35.

29 "Another year is among the years that have passed and this a new one opened in storm and [gloom] how short a period since my prospects were cast by the burning of our lard works in the morning of Jan. 1st, 1850 . . . on that gloomy morning I determined if possible to come to California and so far I have no cause to regret that determination. True I have not made much up to the present time but I have hopes of success during the present year, and with continued life & health, I think with the aid of the last three mos. experience I may be enabled to return to my dear family with means of support and comfort in the excitement of world pursuits let me not forget that life is hastening to its close, that I have one year less to live—and that I have duties to perform of far more importance than the accumulation of gold." Chamberlain Diary, January 1, 1850.

30 There is no record of what fortunes Elisha Douglass Perkins enjoyed after he ended his diary. At the time of his death he was the captain of the steamboat *Marysville*. From this unbroken silence one may conclude that he had a great deal of pride, and refused to go home until he could make a reasonable amount of money. On the other hand he might have become, in three years, enamored of California and reluctant to leave it.

Appendix

FEW COMMUNITIES became so fully involved in the gold rush as did Elisha Douglass Perkins' hometown of Marietta, perhaps because Beman Gates, editor of the Marietta *Intelligencer,* had such a keen interest in the emigration to California. When the Harmar Company, organized in the nearby village of Harmar, left for the gold fields, Gates ran a detailed story of the makeup of the company, its constitution, and its equipment. This account, reprinted here in full, is an interesting adjunct to Perkins' diary, where references to the Harmar Company appear frequently.

Gates also frequently printed excerpts from letters written home by the Argonauts. The letters included here carry to its conclusion the story of the Harmar Company and report, in a warm and informal way, on the California experiences of Samuel Cross, Dr. Hildreth, and others whose names are familiar to readers of Perkins' diary. Among these letters is one from Perkins himself.

In addition to publishing personal correspondence, Gates made arrangements for a series of special reports on the California venture. Below are three of these special reports from Jerome Howard and two unsolicited letters (one from "some of the young men of the Marietta Company" and one from L. M. Wolcott).

These items from the Marietta *Intelligencer,* confirming and supplementing Perkins' diary, show that all Marietta must have gone vicariously to the gold fields, and that the sufferings and disappointments of their men who reached California were experienced also by those they left behind in the town Perkins remembered fondly as "dear old Marietta," "one of the most beautiful places under the sun."

In the items as printed here the date of publication and any explanatory material, sometimes quoted from the *Intelligencer,* are given in square brackets preceding the item.

[April 26, 1849]   HARMAR CALIFORNIA CO.

The members of the Harmar Company started on Monday night last for California. They left here on the S. B. "Hamburg," for St. Louis, where they expect to take passage for St. Josephs. Agents have been sent out to purchase cattle, which will be at St. Josephs by the time the company arrive there. They expect to take their departure from St. Josephs about the 10th of May—taking the usual emigrant route by South pass, Salt Lake, and Humboldt River.

The equipment with which this company started from here was as follows:

Five large waggons, with heavy ducking covers— one of them arranged with a spiral spring bed, for the accomodation of any who be sick, yokes, chains, harnesses, cooking utensils (among which are four stoves), and a large variety of implements. They have 3600 pounds of pilot bread, put up in excellent order by Mr. Hall of this place, and a bountiful supply of beans. The other necessary provisions will be purchased at St. Louis. At St. Josephs they will find waiting 20 yoke of oxen and 4 cows.

Each man is furnished with a rifle or musket; about half of them with revolvers, and all the company with other side arms.

This company is organized under the following constitution:

This shall be known as the Harmar Company.

Article 1. The object shall be to send a party of men to California, to labor for the company.

Article 2. The primary stock to be at least three thousand dollars, in shares of fifty dollars, which shall be assignable, and each share entitled to a vote.

Article 3. At least twenty good able-bodied men shall be selected by the company, who shall bind themselves to go to California, and faithfully labor under the direction of the agents duly appointed, from the time of departure from Harmar in April, A. D. 1849, until the first of April A. D. 1851.

Article 4. The *party* going shall be under the direction of an agent, who, with two advisors, shall constitute a Board of Directors, to be elected by the company; and in case the agent disagrees with both the advisors upon any question, it shall be submitted to the company.

Article 5. The duties of said agent shall be to provide for the party, direct them in travelling, furnish employment, superintend their labors, appoint assistants, keep a just account of work, earning, and expenditures, and make faithful report of proceedings; and at least every quarter after arriving in California, he shall determine the amount of assets, in money or metal, belonging to the company—which he shall forward to the order of the home agent, to be safely invested—one half to the credit of the laboring party, according to the proportions reported by the agent, and the other half to the credit of the several shareholders.

Article 6. An agent shall also be selected with four advisors, to constitute a Home Board of Directors, having charge of the finances of the company. Said agent performing the duties of Treasurer and cashier —keeping a just account with each member, corresponding with the party in California, managing the affairs at home, and keeping a faithful record open for inspection of members.

Article 7. The members shall be allowed to draw upon the home agent for one half of the amount to which they are respectively entitled—the other half to remain until the dissolution of the company as a pledge of fidelity.

Article 8. Either party, upon good cause, may remove their own officers by a three fourths vote— ten days notice of the intention having been given.

Article 9. Vacancies may be filled by a majority of the votes of the parties in which they respectively occur.

Article 10. Each party shall have the power to adopt rules and regulations not inconsistent with these general articles.

Article 11. The expenses of the company shall be paid from the common fund.

Article 12. The party elected to go, shall have power, on the way, and in California, to receive and incorporate other members, upon such terms as may be beneficial to the company—all to be subject to these general rules.

Article 13. All the members of the company may vote either in person or by proxy both in the home company and in the California party, in the proportion of their interest in the proceeds.

Officers

Home Board—Darwin E. Gardner, agent, Henry Fearing, E. W. T. Clark, L. Chamberlain, Asa Soule, advisors.

California Board—Harlow Chapin, agent, A. Hulet, A. G. Hovey, advisors. H. Clark, Secretary.

Pledge

Each member of the party going, shall sign the following statement and pledge:

I am in good bodily health, free from disease, and will, to the best of my ability, promote the harmony and discipline, and faithfully labor to advance the interest of the Harmar Company, and obey its rules.

And furthermore pledge myself to abstain from all intoxicating liquors as a beverage, from all species of gambling and dissipation, and as far as practicable to observe the Sabbath.

And in case of a violation of this pledge, the company may, upon final settlement, make deductions from my proportion—two-thirds concurring therein —after a fair hearing.

The following is a list of the gentlemen who have gone:

Harlow Chapin	Abijah Hulet
Charles Cutler	Henry Clark
M. A. Williams	T. Johnson
Paul Fearing	A. M. Creighbaum
Wm. H. Bisbee	George Roe
Edward Hulet	A. G. Hovey
Almer P. Soule	Gage H. Droun
H. Erhorn	John Mills
E. C. Carter	William Fleck
R. Hamlet	William Irwin
Jeremiah Evans	

Seven of these men are married and leave families.

We suspect that but few California companies embrace so large a proportion of men of business experience and capacity, energy, public enterprise, and correct habits. If California can be settled by such men as make a large majority of *this* company, its population will compare well with that of the other states of the Union.

We trust that prosperity may follow not only the company's enterprise, but that each emigrant may be blessed with health during his journey, with abundant success in his labors at the mines, and with a long life in which to enjoy the fruit thereof.

[November 22, 1849: "The following letter from Mr. J. L. Stephens, who left . . . for California . . . and reached there about the 25th of August . . . was received this last week by way of the Isthmus."]

Sacramento City, Sept. 6, 1849

This morning I have unexpectedly found time to write. I have been too busey in preparing for the mines to write by the last Steamer, which left here on the first of the month. Chesebro and myself have been here two weeks, but have not yet heard a word from the other boys, and, as I wrote you from Fort Laramie, I fear they will see the *"Elephant"* before they reach the mines, though they may be here in a few days. The Harmar Company have not yet arrived, and I think it will be some time before they do; but in this I only judge from what I have seen myself. I had made arrangements to start to the mines (125 miles from this place) this morning, but owing to a subsequent arrangement I am to start in a wagon this evening. Chesebro is not going to the mines with me, but will be on in a day or two. . . .

The trip over the plains and mountains is of such magnitude that it would take quires, instead of this sheet, to give even an *idea* of it; so I will say but little on that head. We started from St. Josephs behind *seven thousand* wagons—some had left two months before, and some only two days—but to say that only 500 of them reached this city before us looks unreasonable; but such is a fact. We made the trip here in 85 days, which is nearly the shortest time on the route. Our reason for such a rush was, that to be at the end of 7000 teams we thought we should never get through; for there is not grass enough for more than half the stock so we thought it better to get through alive, with dead mules, than to perish on the road. Unless something be done to relieve the families which are back on the road, there will be intense suffering this winter. Emigrants here who have friends on the road, are sending word, and are going back themselves, to turn the teams to Oregon, or the Salt Lake (Mormon City) or to have them stop in the valleys where there is sufficient grass to winter. If an early winter sets in they would undoubtedly perish in the mountains; for it snowed on the 15th of August where we were, and it was not a sprinkle only, for it laid in the hot sun a long time.

The sun was oppressively hot in the mountains, but as soon as hidden by a cloud, or below the horizon, it became chilly and begins to freeze. But enough of this—and now something of California.

This "beautiful and fertile valley of the Sacramento," of which Col. Frémont speaks, is an outrageously hot country, and has to be watered to raise anything whatever. Those who have the inclination to cultivate the land in this way may do so—but I shall not. The greater part that you read in relation to this country is *false;* and this everyone knows will testify. The gold is here, but no man ever *dug* a dollar who did not earn it—The gold digging is a lottery sure enough—Some men make a fortune in a short time, but others in ten years would not make a cent. Of those who will leave here, the proportion will be scarcely one in five who will leave with any money; though it is within the power of every man in California to make a small fortune. Any man who has his health and energy can make ten dollars per day at almost anything he chooses to do.—Our clothing costs us but little. We wear nothing that costs more than two dollars.— Vegetables are *tolerably* well up—for instance; mellons from two to six dollars each onions, one dollar each—potatoes one dollar per pound—squashes, one dollar each—everything else in proportion. Substantials are as follows: flour, nine dollars per hundred—sugar, 18 cts per pound—coffee, 12 1/2— tea, one dollar—beef, 25 cents—hams, 60. Mules and horses are worth from fifty to three hundred dollars each, and hay ten cents per pound. Money is nothing here. One hundred dollars is not much more than *ten cents* in Marietta. Some persons in this city are making money as fast as they can count it.—But they are those who had a start at the beginning. . . .

The hardships of the overland route to California are beyond conception. Care and suspense, pained anxiety, fear of losing animals and leaving one to foot it and pack his "duds" on his back, begging provisions, fear of being left in the mountains to starve and freeze to death, and a thousand other things which no one thinks of until on the way, are things of which I may write and you may read, but they are nothing to the *reality*. Still I am in as good health as I ever was, and weigh more by four pounds than I ever did before. I have come to the conclusion that it would be child's play to do as some are doing—to turn round a few times, and then go home, because they cannot pick up gold under their feet as they walk along. This will not be my course. I have spent my four hundred dollars, but I am going to have that back again with *big interest* before I quit the country. Some of my Kentucky friends are at the mines making from $100 to $200 per week. But to stick to the business is the only way to make it pay—at least this is what the old miners tell me. If I hit a rich spot, a fortune can be made in a day, as many have done; but if I do not, it will count up in a year or two. If I am any thing like lucky I will leave this country so as to cross the Isthmus after the sickly season of 1850—if not I may stay another year. . . .

You will almost doubt what I am going to say—but nevertheless it is true. When we packed from Fort Laramie, we started with but little provisions, as we wished to get along as fast as possible, thinking we could buy of the teams on the road. Our provisions lasted us but 1500 miles, and then we had to go hunting. A great many had thrown away their bacon and bread, and had only enough to last them through, and others had to buy as we did. For a hundred miles in places, men would not answer us when we asked for anything—many would not give us a decent answer at other times. And some again would not sell to a *packer,* because as they said, packers could travel faster than they could. Many times I have sat down and eaten dry crackers, in a powdered state, without even water; and often ate raw bacon sides, powdered crackers, and water, for days together. We have been refused a cup of coffee repeatedly—and have often begged men to sell us a pound of bread and could not get it.

But now to give the devil his due. I have seen some Kentuckians that were gentlemen—and others not. From Ohio the same—and the same from Missouri. And indeed the same from almost every state in the Union save Yankeedom. Never have we met the first soul of these that were not as brothers to us —they would always divide with us to the last morsel. The Irish were on a par with the Yankees—and the Germans came next to them. At one time we got out of chocolate and tried to buy from every wagon —but no one would sell us an ounce. At last we came

to an Irish train. I told Chesebro I thought we could get something there. At the first wagon we reached we asked to buy some chocolate. A *lady* said she could let us have some; and the wagon we stopped until she could get at it. She got two pounds and gave it to us refusing to take any pay. Another wagon in the same train gave a tea pot. So much upon this head—but you cannot imagine the reality. —When you are suffering from starvation and ask a man, a brute in the shape of a man, to sell you a meal of victuals, or even a cup of coffee, and he coldly refuses, if the tears would not come then I am mistaken.

We have been on the Humboldt River where the Indians were scattered through the bushes shooting their poisoned arrows at man and cattle. We camped alone several nights where we dare not build a fire lest we should be discovered; and we had to do so because we had traveled as far as our mules could take us, without coming up with any wagons. In such cases we stopped as if it was in a bear's den. To lose our mules by travel was almost certain death to us, and we tried always to be on the safe side. But writing about these things is nothing, when one has had them to *feel.* In consequence, however, of these trials, we have had fine times since we reached this place, where everything to eat is plenty. I do not of course say that all have had as hard time as we met with.

For my part I would not have the feelings of the people we passed on the road, many of whom had their families with them, and who are now 400 or 500 miles behind (with not a spear of grass in their way) for all the gold in this country. Though all of them may get through alive. It is with these persons that we have had comparatively "fine times." . . .

Truly yours
Jos. L. Stephens

[November 22, 1849: "Letters have been received from W. D. Hollister dated Sept. 28th . . . announcing the arrival of Messrs Cross, Perkins, Cunningham and Huntington." "Many of our readers will recognize in the writer of the following letter, a young man from Harmar, formerly for sometime clerk in the post office in this place."]

Coloma, California, Sept. 16, 1849

After having been on the road for months traveling constantly, you can easily imagine the pleasure I experienced at finding myself in good health, by the favor of Providence, and in the possession of buoyant spirits and high hopes for the future.

After leaving the Missouri our road led up the north side of the Nebraska or Great Platte River a distance of about 600 miles. We crossed the north fork of the platte twice. At the last crossing it is very dangerous, being a large mountain torrent foaming and dashing down with maddened fury. Here we saw three men go to their long last home, by the Rafts coming to pieces on which they were crossing the river, after leaving the Platte a short distance we came to the Sweet Water a beautiful mountain stream, which we followed up to within seven miles of the South Pass of the Rocky Mountains.

The South Pass we found very different from our expectations. Instead of a narrow pass, as we supposed, running through the mountains, we found a plain upwards to twenty miles width and from sixty to seventy in length. A person would hardly imagine they were crossing the mountains did he not look to the North and see the Wind River Chain, with its snow clad sumit towering high above them, and the rugged peaks of the main chain of the rocky mountains to the South. We next came to Greene River thence on the Bear River. However, before coming to Green River we concluded to leave the old Ft. Hall road and go by the Great Salt Lake.

After getting to the Mormon City we stopped one week. We don't think the Mormon Settlement here of any importance. They raise nothing except by irrigation, the soil itself being poor. They can raise wheat better than any other agricultural product. Corn will scarcely grow. There are something like ten thousand inhabitants in the Valley. Those are doing best who have turned their attention to herding. After leaving the City of the Fallen Angels we took a northern direction, leaving Salt Lake on our left, and the mountains on our right, for a distance of eighty miles where we again cross Bear River near its mouth, as where it empties into Salt Lake. We then took a western course across the Great Basin—came to Humboldt River, followed it down about three hundred miles to where it sinks into the ground—This point is the commencement of the great desert, so much dreaded by those who go the overland route to the Pacific Ocean. There we took a new route, travelled by about one third of the emigration, and but about thirty-five miles across the desert to Carson River. We traveled up the last named river to its source, about eighty miles to the base of the Sierra Nevada mountains. Here the dusty wayworn traveller is immersed in a dense forest of timber, not a stick of which he has seen for the last six hundred miles until he gets here, the best kind of timber too; Pine and Fir trees from one to three hundred feet in height, and from eight to twelve feet in diameter.

The Sierra Nevada Mountains presents to the beholder the grandest and sublimest Mountain scenery in the world, rising several thousand feet in height and covered with a dense forest extending to its very summit,—our road passing over the dividing ridge at least a mile above the snow of eternal ages. After we came down on the western slope of the mountain, we found the timber of still larger size, and, in addition to the first named trees, the *Arbor Vitae* which is an ornamental bush in the States, grows to upwards of a hundred and fifty feet in height and over six feet in diameter. . . .

We got here a week ago today. Since that time we have been resting ourselves and listening to the news of the country generally. We are about two hundred and fifty miles from San Francisco City, a new and flourishing town at the head of ship navigation on the Sacramento river. At the above named cities wages for common labor rate from eight to ten dollars per day and boarded. At this place, which is in the diggings, from ten to fourteen dollars and boarded.

The digger of gold after understanding his business makes at least an ounce and a half or two ounces, with the chance of striking a good bed and making a fortune in a few weeks though these are rare cases; consequently the inducements to mine are much greater than to work at common labor. The greenest miner makes his half ounce a day and in two or three weeks understands the business, pretty well. Myself and four others are going into the mining business. Those with whom I am to work are men of inexceptional characters with whom I travelled about two thousand miles, and consequently know them well. Tomorrow we are going to commence business. We shall get our supply of clothing and provisions at Sacramento City, as they can be had there at almost as reasonably as in the States; but to show you what labor and especially teaming is worth I will give you the price of some of the articles of food out here but forty miles from Sacramento City and a good road running from that place to this;—flour 25 cents per pound, pork 75 cents per pound, Hard Bread 50 cts. per pound and other things in proportion. The miner pays these prices and makes from ten to fifteen dollars a day besides and feels satisfied. The country at this time is very healthy. The ground here in the diggings is covered with a dense forest of oak and pine trees. I never saw half so much harmony and good feeling among a lot of men in my life as there is here. Everything is order and quiet. Yours affectionately.

P. Howe

[December 6, 1849]

[San Francisco, Sept. 27, 1849]

I am in California, after a passage of 172 days. Long and comfortless days they have been to me. I can assure you, although I was blessed with good health . . . you can have no conception of the sacrafice you make when you set your feet on board a vessel for a long voyage. . . . Besides other bad treatment a man that has been use to your mode of living would find it hard to set to cold salt beef, sea biscuit mouldy and full of living animals, and coffee without sugar, every morning for breakfast. For dinner there is a pan of cold beef set out and every fellow helps himself. For supper you have rice boiled in bags and put into a long tub and pounded soft, by mixing it with water. This is the way we lived on the later part of the voyage. Notwithstanding, I have kept in good spirits, and anxiously looked forward to the time when I could say that I once was a free man, and not only that the hope of being amply rewarded in the end buoyed up and seemed to nerve my drooping energies. I am now in our tent where I have been ever since the 9th of this month, waiting for my goods to be discharged. I think this week will wind up my pilgrimage in

"pleasant valley," as it is called. I am happy to hear that Stephens and Chesebro have arrived safe. There is so much suffering on that route, I was fearful some of them would not stand the hardship. There are a great many thousand on the way, and provision and feed for their teams is becoming so scarce, that the Governor is sending on supplies, from Monterey. . . .

There has been no mail from the States since June. I expect to go to the Mines soon and then I will say something about them. Thousands are flocking to them and a great many are leaving. Some have gold—others have none. Some come from the mines to the city to work by the day. Tell your friends to stay the other side of the Rocky Mountains. The "piles" are awful hard to get.

Yours very respectfully.

D. Hollister

[December 6, 1849]

San Francisco, Sept. 28, [1849]

I arrived here on the 9th of this month, and have been detained ever since, waiting for my goods and provisions, which I expect to get this week. I am then going up to Sacramento City, store my provisions, and dispose of my goods as well as I can. Goods in this place are very low. It would be out of my power to rent a place in the city, as a spot large enough to pitch a small tent for $100 per month. . . . The houses are generally built of wood and covered with canvass. If you want to sleep under one of these roofs a night you must pay $1.50 to $2.00. The City is as stirring a place, as the city of New York and the main streets are as much crowded with teams and business men as Broadway. I cannot say much to you concerning the mines at present. Some are so fortunate as to get enough to go home,—others return from the mines to the cities to work for wages by the day.—Wages for a common laborer per day is from $5 to $8, for a carpenter $12. Boarding is from $21 to $36 per week and sleep on the floor at that. Very few men are able to do hard labor as soon as they arrive. The labor is of the hardest kind.

We have all *Justice* and no law in this country. You may well leave your affects anywhere you choose and they will remain in perfect safety. Thieving is rarely known.—The fact is no one dare molest that which does not belong to him. We are encamped in a place called pleasant valley. This ground was tilled by Col. Stephenson's regiment, and is enclosed. A man was fined one hundred dollars for making use of one rail and had to pay for it.

The U.S. Consul, of the Sandwich Islands arrived here yesterday, bringing news that the French had taken the Islands. He pulled down his colors and left.

[D. Hollister]

[December 20, 1849: "We are reluctantly compelled to omit all that part of the letter which relates to the Journey, and commence our extracts at the period of our arrival in the promised land."]

Sacramento City, October 6, [1849]

When I arrived at Bear River Diggins on the 18th of September. I had one pound of bacon, one pound of crackers [his books and most of his clothing he had discarded soon after leaving Fort Laramie] four pounds of flour in my pack and $2.50 in my pocket. At the "diggins" bacon was $1, a pound, pork 75 cents, flour 40 cents, cracker 60 cents, salaratus $5.00, and everything else in proportion. For the last three or four weeks we had been in company with a couple of Virginians, one of whom had about $20, and the others not a single cent. A council was held, and it was resolved that we would proceed at once to Deer Creek, about eighteen miles distant, where we heard they were digging with some considerable success, and there purchased a cradle with the Virginian's money, or else hire out as day laborers. We reached there in the evening, ascertained there was no demand for labor among the miners, and the next morning went to work "prospecting," as it is called—that is, digging up a pan full of dirt and stone and washing it until you ascertained if there is gold mixed with it. My partner and myself dug and washed all the morning and obtained some minute scales, valued at ten cents—which we threw back again for seed. The other party got about 75 cents, and we concluded—that is, some of us did—return to Bear River.

We were now about four days from Sutter's, and the money and provisions I had would suffice to exist on, until I got there and sold my mule. I sold a common tin pan which I had picked up on the road for *three* dollars, and purchased with the money 20 pounds of damaged flour, at 15 cents a pound, next day started for this place, which we were a week in reaching. We lived on sour flour (mixed up without saluratus) excepting at one meal when we had fresh beef. This sort of living so deranged my system that for 48 hours no nourishment passed my lips except a cup of tea, and during that time I walked 40 miles.

At 5 o'clock P.M., Sept. 27th we reached the Sacramento, opposite Sacramento City, which is situated at the Junction of the Sacramento and American rivers, and two miles below Sutter's Fort. The last dollar I had I paid for ferriage, and arrived in the City with one copper cent (the same D. Hollister gave me when he started around the Cape) in my purse, and a few pounds of flour in my pack.

We went a mile up the American, beyond the City and encamped. Here I found a prize in the shape of a white heron, which apparently had been killed a couple of days. This I picked and cleaned and broiled on the coals, and with the sour dough made as enjoyable a supper as I ever sat down to. The next day, Perkins fortunately met in town a New Orleans acquaintance who offered to furnish him whatever we wanted. A few days after I sold my mule for $125--twenty-five dollars more than he cost in the States, and now I feel as independent as a king.

Hollister and Dr. Hildreth arrived here from San Francisco, four days after we did, and are now encamped with us, near the City, in a beautiful spot among the trees, on the American river, where we have a vineyard of wild grapes within ten yards of us.

Chesebro and Stephen arrived here a month before us. C. remained in the City, sick, until the day before we arrived, when he left for "Redding's Diggins" up the Sacramento. Stephen, I understand, is on the Yuba, 80 miles from here.

We overtook the Harmar Company about fifty miles above the Humboldt, and travelled with them two days—days long to be remembered in my calendar. Shall I ever forget their boiled beans, bacon and slapjacks!—never! "While memory holds her seat, I'll bear them here within the book and volume of my brain, unmixed with any base matter." That company was one of the few that had not been separated. Their cattle were in better order than any I saw on the road, and they had only lost one ox since they started, while others had lost *all* they had, and been obliged to take packs on their backs and foot it.—They have added three new men to their camp, in the place of the four who died of cholera near St. Joseph, and besides their fine ox teams have a team of six first rate mules.—Altogether I think they are one of the most fortunate, if not *the most* fortunate company that has come over the plains. . . .

I wish I could give you some idea of this Sacramento City. It is a perfect New Orleans in mineature. For a mile up and down the river, the banks are crowded with shipping, from the bay scow to the barque of 400 tons. Some of the houses are built of bricks, brought from the States, other of adobes made here, but nine tenths are mere tents—a frame inside and canvas out. The lots rent from $500 to $2,500 a year. Almost every other tent is a gambling hall, where the mysteries of Monte, Faro, Roulette, Twenty-one, Chuck-a-luck &c. are practiced from Monday morning till Sunday night without intermission. . . .

I have not conversed with a man who is not disappointed and ready to return as soon as he can make his passage money, but at present every vessel in San Francisco has her passengers engaged four months ahead. There is gold here, but most of the fortunes are obtained, and what there is left will require hard and constant labor to get. I am also disappointed in the country itself, for it is anything but the paradise represented by Bryant, Frémont, and other speculators. The Valley of the Sacramento, the beauties of which we heard depicted so eloquently, is, with the exception of the immediate banks of the rivers, a barren, flat plain, where the eyes may roam for miles without seeing a single tree. And this is the aspect of the whole country, so far as I can learn. The vicinity of all the "diggins" is generally the most desolate imaginable—high rocky hills, with pine trees scattered over them, and not a blade of grass for miles; but among these I expect

to spend the winter, and perhaps a much longer period. I have given up the hopes of making a fortune here in a few months, but I have not given up the determination of leaving the miserable country as soon as I can hear the Jingle of a few thousands in my pockets. By the help of time and perseverance any man can make money here, but there are thousands who are now making a living. About 40 miles below here is a place called McCollum's Fork, where there are two thousand miners, and only about 500 at work, and I am informed that these 500 do not average $5, a day. But then a man will sometimes strike a *big hole* and raise 15 or 20 pounds a day, and my hopes rest on one of these *big holes.*

Geological knowledge, I have learned, is not worth a dime in this country, and many a miner, with a month's experience will make a better gold hunter than men who have been employed by the Government. The most valuable men in the country are those who can work longest and strongest with a pick axe and shovel, and professional men are not likely to be proficients in such accomplishments. The Harmar Company has arrived here, all fat and hearty. They report a great number of emigrants behind who will need the assistance of the Government, which has sent out relief parties on all the routes. . . .

Now, if you have an enemy you wish to punish, advise him to go to California across the plains. Hollister says, "send him around the Cape," and those who have paid $600 or $800 for passage by steamers will say "send him across the Isthmus." But if you want to be very savage, take my advice and send him across the plains, you cannot punish him more severely. . . .

S[amuel]. E. C[ross].

[December 20, 1849]

Sacramento City, Oct. 15, 1849

I arrived here Sept. 26th. after a journey of a month longer than I had anticipated. . . .

At Fort Laramie our company dissolved, and divided the property. I made a French cart—and a greater humbug never was invented. By the time of my arrival at Deer Creek (60 or 80 miles) my mules were about used up, and to give the animals a chance to pick up, made packs, and pack saddles &c.

Huntington and myself went on at the rate of 30 miles a day, till, we arrived at Green River. The last 76 miles I had traveled in 30 hours—traveling all night in the desert. My mules were so jaded that I had to stop five or six days to recruit.

[At the Humbolt river he lost one of his mules, and] After this walked all the rest of this distance 800 miles—going 25 miles a day, in the broiling sun, over sandy deserts, sometimes without water, and with sore feet, eyes, and lungs, from the vast clouds of dust, and finally arrived as stated above—well and strong, but awful hungry, and without a cent in my pocket—having spent that last I had at Bear River Valley for some flour, which I could get tolerably cheap—20 cents a pound—and on which we had lived for ten days.

[At the city Mr. P. met an old acquaintance Mr. Peyton—from New Orleans who—] insisted on my taking a lot of provisions and his purse. The latter I declined, but the first I accepted to a considerable amount, and packed them up to camp. I was enabled a few days later from the Proceeds of "Old Dave,"—one of my mules which I sold for $121,—to pay Peyton. Wasn't he a noble fellow! . . .

Our little party are at last all safely here, and all in good health. I have gained flesh, weighing now 12 pounds more than I ever did before. This I shall soon work off when I get into the mines, as the work is terrible. I met Conway [a former resident of Marietta—*Ed.*] in the streets the other day, completely broken down. He said it was the result of two and a half months labor in the mines, and few were stronger than he when he went up. He collected together in that time only 1,500. . . .

I have hardly met a man who is not disappointed and dejected, and wishes himself back. Even ——— who is usually cautious in expressing an opinion, shakes his head ominously and with grave face says "twas a very sorry affair."

The fact is, a man in the mines, by the hardest kind of work for 8 or 10 hours per day, can make from 10 to 20 dollars—more seldom the last than the first. The majority now in the mines hardly pay their expenses—making on the average not $5 per day.—The expenses of living are about $3, so you can see what the prospect is of making a fortune by

digging very few have accumulated any considerable amount, and when some lucky individual has hit upon a rich spot, and got out $10,000 or $12,000 it is trumpeted from one end of the country to the other as a fair specimen of what may be found anywhere. . . .

E. D. P[erkins].

[March 21, 1850]

Dead Man's Branch, Dec. 20, [1849]

I have not heard one word from Marietta since I left, until this week, when I received yours of June 17th and August 18th, and one from Robt. Jennings dated in August, which is our latest from Marietta. I arrived in the Sacramento valley in October, after a hard trip across the mountains and plains. I was sick with chills and fever, until I got to the South Pass, from which place I had good health until I arrived at the beautiful Humboldt— or, in other words, Humbug—river. I took the scurvy near the bank of that stream, and from there until I crossed the mountains I had a "good time," I ate grass and willows, and sometimes a little mint. Two of us had this disease at the same time, and the only thing we had to eat was salt bacon, and that we dare not use so we had to "go to grass." I think if I had been detained on the road one week longer I would have been a good subject for a "scratch club" funeral, for when I arrived here I could not ride or walk. I stopped three days on this side the mountains and ate nothing but grapes, which helped me more than anything else.

Since I arrived here I have been well, and have at this time better health than I have had for five years.

Chesebro and Stephens left me at Fort Laramie, and I did not hear from them until I arrived at Sacramento city, where I got a letter from Chesebro, but did not see him. He had gone to the mines before I got in, and my latest news from him is his letter of Sept. 16th. He said Stephens had gone up the Yuba river 100 miles from the city, and he had not heard from him since he left.

Perkins, Cross & Huntington arrived here one week before I did. They saw "the Elephant"—they did. Some of the glorious six (the Marietta company) arrived here barefooted and bareheaded. Ha! ha! ha! Guess who.

I am with the Harmar Company. We shall spend the winter or rainy season here. We have built two loghouses, one 18 by 22, the other 16 by 20. The rainy season having commenced, there will not be much mining done for three or four months.

We are on a branch of the Cosumne river, and in sight of the snowy peaks of the Sierra Nevada. There is a company of 200 men and families back in the mountains, who have sent in for relief from the Government. One man from Ohio came in. He said the snow was three feet deep when he left them. They had killed all of their stock, and made jerk of it.

Dwight Hollister and Sam. Cross have built a house near us. Dwight is well, and we have a fine time. Perkins & Huntington are building about 1/4 mile from us. There are no houses in sight of us. We are six miles from Weavertown, four miles from Hangtown (so called from the fact that 20 Spaniards were hung there last summer without judge or jury) and 50 miles from Sacramento city.

I have not made much since I arrived in California—say about three hundred and fifty dollars, which I have spent for provisions. I shall not say much about the Gold, only there is plenty here, and by a man exposing his health and working in the rain, he can make $16, per working day. As for making any more, I have not seen it yet, but you can hear of some big raises being made. I have not been in the mines long enough to give a full and true account of them. We have not had a fair trial yet. Last week John Huntington was prospecting in a ravine, not twenty yards from our house, but left and went to another place. Two men took the place he left, and in two hours took out two hundred dollars, and in two days $500. So you see it is all a lottery. You make a raise one day, but may be glad to make your board for the next week.

As for boarding, it costs something to live in this country. I have mostly lived on Hall's Pilot bread, salt course and old rusty pork. I will give you some of the prices,—In Sacramento City flour sells from $40 to $60 per barrel; Pork $40 to $50; Dried Apples 50 to 60 cents per pound; dried peaches 75 cents to $1.00 per pound; and Potatoes 75 cents to $1.00 per

pound. All woolen goods sell at high profits. The lowest price boots are $16; water-proof boots $30 to $40 per pair. These are the prices at this time in the city. When we got there they were much cheaper. The prices in the mines are more than double what they are in the city. I have seen flour sold here at $250, and Pork at $200, per barrel, and potatoes at $1.25 per pound. Freight from the city to the mines, this winter, is $60, per 100 pounds. . . .

J.Q.A. Cunningham

[March 26, 1850]

[Dead Man's Branch,] Dec. 23 [1849]
The Harmar Company is no more. It collapsed a flue, and broke the main shaft, on Saturday the 22. I am happy to say that no lives were lost. Through the careful management of her captain, skillful pilots and noble ahem! crew, she was run ashore, and landed on the highest peak of the Sierra Nevada. Her damages being great, her crew thought best to leave her "alone in her glory"

The company dissolved by mutual consent, and from this time every fellow will be for himself, and we hope for good luck all round. The business of the company will be settled by Capt. Chapin. The stock consists of seventy one shares, of $50 each. The laboring party here have pledged themselves to pay the agent $100 for each share—making $7,100, by the first of May 1850. So the laboring party now own all the stock. They have provisions for the winter; two fine log houses; eighteen head of cattle; six mules; three fine wagons; mining tools of every description, &c, &c.

I send you a specimen of the gold. It is some of my own washing. As I wanted to send you two kinds, I have been to the creek and washed out some from there with a pan—the first I have worked from these mines. . . .

Send me some papers from God's country, for I have not seen but three papers since I left home, and they were N. York and New Orleans papers of old dates. The newsboys from New York are crying their papers thro' the mines, and they sell quick at a dollar each. I would give five to see the Marietta Intelligencer—tell Gates, as a hint. . . .

I will state one fact, to show our expenses. Four

of us, who had not heard from home, and were anxious to get letters, sent Mr. Hammett to the city. We got thirteen letters, in all. We paid the agent at Sacramento $13 to get them from San Francisco. Mr. H's tavern bill two days at the city, and two days travelling expenses, in all was $40, not counting his time, which in the mines they average at, $10 per day. I have sometimes paid $2 for a meal, and glad to get it at that. This is the country you read about—it is,

December, 24th.

This Christmas morning I received the office of auctioneer for the Harmar Company. Ha! Ha! Going, going, gone! We are all well.

J.Q.A. Cunningham.

[March 26, 1850: "Mr. Wm H. Bisbee (one of the Harmar company) writes . . . from the mines under date of December 23d, and . . . Dec. 24th. His letters are mostly occupied with details of the difficulties that led to a dissolution of the company. As personal matters seem to have been the principal cause of the disorganization—matters in which the public can have no special interest, and particularly as only one side of the story is told, we find but little in them that we think it proper to copy."]

[Dec. 23, 1849]

At meeting of the Company yesterday, we proposed to pay seventy-one hundred dollars for the stock of the company. This was duly considered by Chapin, Williams and Hovey, and accepted. . . .

Previous to the dissolution we worked 25 days on the Indian Bar, paying in the dust nightly. The amount we earned there was some over $1,500. . . .

I am with a mess of four. We have a fine cabin, 14 by 16 feet, and our winter's flour on hand, and are ready to make money in the spring at the rate of $16 or $20 per day. Paul Fearing is not able to work, and I hardly think he will ever see home again, but I will be ready to help him if he needs it. . . .

Wm. H. Bisbee

[Dec. 24, 1849]

The Harmar company is finally dissolved. The laboring party are to pay $7,100, or double the original stock. By this operation the home stockholders make one hundred per cent on the money they invested, as it is to be paid on or before the first of May 1850, less than one year from the time of investment. . . .

The country here surpasses my most sanguine expectations. The gold is here, and can easily be obtained. Men in the mines average from $16 to $20 per day. The climate is fine, and the weather warm. The rainy season is now upon us, and we cannot work much until about April—from which time we can work without interruption until January. We are now very comfortably situated, in our log cabins. . . .

Wm. H. Bisbee

[March 28, 1850]

Sacramento City, Dec. 24, 1849

My dear Father.—an opportunity occuring of sending letters to the States by private hands, I avail myself of it in preference to trusting the mails, which as yet are badly managed in this country. Dr. Dines, formerly of Athens, kindly took charge of some letters written from San Francisco in Sept.—I trust he has reached home in safety. I remained in Francisco about three weeks, when having received my freight from the vessel, I left this place. I made the passage in a small schooner in about three days; distance perhaps 150 miles. The weather was pleasant, and sleeping on deck in the open air by no means disagreeable. The scenery on the Sacramento River was rather picturesque, and with human habitations would be decidedly handsome. The banks are low, well timbered, but liable I suppose to annual overflows like those of the Mississippi. Containing 8 feet of water at the lowest stage, it is well suited to steamboat navigation, several small boats are now running, though the travel is chiefly monopolized by a fine eastern boat, the *Senator*, which has recently commenced running: a first class steamer, with a richly furnished cabin; make the passage in 12 hours—fare 30 dollars—meals and berth extra. The usual charge for freight on schooners and steamers is by weight, 2 dollars a hundred; by measurement one dollar per foot cubic, a higher rate than from New York to Francisco.

On the landing at this place, which is on the American River, Sutter's Fort being 2 miles distance, I found Douglas with S. Cross and J. Huntington encamped under some trees on the bank of the river. They had arrived from the plains a few days before; were in excellent health, but had been unfortunate enough to lose most of their outfit. The expedition overland seems to have been attended with heavy losses of property, as well as great privation and much sickness and many deaths. The last of the emigrants have scarcely yet reached the settlements, and have suffered extremely. Government relief teams were sent to their aid, at a very heavy expense, but still the amount of suffering has been incaluable. Mr. Stephens and Z. Chesebro came through in advance of their party, and have been in the northern diggings, I have seen Stephens several times in the city. He is at work on Feather River—has made some money, is in good health, and will probably do well.

After some deliberation our party concluded to go out to Weber's Creek, but Captain Chapin's party arriving about that time, we thought it best to remain with them for the benefit of their teams. They were very fortunate in getting through with but little loss; and Chapin deserves great credit for his judgment and good management.

On the 16th of October left for the diggings on the Consumnes, a small stream that is south of this. We reached Indian bar on the fourth day, distance 35 miles. The road running for six miles along the American River stretches off S.W. crossing a barren sandy plain 14 miles without water. On reaching the Consumnes we kept up the bank of the stream without crossing it, about 12 or 14 miles. We passed several ranches, on the route, which at present are not very inviting. They afford good pasturage, but exhibit no enclosures, except such as are made by ditching. The buildings are very rude, and the land uncultivated. The soil is sandy and light, and from descriptions of the country north of this, I am satisfied the agricultural resources of California are entirely overrated. It's true that but little has been done in the way of farming, and several years must elapse first. There is but little time for fencing, and labor is too high for hiring. No surveys have been ordered, and titles are quite uncertain. So that food

must still be imported, and Oregon, the Sandwich Islands and Chili continue to furnish supplies.

The valley of the Sacramento is annually overflowed, which will be an unfailing cause of sickness among its settlers. The Cosumnes has not been much visited by emigrants. Many Indians are found here, but are friendly. The bars are small and contain the fine gold—small scales, very pure. We stopped at Indian bar, so-called from the number of Indians at work. They use pans, and were making three dollars a day. They are miserably poor, were living in brush tents, and subsist during the winter mainly on acorns. The hills are sparsely covered with oak and pine trees, the surface soil reddish from furrugnious [*sic*] clay, and covered with loose broken pieces of quartz. The rock exposed here is a blueish slate, the strata standing on edge, and showing occasionaly veins of quartz.—No gold has been detected in the flint, though this is supposed to be the Matrix.

After making our washers, from hollow pine logs, about four feet long, with a screen at one end, made of copper, with holes punched through it, to sift out the coarse gravel, we commenced operations on the bar. Three persons generally work together; two are sufficient. The machine is placed at the edge of the water, one person rocks it, and dips water to pour on the screen, the others dig and carry the gravel in the pans or buckets. In digging a pointed shovel is generally used, with a pick to loosen the large stones—small *kleets* are fastened on the bottom of the machine an inch or two high, which arrest the gold as it passed through with the sand and water. After two or three hours work the sand left in the machine is washed out through the augur hole into a pan, and the process of separating the gold is completed by dipping the edge of the pan into the water, and as it flows out more or less sand is taken with it, the gold sinking to the bottom. It requires skill, and much gold is lost by beginners. The amount of gold obtained depends upon the richness of the earth, and the number of buckets washed. Where we worked about an ounce a day could be made by the able bodied who could work from morning till night. Few persons unaccustomed to manual labor, can endure the fatigue at first: and after three days digging I was taken sick and confined for a week or more with fever. This ended my mining experience. I made in that time about five dollars. Upon this I retired and by first opportunity came back to this place.

Douglas, Huntington, Cross and Hollister, remain in the mines and intend to winter there. They were expecting to move farther up the mountains into the *dry* diggings, where the gold is found in larger lumps. I have not heard how they are succeeding. Although in California, I perhaps know little more about the success of gold diggers than you in the States who can see in the papers the numerous letters sent home from various persons. But from what I can learn I should think that the majority of miners merely make a living. One hears occasionally of the large amount of gold obtained by a certain person in a short time, but nothing is said about many hundred, who do not make their expenses. Not one in a hundred will make a fortune. The expense of living is enormous, sickness very prevalent and the days spent in moving about, or laying idle from various causes, take away largely from the profits.

Of the passengers that came in the *Samoset* I have heard I suppose of 40 or 50 in the mines, and of that number but two have been among the fortunate. The scarcity of provisions, sickness and rains, have induced a large number of miners to come into this place to pass the winter. Unable to obtain work, they offer to labor for board and lodging. At the hotels and restaurants are almost daily applications of this kind.

I saw Captain Conway [who formerly lived in Marietta], in October. He had made in the mines about 15 hundred dollars, but had left from sickness. He was one of the first to leave the states, and is very sanguine that he shall be able to go back to Marietta and build a ship.

I met Gov. Shannon a few days ago. He has had charge of a party and has spent the season in mining on the Sacramento, about 200 miles north of this. His party have made nothing, but have sustained heavy losses.—They have suffered much from sickness and two have died. The Governor I am told has spent largely. He is closing the company affairs and will probably return home.

I have made the acquaintance of Col. Hagan, a German from the mining region of Missouri. The Col. lived some years at Mime La Motte and formerly was engaged with an English Company in the gold mines of Brazil. He is well acquainted with Mr. Dille and Col. Davidson of Newark, who was speculating in Missouri mineral-lands. The Col. attempted mining here but abandoned it as a failure. He was fortunate enough in the summer to purchase some town lots at a low rate and the rise in value has been so considerable that he will make a good deal of money.

A few men here have made fortunes in trade and by speculation in real estate.—Great facilities are afforded to those who have capital and business tact. Prices are very high for goods, and provisions, and the daily auction sales give opportunities for purchasing at low rates. Supplying the miners is a lucrative trade. The titles to land in this vicinity come through Capt. Sutter, and great doubt is entertained of the validity of his claims. His grants I suppose will be investigated by commissioners and some of them confirmed, but not all at present many are disposed to take possession of unoccupied land and serious difficulties are likely to arise.

Several mass meetings have been held by the squatters, who insist that Sutter can show no title, and that they will sustain each other in settling upon lands around here. These difficulties do not affect the sale of town lots, which change owners freely and at advancing rates.

This city has been surveyed about a year. The lots were offered and mainly sold at $500 dollars each; size of lots of 80 feet by 160.—The custom now is to subdivide them into lots of 20 feet front, and those near the centres of business are sold at prices ranging from fifteen hundred to five thousand dollars. The high price of lumber, 600 dollars a thousand, has led to the erection of tent houses which are very numerous: a frame work of poles, covered with canvass. Without floors. They answer very well in summer, but are unfit for the heavy rains of winter. There are very few dwelling houses here of any other description, and probably one fourth of the population are still in tents. Any thing in demand here sells at a high price. A good cooking stove brings 250 dollars.—Boots of good quality from 15 to 35 dollars. Flour 35 dollars a barrel. Pork $45.

Beef 25 cents a pound, mutton and pork 75 cents, potatoes 25 cents a pound, milk $1 a quart. Eggs $1 each &c. The average charge for board and lodging is 25 dollars a week. The sleeping rooms are fitted up with berths like the old fashioned steam boat cabins and one dollar a night is the charge for lodging alone.

My last intelligence from home was dated in June. I suppose letters have been written but have not yet come to hand. I shall take it for granted that you are all in accustomed health. My own health is at present good, though since living in this boasted Italian Climate I have scarcely been well ten days in succession. The healthfulness of this climate has been overrated and misrepresented like everything else connected with California. Great delusion prevails with respect to this country.

Your affectionate son,
George O. Hildreth

[April 18, 1850: "Several letters received by last steamer from members of the Marietta and Harmar companies. One from Mr. Henry C. Anderson . . . brings the sad intelligence of Mr. Z. J. Chesebro, on the 14th of Dec. . . ."]

Redding's Diggings, Dec. 30 1849
Chesebro's disease was chills and fever, which finally terminated in ulceration of the stomach, throat and mouth, which caused his death after an illness of two months and eight days. He was taken sick on the third day after his arrival in the mines, after having worked but *one day* here. It will be a satisfaction to his friends to know that he had every care and attention during his illness. . . . He never knew what it was to want anything gold could purchase. He was as fine a man as I ever knew, and as such his death is deeply regretted.

[Mr. C. with commendable prudence, obtained an insurance upon his life, the amount of which will, we suppose, be paid to his friends here.]

[March 21, 1850: "Mr. Dwight Hollister writes . . . under date of Jan'y. 8th, on a sheet upon which

Cross commenced a letter in October. From Cross' letter we quote as follows:"]

Well, brother Russell, I have crossed the Rocky Mountains and the Sierra Nevada, arrived safely in the golden land of "Californy" made a very intimate acquaintance with the elephant, and am now recovering from a ten day's attack of fever and ague. While I am writing Dwight is laid up in the tent with a fever, which he has had for two days, and Perkins, Dr. Hildreth and Huntington are down on the river digging for "filthy lucre."

I would like half a day's talk with you, Russell, but I am very glad you are not here to give me that pleasure. You are a lucky man in not having left Marietta, and I am only one of many fools who wish they had stayed at home and been contented with small gains and plenty of comfort. But I am here now, and here I shall remain until I get my pile, and then up stakes and bound for Marietta as fast as wind, steam and horse flesh can carry me.

At the time that I last wrote & designed locating on Weaver Creek, but afterwards concluded to come to this place (Cosumnes River) with the Harmar boys, as we should find it pleasant to be in their vicinity than among strangers. I was attacked by the chills a couple of days before I left the city, and was obliged to be brought in the waggons the whole distance here, and since we arrived I have only worked one hour, making 75 cents. The Harmar boys who have been at work with cradles, have averaged about ten dollars per day and, that is about the average yield at this place. So you see fortunes cannot be made in a week here, and we shall soon move to richer "diggins"

This gold digging is no child's play, but down right hard labor, and a man to make anything must work harder than any day laborer in the States. You must first cut thro' two feet of loose rocks, then shovel out the dirt and carry it to the cradle and wash it until nothing remains but the sand and small gravel. This is then taken out in a tin pan, and washed down until you reach the gold, which invariably sinks to the bottom.

The richest "diggins" in this vicinity have been nearly cleaned out, and the most you can expect is an ounce a day, although occasionally a man stumbles on a "pocket" with thousands in it. Dwight and myself have formed a partnership, and will work together when we are able, which I hope will be in a few days.

Stephens is on the Yuba, 100 miles from here, Chesebro about the same distance up the Sacramento. Cunningham has joined the Harmar Company, and Perkins and myself are in the same camp.

[So far writes Mr. Cross. The sheet is filled up by Mr. Hollister, after Cross' death:—We copy as follows:]

I write in a humble but comfortable cottage, once inhabited by two bosom friends, who came to this country hoping by their energetic labor and economy to obtain money enough in a few years to return to the land of their nativity, and there enjoy, among their numerous friends, the products of their labor. Marietta they claimed as their homes. Circumstances prevented their starting from that place together, or coming the same route.—One chose to come by sea, the other to cross terra firma. After a separation of seven months they met in the land of their destination. When they met, each seemed to the other to have taken a second growth. So well had their respective journeys agreed with them that they hardly knew each other. They formed a partnership, and located in the mines, but the hand of Providence prevented their working together. They had built a snug cabin, and laid in a part of their winter's provisions, when one of them was taken sick, and last night, after an illness of 14 days he died.

Soon after Cross was taken unwell, he accepted an invitation of the Harmar boys to spend a little time with them, thinking that a change will help him. The next day he was taken worse, and sent for medical aid, which was procured immediately. I went to see him as soon as he was taken worse, and stayed with him until he died. I did nothing but wait on him during his sickness, remaining with him day and night, but I could not save him. He was quite delirious, and often talked of his friends and associates in Marietta. You may rest assured

that he had every attention. There were two masons in the cabin with him (Moses Williams and Samuel Hammett) who did everything in their power to make him comfortable, and the other boys also did what they could.

He died without stirring a muscle, or uttering a word. I had a nice coffin made for him, and selected a beautiful spot for his grave.—A pretty little evergreen stands so near its head that its branches hang over the grave.

Dwight Hollister

[March 21, 1850]

Longfellow's ravine, Jan. 8, [1850]

The business of the Harmar Company is not quite settled up. Some of the property has been distributed, and some that could not well be divided, has been sold at auction. One of the Henderson mules, known as Bill, and the Hospital wagon, was sold for $400. The other mules, *Jack, Dina* and *Santa Anna* will bring about $400. more, making $800 for the wagon and mules. We also have nine yoke of oxen on Coates' Branch, about a mile from here, which will be sold in a few days. The weather is so that we are making but little at mining, about $4 per day in average, and that will pay expenses and keep a fellow in a clean shirt, if he washes it himself. . . .

The diggins that we now work in are called the wet diggins, consisting of deep ravines, with the water washing down them like lightening. The man that digs it is in the water over his ankles all the time, and many fall sick in consequence. There has been a great deal of sickness in the mines this winter, and many have died. Our company is highly favored. There is not one on the sick list.

Samuel Cross was taken sick with the fever, scurvy, and weakness in the kidneys, a week ago last Monday. He continued to get worse, and came up to our cabin, where every attention was paid him, by Williams and Hammett (brother Masons) & Dwight Hollister, who set up with him every night, and took every pains to anticipate his wants and make him comfortable. Last night Hollister was very much fatigued, and Hammett and myself set up with him at about eleven o'clock death put an end to his sufferings. Cross had many warm friends among the Harmar Company, and was always a welcome visitor among us. Besides all those mentioned above, I would add that Mr. Chapin was very attentive to Cross, and did all he could to make him comfortable. . . .

W. H. Bisbee

[March 16, 1850: "Mr. A. G. Hovey, one of the Directors of the Harmar Company, writes to his brother. . . ."]

January 6 and 9, [1850]

I have crossed that wide range of country lying west from Marietta towards the Pacific shore, and finally the Rocky Mountains and Sierra Nevada, safe, sound, and unharmed. I have enjoyed perfect health all the time.—I am not sorry I started—on the contrary, I am truly glad. . . .

We commenced digging on Indian Bar. I joined with Hulet, Johnson and Soule, and we made $33 of the "yellow article" the first day. The plan of operation is as follows:—The machine, or cradle, has a box on the top, with a copper screen at the bottom. The dirt is poured into the box, and when the machine is rocked it goes through on to an inclined plane, and the dirt gradually washes down against a ripple, where the gold remains and the dirt mostly washes off. After washing out some 30 or 40 buckets of dirt, the sand or fine gravel, containing the gold, in the bottom of the washer, is taken out in tin pans, and washed out by hand. Sometimes a pan full of sand leaves an ounce, or $16 of gold. The labor of our party was divided as follows: one to rock the machine and pour on water; one to carry dirt and pour it in; one to dig up dirt and pour it in; and the other assisted in digging and washing from the pan. In many places there is the most beautiful black sand I ever saw found among the gold. The largest sum we four made in a day was $51.50. Several of the company have made as high as $20 per day; but an average per day, during the time we were digging upon Indian Bar would not exceed $10. During this time Cunningham was cook for the whole company. After my party had made

$233.25 (half of which belonged to us four, after expenses paid) it began to rain.—Some thought the rainy season had commenced. Chapin went to the city to purchase provisions for the winter, sent Williams to find a place for "winter quarters"—as the rivers overflow all the bars, and the miners have to go into the ravines during the rainy season. Chapin ordered me to get up the teams, and take charge of them on the trip to the city for provisions. Roe, Johnson, and myself went with two teams, and made two trips. Had frequent rains, and out in them often, but the oil cloth suit answered a good purpose. Returned to camp in two weeks, and Dec. 6th went to work and built two cabins. The company divided into three messes for cooking, &c. The large cabin is divided into two rooms by a partition.

.   .   .   .

We have a good deal of rainy weather, so we sit by a good fire and take solid comfort, I assure you, talking of old times in Marietta, Harmar, and other places. Soule brought from the city 150 papers, so we have plenty of reading matter for the present. I wish I could get hold of a Marietta paper, though it might please me too well.

The weather, when it does not rain, is beautiful. This we approve in digging gold, but have not been very successful as yet.—The evenings, in fair weather, are the most beautiful I ever saw. The largest piece of gold yet found in California was worth $244. It was found at Stanislaus. Many have made fortunes who came over the mountains last fall, and others are worse off than when they arrived. Some have made from $100 to 300 a day. We have not made much yet, but Johnson, Hulett and myself think we are equal to anyone of the Harmar company, and that have the merit of being first rate all the way here, and it existed longer than any company that came over the plains last season. We have a good machine, a crow bar, two shovels, one pick, one hoe, two tin pans, a long handled dipper that holds say three pints, to pour water into the machine on the dirt.

Gold is universally kept in buck-skin bags. I have three, and if I had them filled I should leave "instanter" for the states. I intend to spend next summer in the mines, with Hulett and Johnson. We shall go on to some river where we will have plenty of water during the dry season. We have a first-rate tent—better than a house for miners—large enough to hold the effects of three persons easily. It is very light, so we can carry it without much trouble. It cost $24.00. . . .

California has the greatest diversity of scenery I ever saw. Take it altogether it is a beautiful country, and is bound to become the seat of a large and flourishing commerce. Many portions will be settled by farmers.—Portions suitable for agriculture I have not yet seen, but all reports say that the southern and western parts are admirably suited for agricultural pursuits.

Samuel E. Cross is very sick, and I think cannot live but a day or two. Mr. Chapin tended on him, and I think he is as skilfull a physician as any in Marietta or Harmar. . . .

Jan 9, 1850

Mr. Cross is dead. Yesterday I assisted in digging his grave, and at the burial. It seemed hard that he should come so far to die.—Hollister will settle his affairs here. . . .

You probably have no desire to come to California, and perhaps nobody else about Marietta. But after all, the trip is not such a killing affair as some imagine. A trip could be performed with a good deal of pleasure, and in much less time than we did. Small companies are best. . . .

A. G. Hovey.

[March 21, 1850: "Died—in California, at the *mines*, Mr. Samuel E. Cross, formerly of New York, in the 24th year of his age.

For three years previous to starting to California Mr. C. was a resident of Marietta. He has left a widowed mother in New York to mourn his loss, and also a large circle of friends in this place, to whom he was endeared by his many virtues. . . .

He was unwell before he went to the mines: but thought he would soon recover, but in the course of a few days he was taken sick and continued getting worse until the 8th of the present month when he died.

He was tenderly cared for by Dwight Hollister who administered to his wants and scarcely left his bed during his illness."]

[March 21, 1850]

Sacramento City Jan'y. 25, 1850

Dear Father: I have too much to write but feeling that you wish to hear from me as often as it is convenient for me to write, I will again send you a few lines.

Since my last letter there has nothing transpired of great importance except the death of Samuel C. Cross, who died of the intermittent fever after an illness of about six weeks. . . .

Our company are still together and Huntington, Perkins and Hollister are with us, and we shall so remain until Spring.

It rains about two-thirds of the time steady, making bad work for us even when the weather is pleasant.

The roads are so bad that we cannot get away from our present location, and we will be compelled to stay where we are all winter. Great quantities of rain have fallen within the past two weeks which has swollen the streams. The Sacramento has been very high, and the morning after the rise commenced, the water was seven feet deep over the city.—so sudden and unexpected was the rise of the river that people awoke finding their rooms filled and their dwellings surrounded by water. Fortunately no lives were lost.

An immense amount of property was destroyed and the contents of the Post Office was washed away by the freshet.

Gold is plenty, and at present and is found in all parts of the city. From the place where I am now sitting, I can see some twenty persons washing gold in the streets, the late flood having washed it from its original bed into our streets.

Those concerned in this city washing are making from $5 to $20 per day.

I came down to the city the 2d inst. with letters to send home and expected some but I was sadly disappointed & purchased 150 newspapers for twenty-five cents each, which I sold on my return to the mines for $1.00 each.

I saw Jos. Stone a few days ago. He had been

quite unwell but has recovered and intended going to the mines soon.

Phillers House is here at present, starts for the mines in a few days.

Dr. G. O. Hildreth is at present boarding at Mr. Whitney's and is well. I have not heard from Chesebro and Stevens since my arrival.

It is probable I shall stay here until the fall of 1851

Almer P. Soule.

[March 26, 1850: "The latest dates of any letters received here, are Jan'y 30 1850. Mr. D. G. Whitney, formerly of Harmar, late of Quincy, Ill. writes from Sacramento . . . under this date as follows:"]

There is but little doing in the mines this winter, during the rainy season, as miners cannot dig to advantage. The impression is that a large amount of gold will be raised the coming season, and my own belief is, that many years will roll round before it becomes scarce in this valley. Still it will require great toil and exposure to get it, and it is sad to reflect that thousands whose hearts now beat high with hope at the golden prospects ahead, are doomed to disappointment. A great many deaths have occurred this winter, not from any particular disease which prevails, but from general debility and neglect. Mr. Samuel Cross of Marietta died on the 8th inst., at the mines. He was with Hollister, Perkins and others from your place. . . . A young man named Soule, from Harmar, who came out in Chapin's company, told me a few days since that the Harmar company was dissolved, and the men are to work on their own account. . . .

D. G. Whitney.

[April 18, 1850]

CALIFORNIA CORRESPONDENCE

Bidwell's Bar, Feather River, Feb. 12, 1850

Friend Gates: Before leaving Independence, I had commenced a letter, giving as extensive information as I could obtain, in relation to the different companies of emigrants—their origin, their capital, their mode of travel, &c, and whatever might be of interest to yourself or your readers. But the very sudden and unexpected death of my brother-in-law Dr. Hall, of Cholera, placed me in so embarassing and uncertain a situation, that it was rendered very doubtful whether I should proceed upon my long journey, bereft of my companion, or return to those among whom I know would always and under all circumstances be welcome. The love of adventure, and a natural aversion to "backing out" prevailed, and—I am here, as you see by the heading of my sheet. Amid all the confusion attending my departure, your letter was neglected and ever remained a beginning.

To give you a description of my journey—its scenes, its difficulties, its protraction, its dangers—is not the object of this communication. To merely indicate its course, is all I will do.

I left Independence May 15th in company with three others, (one of whom I had joined as a partner,) to overtake the Ithaca, N.Y. pack mule company which took the Santa Fe road, designing to follow a route recommended them by Col. Gilpen [William Gilpin—*Ed.*], and make an unparalleled quick trip to El Dorado. Too large a company, fifty four in all, and a deplorable want of system, occasioned by a difficiency in the early pack-mule education of our Captain, together with considerable sickness, caused a slowness of movement that foretold a late arrival at the "diggins."

Our route, as indicated by Col. Gilpen, was up the Arkansas river to its head waters, thence directly over the mountains in a course N.W. by way of William's Fishery to the Utah Lake [Frémont's route of 1845—*Ed.*]; thence to Salt Lake City and over the desert (Bryant's route,) to Humboldt river and down it, taking the usual route over the Sierra Nevada.

We did go up the Arkansas to Pueblo, some 60 miles above Bent's Fort. Here there was some noise made among the emigrants about the Taos mines; but some we saw upon the return represented them to be any thing but rich. A guide, an old mountaineer, though a young man, by the name of Kinney, formerly from near Lower Sandusky Ohio, was engaged to take our company over the mountains to Fort Bridger, on Green River, which is upon the road to Salt Lake City. Our guide could not be induced to go the route recommended by Col. Gilpen, as it lay through the country of the Utah Indians who were then at war with the whites. We now went up the Boiling Spring River [Fountain Creek—*Ed.*], in nearly a due north course, and along the base of the Rocky Mountains. The snow capped summit of Pike's Peak was in our view several days. We crossed the S. fork of the Platte continuing north and along the base of the mountains, till we reached the Cache la Poudre river, which we followed up into the mountains and on, to the Plains of Laramie and in sight of the Black Hills [Laramie Mountains—*Ed.*], which stretched along far to the north of us. Crossing the Plains of Laramie, which were, thus far south, not more than 8 or 10 miles across from east to west, we continued westerly to the N. fork of the Platte, and two days after reached the waters which flow to the Pacific. Soon after we came upon the head waters of what the hunters call the Snake river, but which is, if I remember rightly, called by Col. Frémont, Bear River.—We followed down this several days, until we came within a days drive of Brown's Hole upon Green River, which was a distance of 35 miles, without water. Brown's Hole is the only place where the Green River can be crossed or approached, for several hundred miles above or below. Here we had to raft our things over, and swim across the mules, which detained us two days. Thence our final course was south of west to the Utah Lake; our company deciding to go there instead of Fort Bridger. We entered the Utah valley, which surrounds a beautiful fresh water lake of the same name, on Sunday, July 29th, and the next day crossed the Salt Lake valley, encamped 12 miles south of the "City of the Great Salt Lake," near the houses of some Mormon settlers. To fully appreciate the high gratification afforded us by a sight of civilized men, women, babies, houses, barns, fences, cattle, swine and sheep, you will be under the necessity of travelling for two and a half months upon the wild plains and mountains of our almost unlimited territories. We had plenty of game upon the way, thus far; hardly ever without buffalo, elk, deer, antelope, birds or fish: but there were some luxuries to which we had been habituated and which were never forgotten, entirely beyond our reach. Milk, cream, butter, cheese!—what meal can be perfect without them! I believe I recalled the

honest countenances of every cow my father ever owned, since I was large enough to drive them to pasture; deeply regretting every stone I ever threw at them, to get them started from the field and every kick I ever inflicted to make them "So!"—Mr. Editor, have you a cow! If so, give her in summer good pasture; and in winter plenty of good hay and the choicest cabbage stumps you can command; for she will well repay you: and if you ever take an overland trip to California, you will, while craving her precious products, receive a special consolation for so doing!

Having arrived at the Mormon Settlements in my hasty sketch of our route, as I remained three weeks among that persecuted, but enterprising and energetic sect, I will detain you a few moments in reading what I may say concerning them.

A mud fort, enclosing about three quarters of an acre of ground, constituted all the shelter of the whole Mormon population of the valley, one year previous to the time I was there. Ten months previous, they had so conciliated the Indians, that they determined to lay out their city, and commence separate residences. Their city was laid out regularly, the streets being at right angles eight rods in width, and some three hundreds yards apart. Each head of a family was entitled to a city lot, free of charge, of one and a quarter acres. In a few months, every family was under the shelter of their own house, and cultivating their allotted ground. Besides a city lot every one may cultivate as much of the surrounding valley as they choose. Many, vacating for the summer their city residences, moved up the valley and commenced more or less extensive farming operations. Two flouring mills were built. Nothing in short was left undone that tended to promote their convenience and comfort, and render them independent of all the world.

All their crops require irrigation, and no spot on earth affords greater facilities for it. The valley, surrounded by mountains covered with perpetual snow, is by the Utah outlet, which they have piously named the river JORDAN. There are numerous streams flowing into this from the canons of the surrounding mountains. As the plain slightly inclines towards "Jordan," they are easily turned in any direction, (except east,) and are made to ramify the whole cultivated portion of the plain which at present, lies east of the river.

To give you as distinct an idea as possible, of their mode of irrigation, we will suppose a range of fields, some of corn, others of wheat, pease, beans, barley and what you please, extending in a line north and south. Along the east sides of these fields is turned a stream either north or south, the natural course of which is west. The rows of corn run east and west, and fields which are sown broadcast are furrowed into narrow spaces in the same direction. These furrows and those between the rows are made a little lower than the surface of the water in the artificial channel upon the eastern border. The fields slightly incline to the west. When they wish to irrigate, a communication is made with a shovel, between the stream and each furrow. The water flows slowly along the furrows, until the earth becomes sufficiently saturated, when it is shut off by throwing a shovel full of earth into the head of each furrow. Once a week is usually sufficient to water the crops so that when the preparations are once completed, which is attended with considerable expense, the labor of irrigation is slight. It ensures a more certain crop than where rain is depended upon. They were in the midst of the wheat harvest when we entered the valley, which was sufficiently plentiful to ensure their 8000 inhabitants food for the year and the emigrants all they wished to purchase. We arrived in the valley nearly destitute of provisions, and before entering it, it was with much solicitude that we considered whether they would have any to spare us. The price of flour was 10 cents per pound—bacon (obtained entirely from the emigrants) 12 1/2 to 18 3/4 per pound. They paid for laborers in the harvest fields $5.00 per day.

The farming portion of the valley, at present, extends from the south border of the city towards the south 8 miles. Here, a fence running east and west, separates the cultivated fields from pasture lands beyond. There are farms, however, twelve miles from the city but each man's fields are separately enclosed. The pasture is common property. Those in the city owning cattle, drive them towards the river northwesterly from the town; and some drive them "beyond Jordan," (a common expression) and herd them during the day, driving them in at

night and penning them up till morning. Cattle are quite plenty among them. There are but few sheep and less swine. Horses and mules were very plenty. An Indian, by the name of Walker, with his band, make a regular business of going to California, stealing droves of horses and taking them back to the valley for sale. The Mormon interpreter, Dimmock [Dimmick—*Ed.*] Huntington, holds a great influence over him, and acts as his agent in the disposition of the horses. Many of the emigrants got fresh animals of Walker, Huntington receiving a per cent for his services. I witnessed some of these bargains. The Mormons count upon Walker's tribe [the Utes—*Ed.*] as allies, should they need their services in any emergency, and no doubt they would prove faithful ones.

The most careful cultivation, of course, is in the city, which is emphatically a City of Gardens. Every street has a running stream of clear water in what would be, in eastern cities, the gutter. Branches from these run through the gardens supplying the place of rain which seldom falls there. The houses, built of unburnt brick or adobes, are small, scarcely any being more than one story high, and are rather miscellaneously situated, some near the street and some back in the gardens, and present little the appearance of what we would call a city. They had commenced a large public building or church, of stone from the mountains, where there is an inexhaustable supply for building. A large and commodious house was building for the President, Brigham Young. Whether it was for him individually, or for their presidents I did not learn. They design, in a short time, to commence a splendid temple, which shall equal Solomon's, in magnificence.

The general government consists of three presidents, which they denominate, first, second, and third presidents, the First President having the title "the President," and a council of twelve, (corresponding to the twelve apostles) which constitute a kind of Court of Appeals. They are also the legislators of the sect. They have the city divided into wards, (thirteen at present, I think,) over which presides a Bishop whose jurisdiction corresponds, temporally, with a justice of the Peace. He has also a spiritual jurisdiction; but as I did not enter into the spirit of their eternal affairs, I cannot say how extensive it is. An appeal may be made from their decision, to the presidents and Council, and from the latter to the voice of the assembled congregation, whose decision is final. So upon the whole, it is rather a democratic government. They have also a "Seventy," who I believe have license to preach, but have nothing to do offically with temporal matters.

From what I could observe of their social habits, the Mormons differ very little from the "world's people" in the States. They are very friendly in their intercourse with one another and with strangers. They are hospitable and obliging. Perhaps some Illinoions and Missourions who passed through their city would take exception to what I say, and well they might, for the Mormons hold towards them a special hatred. None of those engaged in the mobs which attacked the Mormons in Illinois dared go through; nor were their fears unfounded; for, had the Mormon's recognized any of them, they would surely have been tried and executed. This I was assured of by one of their presidents, from whom I received very hospitable and kind treatment.

I had expected to find them very fanatical and superstitious, but in this was totally mistaken; at least as to the kind. Their fanaticism is more of a worldly one than else; delighting rather to dwell upon the future prospects of their success as a sect than upon their individual future state; to recount their persecutions and sufferings, rather than reflect upon the consolations afforded them by their religious belief. Their superstition instead of inspiring them with a fear of future punishment for indulging in the worldly amusements proscribed by orthodoxy, manifests itself in their accrediting to their leaders, such as make the pretensions, power to work miracles as the laying of hands on the sick and healing them, believing in their inspiration, and in special interposition of Heaven in their behalf. They seem to care less for their religion than their nation.

As an instance of the latter kind of superstition I will refer to the cricket miracle. The first summer they attempted to cultivate the valley, their crops were very much injured by a kind of large, clumsy, black cricket, which sometimes perfectly blackens every bush and bunch of grass for miles in extent. They were plenty for hundreds of miles before we entered the Mormon Valley. The second year their crops again seemed to be in the same danger of destruction, from the multitudinous appearance of the crickets. But mark! Heaven sent a colony of gulls, a bird never seen before in the valley by the oldest hunter, nor within hundreds of miles of it, and these angels devoured the destructive insects, leaving the crops free to yield their thirty or sixty fold. Not a cricket was left in the whole valley! This they believe to be as truly a miracle as that of the quails being sent in such larger quantities to the famishing Israelites. Not being myself very superstitious I could not be gulled into a belief of the miracle, more especially as I saw no gulls but plenty of crickets in the valley; though they were, from some cause, confined principally to the more uncultivated portions.

None of the austerity of manners which usually characterizes new sects, but which in their prosperity invariably wears off, as for instance the dress and rigid customs of the early Methodists and Quakers, is apparent among the Mormons. Cotillion parties are held almost every week under the sanction of the leaders of the sect. I had an invitation from the Clerk and Recorder [Willard Richards—*Ed.*] to attend, which I did not find it convenient to do. Their Sunday is as much a business day as anything else; that is business pertaining to the general welfare of the state. It is a court day—cases being tried and decided after sermon. The sermon itself is devoted to advising the people, as to their crops, fences, &c. Pomeroy from Lexington, Mo. who took a large quantity of goods there for sale was tried on Sunday after sermon for participating in the mob against the Mormons. He proved himself clear, and rather a friend of the Mormons, whereupon the President advised the people to patronize Pomeroy, which was done very freely, his accusation and acquittal being several hundred dollars in his pocket.

Their laws concerning property differ little from the laws in every state. Each one "gets what he can and keeps what he gets." There is however this improvement upon our laws: no one is entitled to any more land than he cultivates, or causes to be cultivated. All their contributions to the church or rather to its individual ministers are voluntary.— strictly speaking the church and state are about the same thing. How the tax is raised for the support of the state I did not learn. I spoke of each one's being entitled to a city lot. When the lots are all taken up within the present limits of the city it will be enlarged, so that every one who may become, hereafter, a Mormon shall have his city lot of an acre and a quarter.

In their religion the Mormons profess to be Christians in the strictest sense of the word. They admit the truths of the Bible and found their creed upon it. They call themselves the "Latter Day Saints," averring that the New Testament speaks in unmistakeable terms of their rise and spread over the earth. Joe Smith they consider as a Prophet from the Lord, and the Book of Mormon, of which he was the special interpreter, a Revelation from Heaven, throwing light upon the dark and mysterious passages of the Bible. They say there is no declaration or intimation in the New Testament that the performance of miracles shall ever cease, but the contrary that it expressly declares that they shall follow as a sign of their believers. I recollect one passage referred to by a very bright and intelligent young man who spent an evening with me at my camp, to prove this. It is I believe in the last chapter of Mark where it is said (I quote from memory,) "He that believeth and is baptised shall be saved but he that believeth not shall be damned. And these signs shall follow them that believe. In my name, they shall take up serpents and if they drink any deadly thing it shall not harm them: and they shall lay hands upon the sick and they shall recover." That after Christ had been received up out of sight "they went forth every where, the Lord working with them confirming the word by signs following." He held that these declarations concerning those that believed was as unlimited as the command to "Go forth," &c. He informed me that the reason he was there, a Mormon, was the fact that a sick brother, given over to death by his physicians, was cured by the "laying on of hands" of a Mormon preacher! Out of respect to the young man's enthusiasm and apparent sincerity I did not venture to suggest, that had the doctors not given him up and stopped their dosing, it might have fared hard with his brother, notwithstanding the laying on of hands.

In answer to a question as to why miracles had never been performed since the times of the apostles,

he answered that they were performed for many years after and that it was only when the Church became corrupt that the power ceased; and that it might now be made manifest that *they* were the true Latter Day Saints, the power was given them from on high the same as to the saints of old. The Mormons have a most profound veneration for "Joseph," comparing him with Christ and calling him the greatest who has ever appeared. In London (there are many English among them) . . . which contained numerous hymns in his praise. I recollect particularly one parodied from the celebrated song "The Sea! the Sea! the open sea!" which was ludicrously metamorphased into "The Seer! the Seer! Joseph the Seer!" &c.

Their worship is not characterized by what we would consider devoutness. They seem to think that no kind of musical instrument can commit sin. Every Sunday at meeting the Brass Band performs marches, waltzes and quick steps to the edification of the congregation. One Sunday while I was there they played the "Love Not Quick Step" with considerable taste. They have vocal music also.

I have said thus much concerning the Mormons that you may the more distinctly understand what kind of people they are, who are knocking for admission as a sovereign and independent state into our Union. I have endeavored to represent them in as fair and impartial a light as possible. I have said their Church and State are about the same thing. It is natural for men who profess any religion to endeavor to gain as many converts to it as possible and when they are a large majority of the people it is equally natural for them to make converts by law and force. The Congress of the U. States should not admit a state into the Union where every one is not entitled to think as he pleases in regard to religious matters. The Mormons are and must ever be an isolated community, since there is not a section of country fit for cultivation within many hundred miles of them. Whether any will ever settle on the Salt Lake Valley besides Mormons is a matter of some doubt, even should all religions be tolerated by them; so it may remain forever a Mormon State. Whether such a state would be profitable to our Union is problematical. Many of them profess allegiance to the U. States while some leaders seem desirous of independence. The subject of their admission should be viewed in every light, and well considered by Congress, before making them a State. Previous to the cession of this portion of Mexican territory to the government of the U. States, they without doubt intended to establish an independent government and this desire still clings to some, especially to the English portion of them. I will close this letter and resume my trip to the next.

JEROME B. HOWARD

[April 25, 1850]

CALIFORNIA CORRESPONDENCE

Bidwell's Bar, Feather River, Feb. 16, 1850
Friend Gates: To resume my journey; we left the City of the Great Salt Lake Aug. 18, my company being entirely new with the exception of my partner. The Ithaca joint stock company dissolved at the Salt Lake Valley, and the majority of these determined to remain there till October and go through by a southern route, that being the last recommended to them. Four constituted our company, but we were joined in a few days by four others, and with this small force travelled up the east side of the Salt Lake, leaving the last Mormon settlement, forty miles from the city; where our route was again to be through an uninhabited region (except by Indians, which were seldom seen,) until we should enter the Sacramento Valley. We went around the North end of the Salt Lake, and struck the main travelled road 200 miles from the Mormon City. We continued on, passing hundreds of emigrant wagons, till we reached the Humboldt River. On our first reaching it, who should we meet but fifteen of the Ithaca Company, among whom was Dr. [Elijah —Ed.] White formerly of Oregon who had come across the desert at the south end of Salt Lake, a shorter route by 200 miles than we came [the Hastings Cutoff—Ed.]. They started from Salt Lake City the day before we did.

Joining our old friends we proceeded down the river passing multitudes of emigrants and after some days of hard travel reached the sink. The Humboldt River runs west thro' the "Great Basin" without a tributary for some three hundred miles and consequently it grows smaller until it becomes entirely lost in the sands of what is called the "sink." It is the most serpentine of all rivers I ever saw being made up of an unbroken succession of bends. If some power could get hold of the end of it at the sink and straighten it out, it would reach to the Pacific. From the sink we had a desert to cross without grass and with water at but one place, of some sixty miles. The water was boiling hot and brackish at that, so that when it was cooled some of the animals would not drink it.—There were several springs and the ground upon which we encamped appeared to be, as it no doubt was, a mere crust rock. The tread of men or animals upon it produced a hollow sound which with the rumbling and puffing below, together with the heat of the shell rock, rendered it rather a frightful resting place. Steam was constantly rising from the springs and at intervals it came in puffs. What they were doing down there I could not imagine.

The next day we reached Truckee River a fine bounding stream of pure cold water from the Sierra Nevada. The mules appeared as pleased at the sight of it as their riders, for they had been three days without water, and grass except what we cut and dried for their dinner the first day. We found a pretty good camping ground and grass for our animals. After a days rest we continued up the river towards the middle pass of the mountains.

There are three routes from the Great Basin to the Sacramento Valley. The one we took was the old and middle route. A hundred miles or so back of the sink a new route had been discovered striking the head waters of Feather River, and this was recommended by a Mr. Lawson [Peter Lassen—Ed.], who has a rancho in the upper part of the valley where they would come in. Many took that route and everyone I have seen curses the route and Lawson too. It was feared he would be assassinated at one time. The other is the Southern [Carson—Ed.] or Mormon route which is farther than the middle but a much better one for wagons. That leaves the old route two miles west of the sink, and comes in upon the head waters of the American River.

We continued up the Truckee river three days, when we crossed the first mountain, and came to a very pleasant valley, where we encamped. We nooned the next day near the log houses of the unfortunate Donner party, so many of whom perished there of hunger and cold.—(see Bryant.) This was the 20th of September, and the afternoon of the same day we began the ascent of the summit. While I was upon the point where a bucket of water turned upon the ground would divide, one part running toward the great basin, and the other towards the ocean, I witnessed a sunset whose grandeur was at most inspiring. Before me were peaks of mountains each side of the Yuba, becoming lower and lower as they receded to the west. They were in the Sacramento Valley! How many slow and tedious hours, and days, and months of toil had I undergone to enter it!— Here it lay before me. The parting rays of sunset gilded the craggy mountain peaks of this golden valley, imagination converting their fantastic crags into castles and turrets at pleasure. DURAND is the only artist living who could represent it upon canvass.—the warm—almost hot—misty atmosphere of his picture, which I never before thought quite natural, and which is so unlike the clear pure air that pervades the pictures of Thomas Cole, enveloped the whole landscape. After feasting my eyes with this enchanting scene, I began to descend into the Yuba valley.—The sun sank below the western horizon; the damp chilly atmosphere of night admonished me to invest myself with my poncho; darkness soon enveloped me; the path was steep and rough; but trusting to my sure-footed Billy, and conscious that I was in the Sacramento Valley, and within two days of the "diggings," I proceeded with a light heart, making the woods re-echo with song. I came up with the main company, who had encamped far down the valley, about 10 o'clock, and after supper, spread my blankets upon the frosty ground, and was soon locked in the embrace of Morpheus.

There is a great disparity of temperature between sunshine and darkness in the mountains. After riding all day in the scorching rays of the sun, the late autumn night would follow. I slept in the open air during the whole trip, and frequently would find my buffalo robe in the morning thickly coated with frost. The change seemed almost instantanous upon the appearance or disappearance of the sun.

Crossing from the Yuba into the Bear valley we continued down it until we arrived at Johnson's (now Gillespie & Co.'s) Rancho, which is upon Bear river, at the foot of the last hills of the Sierra Nevada. On our way down the river we witnessed the first mining operations, and saw the first gold. We passed by the encampment of the Zanesville company, whom I had not seen since they left Independence. I learned with pleasure from Dr. Kernian, of the safe arrival in the valley of your townsman E. Douglas Perkins, and regret to say that I was unable, notwithstanding dilligent inquiry, to meet him at the city—Sacramento City, at the junction of the Sacramento and American rivers, and near Sutter's Fort. I have also learned of the arrival, by sea, of Dr. George Hildreth.

Remaining at Johnson's two days, we started for Sacramento City, which we reached Sept. 30th, having taken it leisurely through the mining district. We remained a week in this city of tents. It was laid out by Mr. Burnet [Peter H. Burnett—*Ed.*], now Governor Burnet, and was rapidly settled, lots soon selling for several thousand dollars. There being little lumber in the country, the houses were built of all kinds of cotton and linen stuffs stretched upon frames. There were two or three buildings of wood and one of sheet iron and zink. But notwithstanding its ragged appearance, there was a great deal of business done.—Ships, brigs and sloops from almost every port in the States, were lying at the dock. The Sacramento river here is a noble stream. The fact that at low water season vessels drawing nine feet of water can ascend to this point, which is 150 miles from the ocean, sufficiently proves this. The Hudson does not compare with it. The largest ship-of-the-line in the navy can go above Sacramento City in the winter.

There have been several cities laid out above Sacramento. Twenty miles above, (25 miles by the river,) are Vernon and Fremont—the latter opposite the mouth of Feather river; The former upon the left bank of both Sacramento and Feather rivers. Up Feather river, near the mouth of the Yuba, are Yuba city and Maryville. At Johnson's, since I was there —there has been a city laid out called Kearney. There are post offices at these places established independently of the government; though I suppose the mail agent will devise some plan by which the mail shall be carried at the expense of the government.

This should be done. There is a large and rapidly increasing population here, who will demand a better, regulated arrangement for obtaining letters. I received five letters from home the last mail, which cost me, besides the postage, ten dollars. I have to pay fifty cents for every one I send to San Francisco unless some one goes especially for them, and presents an order signed by each one expecting them. No one can go to and return from that city, from here, at an expense of much less than $150.

There are some new towns below Sacramento, one of little importance at the junction of the Sacramento and San Joachin*, rivers, ridiculously called New York! The man who had so little taste as to give it such a name, ought never to sell a lot for over ten dollars. Up the San Joachin one hundred miles is Stockton. Benecia is on the bay, 35 miles from San Francisco.

Nearly all the articles purchased for the mines of the upper country are obtained at Sacramento. Transportation is very high, and especially after the rainy season commenced. During the hardest of the rains, the roads were entirely impassable. During the fall, while it continued good going, teamsters charged 30 cents per pound for every article they brought to this bar. After the rains set in, it raised one and two hundred per cent. The prices of provisions and tools here were enormous. The following is a list of prices while they were highest:

Flour, per lb.	$1.75	Frying Pans	$8.00
Mess pork	1.25	Mack, Blankets, pair	$40
Potatoes	1.25	Horse do. „	$16
Onions	1.50	Coarse Stoga Boots „	$64
Cheese	3.00	„ „ Shoes „	$16
Butter, poor	2.50	HICKORY SHIRTS	
Dried apples	1.50		$3 a piece
Sugar	1.25	Red Flan. do. $10 do.	
Tea	3.00	Good rifles have been	
Coffee	.75		sold for $5.00
Saleratus	5.00	Shot guns for	$1.00
Rice	.75	Wool Hats	7.00
Beans	1.00	Sail-duck wall tents,	
Pickles per			10x12 feet $175.00
	quart $5.00	Small drilling tents,	$50

* Pronounced San Wah-keen, a like a in what.

Molasses	3.00	Boards, square foot,	$1.50
Brandy	7.00	Shovels	$10
Vinegar	2.00	Pick-axes	$6.00
Board per day	5.00	Tin Pans	8.00
Tol Candles 50¢ a		Crowbars	$12.00
	piece	Sheet Iron, square foot	$6
Tin Coffee Pots $6.00		Tin, per sheet	$5.00

The above is a list of prices at this bar. Above as they are still higher. When we left the city, which was Oct. 6th, we designed to go to the Upper Sacramento mines, or upon Trinity river, which is west of the coast range of mountains which separates this valley from the Pacific. We continued up the valley within a day's drive of Redding's Diggings, when we met so many returning with such unfavorable accounts that it was impossible to disbelieve their united testimony. We had met persons all along the road who gave us very unfavorable accounts of the Upper Sacramento, but as there are so many and contradictory accounts as to mines generally, we did not feel inclined to follow the dictum of every one we met. After deciding to return, our only course seemed to be to Feather River. We arrived at this bar October 20th.

There are some things concerning the country and its inhabitants, that I learned from observation, with a description of which I will close this letter; in my next giving you some items respecting present and future prospects of mining in this country.

Almost every one who has had a desire to become acquainted with California, has read Col. Fremont's description of it; to which I refer them for correct geographical knowledge. With his description of soil, climate, &c., of the valley, there is scarcely an immigrant to agree. They are unanimous in pronouncing it a country unfit for settlement or cultivation. This seems an impeachment of Col. Frémont's veracity. I recollect while crossing the mountains into the Salt Lake Valley, our guide, and many of the company following his cue, were unceasing in their condemnation of his map of Oregon and California, as being worthy of no dependence, and more incorrect than some little two penny maps that were in the company. Now the truth is, that Col. Frémont's map is the only map existing that approaches correctness, and wherever his route lay, every thing was perfectly represented. Of course, if he crossed a branch, continuing on to the main river without following down to the junction, which of course he could not do in many cases, the point of their junction might vary from the truth. This must of necessity be the case with all explorers. They cannot take an observation from each square mile, without which it is impossible for a map to be perfectly correct in every minutia.

Now let us inquire the cause of the discrepancy between Col. Frémont and the united impressions of our gold seekers. It all resolves into this; not one in five hundred of those who almost curse him for his glowing description of California, has ever seen the country he described. They apply to the valley of the Upper Sacramento his description of the country around Los Angeles, Monterey, the Tula[re—*Ed.*] valley, and the valley of the San Joachin! When I go to San Luis Abisbo and find no fruit but acorns, and nothing but barren red soil of the hills I am now among, I will join the general clamor against Col. Frémont.

As will be learned from the authority referred to, the valley is bounded upon the east by the Sierra Nevada or Snowy Mountains, and upon the west by what is called the coast range, though it is not so, strictly speaking, since there is between it and the coast the San Juan valley, and another range of lower mountains, which form the coast itself. The range between the San Juan and Sacramento valleys, which is generally about 20 miles from the ocean, is sometimes called merely "The Sierra," that is, *The Mountains.* They are at present covered with snow, and present a beautiful appearance, from the hill tops around us. The valley is divided into the upper and lower valleys by the "Three Buttes," an isolated group of mountains which extend partly across the valley, east and west.—These Buttes (pronounced as a monysyllable Butes, "u" long.) are in a course south of east of this bar, so that we are in the Upper Sacramento valley. The valley becomes more and more narrow until the mountains approach the Sacramento on either side, which they do at Reddings, that being the extreme upper end of the valley. It is, from the foot of the hills of the Sierra, west of the valley, from 40 to 50 miles, opposite where we now are. For this distance the country is

one unbroken plain, in the summer destitute of grass and trees, except up the borders of the river and its branches, and some Indian villages, upon the water courses.

The Indians of this country more resemble in features the inhabitants of the Islands of the Pacific, than the aborigines east of the Rocky Mountains. In their habits, they resemble neither. They are small in stature, and extremely filthy in their habits. They have a large mouth, thick lips, small black, sunken, treacherous eyes, a short flattish nose, low forehead, straight, coarse, black thick hair, and rather high cheek bones.—Many of them are tolerably well formed, and active. They are fond of dress, and will work, which is a distinguishing trait between them and the Indians east of the Rocky Mountains.

Those who cannot obtain from the whites clothing, make a kind of breech cloth, or rather a substitute for it, of willow bark. Sticks of willows are selected which are at least three feet long, the outside bark scraped off, and the inside pulled off in shreds the length of the stick, and a bunch of these tied together at one end in the form of a wisk, is tied around the waist, one before and another behind, and hanging to the knees. This is more peculiarly the dress of the women. In summer, many of the men, and all the children, are entirely nude.

Their houses, if houses they may be called which are not houses, are formed by digging a hole in the ground, in a circle of some fifteen feet in diameter, and roofing it over with poles, bark, and lastly earth, leaving an opening upon one side for crawling in at.—The roof presents a dome-like appearance. These are promiscuously situated, there being no attention paid to streets, alleys, and public squares.

Their food consists, principally, (besides what fish they catch and animals kill,) of acorns. Nearly or quite every hut has a crib filled with them. The cribs are built of small sticks wove together; are circular, about four feet in diameter, and sometimes 8 or 10 feet high. After the acorns are dried, they have little of the bitter taste, though somewhat astringent. Their mode of preparing flour from these acorns is quite ingenious, and somewhat complicated. I witnessed the operation near Potter's rancho.

The place selected for operation is upon a gravelly bar of some stream. The acorns are shelled and pounded into small pieces, probably in a stone mortar, this part of the operation being performed at their village—I not witnessing it. Into a bowl like basket, they boil water by means of heated cobble-stones of from two to six pounds weight.—Into this they throw the pounded acorns, which are ground fine by the agitation of the stones. They then scoop out a concave hole in the gravel, or coarse sand capable of holding about six gallons, and after patting the bottom smooth, they pour in the contents of the bowl; or several of them, after taking out the stones with a paddle which they do very dexterously. The acorn meal is held in suspension by the water. After the meal settles to the bottom, in the form of a cake, they pump out the water, which remains, a portion of it at first draining through the sand, by means of a most ingeniously constructed pump. A woman (the women do it all,) bends over it upon her hands and knees, and with her mouth sucks up the water, squirting it out by huge mouthfulls. They then, with a sharp stick, divide it at right angles, into squares of about five inches, cutting through to the sand. This drains off all the water. Then each square cake is taken up in the hand, and the gravel adhering to the bottom washed off in a bowl of clean cold water.—Thus prepared it is ready to be made into mush or bread. They took their supper from it while I was with them. Thinning it with boiling water into a kind of soup, they ate it, using their hands for spoons. Bending their fingers in a cramped manner, and hugging them close together, they would dip into the soup, and hiding the larger portion of their hands in their enormous mouths, swallow every drop it contained. The little children ate in the same manner, with surprising facility. One little fellow of four years old I noticed particularly. A grunt and a snuff were regularly heard between each mouthful, which followed one another in rapid succession. Occasionally the little brat would get a gravel stone in his mouth, which he would spitefully spit back into the dish not breaking his time in the least.

Two of the women thus engaged, observing me watching them with such intentness, charitably supposed I must be hungry; and very hospitably beckoned me to seat myself between them and partake of their meal.—Fearing to violate the etiquette of their style, from my inability to use their spoons,

without extreme awkwardness, and not being in a famishing condition, I intimated to them that "I had no occasion." Out of curiosity I did taste a bit from the portion settled in one of their draining holes, and found it quite palatable.

Some baked cakes for their supper. The same heated stones used for boiling the water were used for baking. After heating, they would cover a layer of them with vine leaves, upon which they put the dough, and after covering with leaves, would pile on hot stone. I did not eat any of it, for after my first taste, I saw a woman catching vermin from the head of a little boy, with which she pieced out her supper, which fully satisfied my curiosity as to their epicurism.

Most of the Indians in the valley belong to some rancho, and are little less than slaves. They cannot be sold, except with the estate however, and this is about the only difference between them and the state of American slavery.

The ranchos are the dwellings and lands of such as were obtained from the regular Mexican or revolutionary governments which existed before the American conquest. Many of the real or fictitious titles have changed within two years. Some of the estates are very large, embrasing all the arable land upon the rivers for one, two and even five leagues up and down the river. Indeed almost all the land fit for cultivation, in the valley, is covered by these grants, frequently ecclesiastical grants. The Congress or the Supreme Court should not recognize their validity, I have not the least doubt. The treaty with Mexico would not be violated, in most cases, should they be declared void; for from the best sources of information I can reach, they, in but very few cases, emanated not from the regular Mexican government, but from the ephemeral independent governments, so many of which have arisen in this country. If they are recognized, Congress should admit it as the aristocracy of California, and not as a free state. Under such regulations, it never can be generally populated, nor profitably cultivated.

As my letter is already too protracted, I will close with an account of an accident that occurred here the day of the date of this letter. Six men attempting to cross the river in a small skiff, were carried over some rapids below and upset. Three were drowned. Their names were Wm. Sheay, of Lenowee Co, Mich., and Dr. McConnel and Wm. Feagles from Orange Co., N.Y., but latterly from Iowa. Mr. Thomas Edsall, and his brother-in-law, Jas. B. Jones, both from Orange Co. N.Y. and Mr. Easterbrook, from R. I., were saved. The escape of Mr. Jones, with whom I was acquainted in Albany, N.Y., was very strange. He could not swim at all. Carried by the current, under water, he was hurled down stream, and nearly suffocated, when he happened to find his head above water and took a few precious breaths. Taken under again he endeavored to hold his breath, and considered with calmness his situation, deciding that he must be drowned. Suddenly he felt something thrike his left forearm, which grabbing he found to be a rope! Pulling hand over hand, his head under water seemed as though he would never reach the end. At last his head came up in contact with a rock, up which he pulled, till coming out of water, he found he had the boat rope, and that the boat had caught upon a craggy point and remained fast. Hardly had he pulled himself upon shore when the boat left, and was carried three hundred yards farther down the river! The other two swam ashore. The determination to spend that forenoon in writing to you, was all that hindered me being in the boat as I was engaged with these men over the river. The evening before, I crossed at the same place, and in the same boat.

Returning from down the river, whither I went to see if I could find the bodies. I observed some flowers in bloom upon the hill side. As an evidence of the mildness of the climate among the first hills of the Sierra Nevada, I enclose one of them, which I request you to present with my compliments, to Dr. Hildreth. I hope it will not become so mutilated by its carriage as to lose with him all botanical interest. There are now about eighteen species of plants in bloom.

Truly Yours,
JEROME B. HOWARD

[May 23, 1850]

CALIFORNIA CORRESPONDENCE

Bidwell's Bar, Feather River, March 17, 1850
DEAR GATES: The "jug-handle" climate of California deserves a few remarks. An Irishman would say,

that first it never rains, and after that, it rains forever. When I arrived in the valley, last September, the ground was as dry and parched as though it had never recieved a shower. The roads were dusty in the extreme, so that the constant travel kept up, had covered the grass with a thick coat, making the whole plain appear even more barren, than it really was. Some sections of the plain, and for miles in extent, where the water had overflowed in the preceding winter, upon being thoroughly dried by the sun, contracted as to leave fissures in the earth of from three to eight inches wide, and of a depth of over twelve inches. Our mules walked over them with great circumspection. The surface was so hard and baked in such places, that no trace of the animals would be left.

The first rain that fell in the fall was upon the 10th of Oct., and it rained some the following day. The night of the 11th, it rained almost incessantly. When I forget a thorough soaking I got in my bed, I shall forget that night. From this time, till the first of November, it was fair. I think it was the last night of October, that it rained again, the water raising in the river, at this place, one foot. The next night, it rained again; from that time, an almost unceasing rain poured down for three weeks; not a day passing but what it rained. It was during this season, that every thing rose to such high prices. The wagons which were upon the way from the city, (This unlimited expression is always applied, here, to Sacramento City,) were forced to stop. The entire load was, in many instances, ruined, where the wagons were not prepared for the rains. Mules, horses or cattle were equally unable to walk over the miry earth. Many animals were mired and lost. A scarcity of provisions, especially of flour, followed.—There passed one week, in which it could not be obtained for any price. It was feared, by some, that a famine would follow.

Just before any one starved, however, the rain ceased, and some two weeks of fair weather followed; during which time the roads, except in a few places, became quite passable. Provisions arrived of all kinds, and in every manner. Packmules and wagons, pack-men and boats, were in active operation, prices gradually lowered, until another rainy spell followed, during which they were stationary. Things continued to arrive by whole boats, which could approach within nine miles of here, whence they were packed over a rocky path running near the river bank. Another still longer season of dry weather succeeded, which nearly ruined some who had, at a heavy expense, engaged in boating during the rains. Those who brought provisions by land during the dry interval, could afford these much less than what they cost the former. As at home, speculation was confined principally to flour.

This article is obtained almost entirely from Chili and Oregon. Instead of barrels, it comes in two hundred pound sacks. If not impared from dampness it is a good article.—The speculators last fall bought up all the flour in the city; and from $16 per sack or barrel, it suddenly rose to $40. Reports were circulated that the Chilian government had prohibited its exportation. It remained at this high price until lately when plentiful arrivals forced a change.

Ah! "oasis" in a desert! The sweets of life mingled with its privations! What think you I have been doing, since penning the above paragraph! It was too good to be passed over in silence. I have just been roasting a fine, large, sweetpotato!! Ten thousand blessings upon the head of the Yankees who thought of bringing it!

By the by, Mr. Editor, you cannot imagine how good everything tastes here that *is* good. The principal living is "slap-jacks," with pork or beef. Some use, almost exclusively, for bread, sea bisquit, while others make their flour into short-cake, using the fat from the fried pork for shortening: others bake it in the same manner without the fat, making what I would call a long-cake. A few, as though it were a premium upon cruelty, eat nice light bread made by their wives. After all the permutations possible have been performed with bread, pork and beef, and reported sundry days, how refreshing is a pickle! how grateful a piece of cheese! how invigorating, a potato! and above all, far more rare, how ecstatic, a sweetpotato! Seriously you can know nothing of the zest, which long privation gives, to "a feast of fat things."—Most of the miners here live in tents at the end of which they build a fire place and chimney. Some are fortunate enough to have iron stoves. There are some cloth houses after the style of those at the city, some shingle shanties, some half subterranean and some "regular" built log cabins, enough

perfect for electioneering purposes. These are the most comfortable of all, but around here a scarcity of suitable timber limits the number. Up the river, where timber is more plentiful, there are many good log houses. I had nigh forgotten to mention another type of house, made of raw-hides stretched over a frame.

More sleep with their blankets spread upon the ground than otherwise, while some make a frame, stretch canvass upon it and raise it from the ground with four legs. This is a much healthier manner of sleeping than upon the damp ground.

On the character of our population I must say something. My remarks are of recently almost entirely limited to those who live upon the river, since it is here I have spent most of my time. Although we find nearly every state represented, still there are three from Missouri where there is one from any other state. The reason of it, is this: that most of the later emigrants took the northern route which comes in at the head waters of the north fork of this river. They arrived so late that there was no other alternative than to come here. Some were so late as to be obstructed by snow in the mountains and leaving oxen, wagons, and all, footed it in to the valley. Most of the Missourians, and there were thousands of them, came with oxen. They could get ready at their own door to immediately commence their journey; and, already owning cattle and wagons, their fit out cost them little of money, which induced thousands to start who if similarly situated in Ohio would never thought of starting.

They are generally open and rather rough in their manners, most of them addicted to drinking some, and occasionally having a "spree." Nine in ten of them gamble either upon a large or small scale. The same is true of two thirds of all the miners. Large prices for all necessaries of life soon habituate them to look upon three dollars as they would upon a quarter in the states, and when they get to gambling they will lose fifty or a hundred dollars with little concern. I have heard them say when witnessing their games, "whats a hundred dollars! We often lose that in states," and another, a less of a gamester, I can afford to lose an ounce, for I can dig it out of the bar again to-morrow." thus verifying the proverb "light come, light go." Few take the pains to think that every half dollar saved here to be carried home, is just as good as a half dollar at home. —Frequently, when the players become drunk, a quarrel follows the game. About the only games played are the Mexican game monte, and poker.

Among the Missourians are a large number of the late volunteers to the Mexican war. Many of Doniphan's regiment are here. They are generally good fellows and good workers. Our "Army of Occupation" and Army of Invasion are well represented from every state. The soldiers were paid off in about the right time for them to start;—having just returned from one adventure of danger and blood they were ready to engage in another for gold.

Little can be calculated upon the future permanent population of California, I do not know of one in fifty but what intends to return, most within one year and the rest within two. No doubt there will be many who will settle in the southern part of California—enough to make a population sufficient to support the dignity of a State. But the talk about the unparalleled increase of population we read of, especially in the San Francisco papers, is all nonsense. To base the future character of the country and its claims for admission into the Union as a state upon this "unparalleled increase of population" is as absurd as it would be for the village of Saratoga Springs to apply to the legislature for a city charter because of its "increase of population" during the hot season. There are already thousands going home every month, some having made their fortunes, some discouraged and without gold, some from fear of sickness. Many came who were rich at home and accustomed to a life of luxury and ease. They—many of them—go home almost immediately after their arrival. Rich men's sons, such as led a life of dissipation at home, find nothing here worth staying for. Such as can turn their hand to any kind of labor that offers, which will pay well, are those who if prudent will surely go home with a few thousands. The opportunity offers itself to every one of doing this much.—Even those who work by the day get Congressmen's wages, including board. Nearly every one seems to be doing better than this, for hired laborers are scarce.

From what I hear of the immense fleets that have sailed from the different parts of the United States,

filled with gold seekers, I auger some considerable disappointment in their expectations. The immense emigration which arrived overland, last fall, together with a large increase from the vessels that arrived has lined every river and branch where gold is to be found with those who have laid their "claims" to a certain portion of the river "for the purpose of cutting a race and building a dam," as their notices read. Every bar that has not been worked is claimed. Mining as it has been and is at present carried on will cease after this season. No more fortunes will be made single handed. This leads me to speak of mining as it has been carried on thus far. There are two kinds of "diggings" the wet and dry.

Most of the dry diggings are in the southern part of California, in the neighborhood of the San Joachin and its tributaries. They are called dry because of being in ravines which during the summer contain no water: in the winter however they are wet diggings. The gold of the dry diggings is very different in its appearance from that taken from the beds of rivers. It is in an irregular form somewhat spherical as though it had just been melted and thrown down. Whether the volcanic action which threw it from its more primitive bed, took place while yet this portion of earth was covered with water, or after the "dry land appeared," is for future geologists to determine. From the appearance of the lumps I would judge that water received them first. The lumps vary from a pennyweight to a quarter of an ounce and occasionally they weigh eight or ten pounds troy. The gold is of a reddish hue.

The manner of working dry diggings is to dig down to the base rock, in the lowest part of the ravine, and removing the earth look for the lumps in the crevices. They then follow up the side of the ravine making bare the base rock, which is almost always of talcose slate with a perpendicular dip, that is the strata are perpendicular, so that their edges form the surface of the bed rock. The rock being thus situated affords innumerable fissures, or, as miners term them, crevices, in which gold has lodged. Occasionally dry diggings are found upon slightly elevated table land when it is obtained by sinking a shaft to the base rock of talcose slate and is found in the crevices before mentioned.—There is no washing for gold in the dry diggings. The dry diggings are more uncertain than others. Occasionally, fortunes are made in a short time, and others spend months without striking a lead and scarcely make their expenses.

I have never seen any dry diggings, and am indebted for what I know of them to Mr. Clay Taylor of St. Louis, a very intelligent young man who spent the winter before last in them. I have seen specimens of gold from them which are as unlike the gold I am soon to describe as is the manner of obtaining it.

When gold is obtained upon the banks and in the beds of rivers the diggings are called wet. The gold is in thin scales. To give you an idea of their average weight I have just weighed five grains, worth at our rates of $16 to the ounce, 16 1/4 cents. I counted 67 scales and particles of gold, in the five grains averaging about 13 to the grain. I put the largest scale in the ballance and it weighed as much as 32 of the smaller particles. There is much real gold dust in the sands of the river, so fine that if there is a particle of grease upon the bottom of the washing pan this dust upon touching it will float upon the surface of the water, and if plenty perfectly cover it. I have some way the knack of panning out the black sand from the gold, so as to save a great share of this dust, which in most hands is usually lost.—Last winter I bought 36 pounds of black sand which had been run through a fine hair sieve, the meshes just large enough to let through the fine black sand, particles of gold as fine or finer going through of course. I panned it out (a process to be explained) and obtained almost two ounces of the most beautiful gold I ever saw. It was finer by half than the finest rifle powder. I gave ten cents a pound for the sand, and panned it but in little over a half day.

Of the different processes of obtaining gold I will first name "panning," the simplest of all. It pays well only where the earth is very rich. A pan of gravel sand and small stones is agitated, under running water, by shaking the pan with a circular motion so as to loosen up the earth, the lighter parts of which are carried off by the water, the gold and black iron sand, precisely like that used upon writing desks, and which is invariably found with gold, settling to the bottom. The shaking may be more violent as the pan becomes less full. When but about one eighth

of the earth is left, it will be found to consist of little stones and gravel of the heavier kind, the black sand, a kind of stone as small as the black sand, much resembling a ruby, and lastly the gold. The small stones and gravel are washed and picked out very easily, but to separate the black sand from the gold requires an indescribable and hard to be acquired twist of the pan which I must pass over in silence. Some perform it very dexterously and savingly, while others waste much gold and are a long time about it. Those who follow panning usually obtain their earth from crevices which they pick loose with a butcher knife and take out with a spoon.—They wash but a few panfuls in a day but those of rich earth. I know of a little Dutchman [German—*Ed.*] who made ninety dollars in three days from a small hole near the river's margin.—During rainy days, when nobody else would work, he would pan his ounce a day, upon a spot where any one else in fair weather would not make an ounce a week.

By the by, Dutchmen, in the mines, are usually very fortunate in their labor. They are industrious and saving. Some of them seem to find the gold by instinct; especially those of that class which we sometimes see on their way to their new home in America. I mean one of the miserly looking fellows 4 feet 11 inches and three quarters in height, that can live a week upon two Boston crackers and a glass of water; with small black dreamy eye, a low forehead prominent above the eyes; a small sharp nose contracted into wrinkles, which looks as though it could track a half dollar rolled across the floor by smell; one who would beg his dinner with a thousand dollars in his pocket, who would mourn the loss of an half eagle, as he would that of a child. With such an one for my partner I would feel sure of a fortune. He would point on coming upon a rich place, as surely and certainly as a setter dog would on coming upon a woodcock.

Washing out gold with the "rocker," I will next describe. The rocker consists of a bottom board 14 inches wide, and 3 feet 6 inches long; of sides which are flaring or wider at the top than the bottom, of 8 inches in height and the length of the bottom, the ends are made so as to give the sides the proper slant. Say 14 inches at the bottom and 18 at the top. The sides are not of uniform height, being highest near the middle and sloping towards the head and foot of the rocker. The half of the rocker towards the head is covered with a screen of sheet iron, perforated with holes half an inch in diameter. It is furnished with rockers and when in operation is placed where water can be handily dipped and poured upon the screen, around which are sides 4 inches high so that, it may contain a pan full of earth, or more. The water is poured upon the earth, which with the motion of the rocker carries through every thing smaller than the holes of the screen. There is an apron under the screen which runs to within an inch or two of the head of the rocker, and inclines downwards as it runs toward the head. All the earth falls in one spot close to the head of the rocker. The gold, from its greater weight settles and remains at the bottom. A clete an inch in height is nailed on across the bottom immediately under the lower end of the screen, which stops the gold and over which the gravel &c. washes. The foot of the rocker is mostly open, and has a clete like the former which stops the gold that may wash over the first clete. There are many variations of the model I have described; some having a narrower bottom board and a piece each side 5 inches wide at an angle of 45 degrees from the bottom and then side pieces 5 inches wide. Some are 4, 5 and 6 feet long, in proportion to the number working it. The prices of rockers are from $150, for the larger ones to $40, for the smaller.

Before setting the rocker, the owners "prospects" for a suitable place. If the earth upon trial yields 25 cents to the pan-ful, they may calculate upon some $16, to the man, there being fair men with a medium machine.—Some places are found where 75 cents and one dollar to the panful is obtained, which is very rich, and will soon fill the purse. The first case supposes a spot where there are so many stone to remove that not more than 64 pan-fuls of earth can be washed. There are many places where earth is easily obtained and plenty of it. I worked last fall in a place where two of us washed 180 pan-fuls,—a large days work. We made $24.40 apiece, the earth yielding only a fraction over 15 cents, to the pan-full. Most of the mining has been done upon the bank, and bars of rivers by use of the above described rocker.

There is another kind of rocker, far more complicated and expensive, which saves *all* the gold contained in the earth washed. The common washers, as usually worked, lose all the gold dust and most of the small particles which constitute a large proportion of all the gold. They are usually set so slanting that the water runs off rapidly, carrying in its current those small particles, and frequently large ones entirely out of the machine. The Quicksilver Machine obviates all other losses. It is constructed as follows: though as in case of the common machines their forms vary some. The quicksilver trough is 5 feet long and from that upwards in proportion to the number to work it. The bottom is 12 inches in width, with sides some six inches in height, and spreading out as they rise to some 15 inches at the top. There are a large number of cletes across the bottom, about 2 inches high in the center, and as they approach the sides of the trough rise in a circular form, so that they are four inches high at the sides. The cletes, or "riffles," as they are called in the machine, are about three and a half inches apart. This trough, or "riffle box," slides under a sheet iron screen extending the whole length of the box, sitting upon a kind of frame work under which are rockers. There is a handle some two and a half feet long extending out horizontally from the middle of the machine, and a corresponding piece, but shorter, opposite which strikes a firm post under it, giving the machine a jolt when it strikes which stirs up the earth and gives the gold a chance to pass through it. Attached to a frame foundation upon which the rockers work, is, at the head of the machine, a wooden trough 10 inches by 20, the bottom perforated with small holes. The water which supplies the machine is poured into this by the pail full and the earth thrown upon the head of the screen under the water trough and is washed through, the larger stones working off at the foot of the machine which is set with considerable slant. The quicksilver is put in the lower partitions, three or four of the upper ones retaining the coarser scales of gold. The finer ones are instantly licked up on coming in contact with the quicksilver and form an amalgum with it. When the quicksilver has taken up such a quantity of gold that it begins to assume a solid state, the amalgum is taken out, strained through buck or doe skin (the last is better because thinner,) which leaves a lump of tolerably solid quicksilver and gold. This amalgum is then put into a cast iron retort, twelve inches deep and four in diameter, with rounded bottom and ground rim to which is fitted a ground cover so as to be air tight. It is fastened on by a clamp, which is tightened with iron wedges like those used upon the scythe heel. In the top is screwed an iron tube resembling a musket barrel, which is bent so as to go under the surface of the water contained in a tub for the purpose of condensing the quicksilver vapor, which passes through the tube. On coming in contact with the water the quicksilver assumes the form of minute particles, like dew, which collect in larger drops on reaching the bottom of the tub. When the quicksilver is all driven off the gold is left in a conglomerate mass! Pure except what little quicksilver it may contain, and such alloys as are combined with the native gold.

These machines have been in operation here only about two weeks. Six men which is the number required, to work the size I have described, have obtained, in places where a common one would never be tho't of being worked, two ounces or $32, to the man. There are now but four upon the bar, but are rapidly being introduced as orders for several have been given. They have been worked with success upon the Yuba a branch or this river south of here.

                    Truly yours,
                    JEROME HOWARD

[May 30, 1850: a continuation of the preceding letter]

I believe I have described as minutely as is necessary and as intelligibly as I could, the different methods of separating the gold, from the earthy matter which contains it.—Next I will proceed to describe the different methods used to obtain the earth.

It may here be the proper place, to speculate upon the probable origin or more primitive state of the gold found in our rivers.—Some of these particles are partially enclosed in white quartz. Upon examining the ledges of talcose slate which form the bed of the river and its banks, in some cases to a height of twenty or thirty feet, occasional thin strata of white

quartz are found of a vertical dip, with the strike, or horizontal direction of the strata, at right angles with the general course of the river, or in other words north and south. I have examined many of these but never found any gold. Upon the top of the surrounding hills I have discovered ledges of white quartz cropping out in places. I have never found gold upon heating up this quartz. White quartz is also one ingredient of granite, of which there is more than of any other kind of rock, in these mountains. I have never discovered gold in granite. Now where does it originate and in what company? A hard question to answer. Whenever found attached to any other mineral it is always with white quartz.--My opinion is that it is contained in the quartz of either the strata running through the talcose slate, or in the ledges of quartz or both. Possibly it is in the granite, though the quartz is not of that pure white which encases gold. The action of water, heat and air, upon the once bare rocks of this region, operating for ages has converted their surfaces into fine earth, which rains have washed into the ravines and thence into the rivers, as well as that washed from mountain banks of the rivers directly into them, the finer parts of which have been washed into what might of been a lake or sea, filling it so as to form a plain—the "Valley of the Sacramento." The particles of gold in a globular form corresponding in weight to the scales now found in the rivers, were washed down with the other mineral & earthy matters into the rivers. The action of the water impelling downwards the rocks with which the bed rock is now covered with three or four layers, beat out the globules of gold into the scaley form they now present. It will be seen upon examining one of the larger scales that the edge is rounded in the same manner as the edge of a lead shot beat out with a hammer would be. The same is true of the smallest scale, as will appear from examination with a microscope. The gold which I have supposed to be gathered, if I may use the term, from an extended surface into a much smaller one, namely: the river beds and banks up to high water mark, is plentiful in direct proportion to the nearness of locality to the bank. If so, how much less near must have been the proximity of the original globules than is that of the scales in the river. In a handful of earth containing gold from the river bed or bank, you would be exceedingly successful if you ever discovered the presence of gold, though it might contain 50 cents to the panful. The ounces of scales which upon the river bank may be collected from 50 panfuls must have been distributed through as many, and perhaps fifty times as many, wagon loads of earth or rock. If my theory is true there is no prospect of ever finding mines in the mountains. The higher up the river we go, the coarser the gold, though its form is globular or I may say that in general most of the surface of the lumps is curvilinear, which shows that they are not detached from masses of gold which many suppose exist in the mountains, but that they are of the form they originally possessed when they were, with the rock which enclosed them, thrown out from the melted interior of the earth.

There is a seeming exception to the fact that the flattening of the scales was caused by the motion of the rocks in the river.—Scales have been found upon the tops or nearly to the tops of some high hills. How account for their being flattened? Indications of high water mark have been discovered near the tops of high hills. The young man who noticed the fact was led to look for a cause. In a canon below he discovered unmistakeable evidences of a vast land slide which would fill up the canon and raise the water to his indicated high water mark, which was several hundred feet above high water mark of later years. The broad thin flattened scales of gold are carried along by a current of water with great facility. They are very easily thrown out of the pan by the agitation of the water, as every day experience shows.

That the courses of rivers and their quantity of water, have undergone great changes during the long period they have existed, no one can doubt. What has been highest water mark no one can tell. But as I might write enough of these speculations for two letters, I will stop here and finish this, which is already enough protracted.

The earth which contains the gold, is different in different situations. In the beds of rivers and in the bars it is washed out of coarse gravel. Upon the banks occasionally in gravel but more frequently in a reddish clay. This earth is invariably covered with rocks from 50 pounds in weight to three tons. These must be removed, which for three or four men is no play. After clearing a place five or six feet square of these rocks, then the earth [which] is intermixed with smaller rocks is "picked up," and carried in pans or buckets to the machine and washed out. After all the earth is removed so as to leave bare the bed rock, and the base rock "creviced" into this and so on till the whole is dug out. The depth of earth fit to be washed averages less than one foot in depth.

The miner during his labor treats only upon rocks. The idea of many in the states that the bars and banks which contain the gold is obtained without rolling and tumbling about heavy rocks. Upon the island bars of the river the same is true. The surface is covered with rocks and they with several feet of gravel and rocks must be removed before coming to the base rock where the gold is found. The same may be said of the drained river bed.

The gold in the river, or under it is obtained by draining a portion by means of daming. There are two kinds of dams; "wing dams," and dams extending entirely across the river. Wing dams extend partly across the river and thence down it, thus throwing all the water upon one side—draining the other below the dam, and between the wall extending down the stream, and shore. The place selected is at the head of a rapid so that back water may not prevent the draining. The earth obtained from the river bed is usually much richer than that upon the shore. A company of six men can build a wing dam in some ten days. The rocks being removed till the earth immediately above the bed rock is revealed, the process of washing is carried on the same as before described.

Dams entirely across the river must be accompanied with a race for the water to run in. There is one being constructed on this bar. There is a bend in the river, and a rapid at the commencement of the bend—the same the boat was carried over, as mentioned in my last letter. The race or canal commences just above where the dam is to be placed, and runs into the river below draining about 300 yards of the entire river.

Instead of cutting canals, there has been introduced a kind of aquaduct of sail duck.—It is sewed together so as to make when distended a pipe of 3 1/2 feet in diameter—some are more than that.

How they will work I cannot tell as they have not yet been put in use. Damming cannot be commenced before the month of August, as the river is kept up by the melting of the snow upon the mountains until about that time.

In my next I will say something of the future prospects of mining in this country, and make such miscellaneous remarks as may occur to me concerning the country generally.

I must make a remark concerning March in California. Feb. was a fine month—sunny and pleasant. When March came, rain commenced and it has copiously maintained its reputation for unpleasantness as ever I knew a March do. It snowed hailed or rained every day until today which is a fair day.— May many such follow. Truly yours.

JEROME B. HOWARD

[May 30, 1850: "The following letter will be recognized . . . as from some of the young men of the Marietta Company."]

Longtown, Eldorado Co. California, April 9, 1850
Friend Gates:—We have been desirous of placing in your hands some items of a reliable character in relation to this new and interesting country—some of which might be interesting to your readers—but when we consider that nearly all the papers of the States are teeming with letters from talented correspondents in the country to their friends in the States—which are more favorably located than we are for the acquisition of authentic information—we feel fully impressed with the idea that we can send you nothing but what will appear stale and uninteresting, either to yourself or readers. Having a little spare time we conclude to send you a few notes from which you are at liberty to extract any facts you may think proper for publication. You are all ready in possession of all the details of the organization of the companies to which we belonged, of their departure from home of their progress of their irreparable loss by the death of several of their members, of their toilsome journey over the plains and mountains and their safe arrival at their final place of destination—a little before the commencement of the rainy season, which has continued up to the present time. The rains commenced about the first of November and continued without intermission, with the exception of a few days at a time—for four weeks, and then only for something over a week. The rains were then almost constant until Feb. during which month we had scarcely any. March was a very stormy month—now however the weather seems fairly settled. The month of Feb. was as pleasant as May in the States. Those that spent the rainy season in the mountains and provided themselves with comfortable quarters and provisions, as most did, spent the time rather pleasantly than otherwise. The game has been in fine order, and during the winter was abundant— which furnished the miners with fresh meat. Of late the mines have been supplied with fresh beef, at from 30 to 40 cents per pound. Some grisly bear have been killed here during the winter. Grass and other vegetation has been up for some weeks, and stock of all kinds looks fine.

The market at present in this country is more fluctuating than in any part of the world. Articles that a few weeks past would command high prices at this time bring but little more than first cost. On the other hand articles that a short time past would bring but little now command fair or high prices. The great scarcity and high prices of lumber last fall roused the sturdy Oregonian, the sleepy Chilian, and the speculating Yankee. Each has contributed largely so as to completely overstock the market with lumber and building materials of all descriptions.—Lumber was selling as high as sixty cents per foot last fall, it can now be had at from 6 to 12 cents. Buildings that were framed and set up last fall commanded a very high price—now they are sold at auction to pay expenses. There are some leading articles, exports from the U.S. almost exclusively—that are now uniform in prices, some of which are Mess Pork, Boots, No. 9 and 10 of best quality Dried Apples, Peaches, Butter and Cheese well put up. Cider Vinegar, Sheet Iron, Sole Leather, Shoe Nails and Tacts. The above named articles have been generally high and as they are indispensable to every miner must continue to bring a fair price.—Most kinds of groceries are in good demand—liquor especially, sales of which have made fortunes. Socks, wool hats or Buena Vistas, and woolen shirts are in good demand in the fall. Ready made clothing and dry goods are low. There is a moderate demand for hardware—prices generally low—except cooking utensils.

Government of the country. The legislature has held a session—passed a code of laws, and dividing the Territory into counties, constituted the requisite number of officers made the first Monday of April the day for the citizens of the county to meet at some appointed place in the several counties—and then and there to elect their respective public servants. As dutiful citizens they meet and elect their Judges, Attorneys, Sheriffs, Recorders, Justices of Peace, and we presume these worthies are regularly installed in their several offices, and in the regular [performances—*Ed*.] of their respective duties—and we are now enjoying the advantage of a well organized republican government—although we can hardly appreciate their value.

The mines.—On this subject our information is mostly derived from others—as our late arrival in the country necessitated us to immediately seek winter quarters, consequently we have travelled but little. One of our members however went on an expedition to the southern mines and into the mountains and had by the way rather a hard time of it, camping four weeks in from three to five feet of snow, and seeing in close contact something worse than the Elephant. Another went north during the pleasant weather in Feb. and between the two we think we understand the condition of things tolerably well. The great rush has been in the South to the Stanislaus, Mariposa, Towalumn &c, in the north to the Yuba, Feather River and the three forks of the American. These northern streams are already crowded to overflowing with anxious gold diggers, although the high water in them occasioned by the melting snow will not admit of much being done on their bars before June. These streams are considered the richest in the country and no doubt are —still thousands of miners are scattered through the country on ravines and streams of less note. The average proceeds of mining labor is small, being as stated by a gentleman of San Francisco who took some pains to acquire information, not over 4.00 per day since the mines have been worked. The fortunate exceptions to this are of course quite a minority of the population of the mining districts, and the chances for individual success daily decreases as the

number of laborers increases. The gold varies much in fineness, being worth from 18 to 23 dollars per oz. according to the number of carats quality. That of the Cosumne, Yuba and Feather River is considered the best, and worth the most to take home—here all sell alike—in payment for goods or debts passing at 16.00 and for specie or exchange in the States at 15.50. We suppose you are well posted as to the items of a miner's life. His wants are not few, but many, for little in the way of comforts or luxuries can be had—the fare generally being pork mess beef and beans and the various preparations of flour that his ingenuity or taste and means can produce.—A rough shanty in winter and a tent in summer is his substitute for home. The day is spent in hard knocks among hard stones, and the frequent result of a hard days labor is hardly any gold. At evening he returns to his hard looking home, has hard fare for supper, smokes a pipe and retires to sleep upon the hard side of a clap board and we can from experience say generally sleeps hard and sound. This way seems like a hard picture but we are notwithstanding hardy, healthy and in good spirits, and as fat ragged and saucy as our fond mothers could wish to see us.

A few words as to the prosperity of our amicable and respectable order in this land, and we are sorry to be obliged to confess to our breathern at home that the love of filthy lucre has tended to make us forget our duties and that in the acquisition of the same our time has been so occupied that we have found none to spend in making proselytes—consequently we remain in number—excepting one death —as when we started. We often think however of the good old times in the "Hole in the Wall" and hope our breathern as they occasionally assemble will think of their representatives in this heathen land. We hope the order at home still flourishes and supports its dignity, and we all unite in the best of wishes for its prosperity. We add this notice for the benefit of our fellows in your town—not because we suppose it will interest the public generally—though our mysterious notices, as they appeared in your paper must have attracted some attention and excited some interest. We submit the above remarks upon this country to you to do with as you please, and hope you may find something in them of interest, and we think we have included in them most of the topics inquired about in the States. We subscribe ourselves, yours most truly.

[February 20, 1851]

CALIFORNIA CORRESPONDENCE

Cullima, Eldorado Co. California, Dec. 27, 1850

B. Gates, Esq.—Dear Sir—Though personally unacquainted, yet I would beg leave to communicate through the medium of your columns, with my numerous relatives and friends in Washington and adjoining counties regarding my California life and knowledge; and I trust by avoiding all exageration, little may not be without interest to the mass of your readers.

After a hundred days trip across the Plains, and enduring hardships that but "one in a thousand" would a second time enter upon, I found myself in the world's focus, in the midst of the greatest of all modern humbugs, August 28th. Myself in ill health from the excessive hardships of the trip, surrounded by the dead and dying on every side, and hundreds offering their labor for board,—certainly left my mind in no condition to recieve a favorable impression regarding California.

For six weeks I was ill, with the prevailing disease, dirahea, and unable to make my board, but I then recovered, and located here and during the last ten weeks have literally "dug" twenty eight ounces of the ore.

Many may think this doing well, and indicative of any thing but a humbug country. But allow me to call the consideration of such, to the time, detail, expense, and the very great risk of health, life, fortune, in getting here,—particularly of the overland emigrant, and the life of severe labor and great exposure of the miner—& such wages amounts to a mere pittance. Besides, not a person near me has done as well as myself. Of the company of thirty five men I came out with, but eight of them can show but two or three hundred dollars each—and the best mining season is over.

A large proportion of the miners, left the streams in the fall, for the purpose of throwing up dirt in the "dry diggings," which they are now washing,—

the rainy season having just set in, and it is proving a general failure; so they are obliged to change their quarters after expending several months valuable time, and one and two hundred dollars in building, provisions, &c.--taking all the dust of many. A dozen persons just returned to the bar on which I work, (one mile above town) with not dust enough to board them a week, and they left it two months ago, with two or three hundred each.

Provisions are low, in comparison to last years prices, and the great competition will keep them low, so that the suffering and starvation in the mines this winter foretold by many, will be prevented. Flour 18c., pork 25c., potatoes 25c., and prices can't advance but a trifle, if any.

I tell you that every kind of business here, is filled to satiety, and every mean trick and invention is resorted to make money.—In this respect the whole country is a facimile of our most thronged cities. Merchants have made but little this season, and many have now closed their doors for the mines.

Gambling is still the best business, but the diminished yield of the mines, causes it to suffer proportionally with every other. Speaking of gambling, E. M. Crommic, a prominent one of the order, and proprietor of several gambling saloons here, recently vamosed for France with the monte banks of several of his Brethren, amounting to $50,000 or $75,000! The blessings heaped upon him are not proper for christian ears to harken to.

To change the subject, I can assure you, that the general sentiment here is—home as soon as I can get there. The "Home Fever" (I don't know whether the desire is confined to this country or not.) prevails here to an extent equalling that of the "California Fever" in the States at any time. In the strong desire for home, fortunes are forgotten, and to get even, or to get enough to take them home, is as much as is generally expected.—Hundreds of this years emigration, were so disappointed, that they sold their outfits immediately on their arrival and left—and thousands would have done so if they had the means. But here we must stay and pack mud till we can get away.

Miners have not averaged more than three or four dollars (per day) this year. Last year wages was $16.00 per day, this year $5.00, and next year $2.00 to $3.00, will be the highest.

Oh! the poor emigrants of 1851, for a few—perhaps many—favored ones are still behind, incurable by reason or truth, & by nothing but experience. Franklin's school for fools. To such I would say, come by the Isthmus and no other way, and to avoid disaphaintment [*sic*], anticipate nothing above 2 or $3.00 per diem, and an amount sufficient to take you home in a year, secured at the expense of health and every comfort. Thousands have become grey here, by hardship and sickness. Multitudes have grasped fortunes, and perfectly careless about money—thinking it as easily gotten again, have lost all at the bar and monte table. Very, very many gamble here and practice all the accompanying vices which they would shun at home. The reasons are evident. There is a bar or gambling house at every step, and no law to restrain these things. Every attraction is offered and every trick practiced to get the poor miner's dust. We are thousands of miles from home, and comfort ourselves by thinking that a knowledge of our indulgence in vice will never reach them. Here, there is no parents eye to guide, no wife to warn, no sister to entreat, no church, no sabbath, for Sunday is much the busiest day in this region, and in short, all the animal and vicious passions are let loose, and free to indulgence without any legal or social restraint.

You doubtless find it difficult to aim at any definite conclusions regarding California because of the conflicting accounts you recieve. These accounts are as varied as human character. A person's health, success, or disappointment, has great influence over the character of his report. It is to the pecuniary interest of land jobbers, traders and many others whose interests are fixed here for a season, to continue an excitement and rush to California, which they will do thro' the influence of their published lies and exaggerated tales.

Fortunes too, are made here every day, (& lost,) & their possessors being blinded by their glittering dust, see not their thousand less fortunate neighbors, and therefore from such, you recieve the light side of the picture, highly magnified. Many come here to get a fortune immediately and return, many loose fortunes, or are sick much of the time, many spend their time, money and strength in prospecting the country, and finding no places answering their anticipations, quit the mines in disgust, and pronounce

them all humbugs. From these you get the dark shades of the picture exaggerated to a degree equally incorrect.

But notwithstanding all I have said, I believe, with health and contentment, an energetic man can make at least a thousand dollars a year, and stands a chance for making more.

But these blessings are secured to precious few, for the multitudes of reports constantly float, and the dissatisfaction among miners, maintains a continual excitement and passing to and fro in every direction. As for health, it seems that every one must be sick here. The sickness of this year far exceeds that of last. Without reference to the Cholera in Sacramento, Missouri has lost about one eighth or tenth of an emigration of young able men. We have no certain knowledge regarding the balance of the emigration, but have no reason to disbelieve that an equal fatality has attended it.

Most of the mining towns are temporary affairs, thrown hastily and slightly together but Cullima is built upon a more permanent scale, as her mines are more permanent.—Here is Capt. Sutter's mill, in digging the race for which gold was first discovered, June 1848. Some writers, and among them S. A. Mitchell, in his recent map of California place this mill on Feather River, instead of the South Fork of the American. To give a fresh impetus to business, the citizens talk strongly of tearing the dam away, as it backs the water over one of the richest bars in California. They propose remunerating the owners (Capt. Sutter has sold and is living on his Hawk [Hock—*Ed.*] farm on Yuba) with a certain per cent of the proceeds. The rainy season commenced last year the first of October, but thus far this year I have not lost a weeks work by the rain, and as for winter, it is nothing in comparison to spring at home.—while you are muffled in your cloaks, we are comfortable in our shirt sleeves, and I have not labored a day here without perspiring freely. We may find something different ere it is over, but—nous verrons.

Ex-Gov. Shannon is among the disappointed ones, I understand. I shall probably leave here next spring.

Very Resp'y., Yours &c.
L. M. WOLCOTT

Bibliography

Unpublished Journals and Papers

Bruff, J. Goldsborough. Journal of J. Goldsborough Bruff, August 28 to November, 1849, and March 17 to October 9, 1850. 2 vols. Henry E. Huntington Library, San Marino, California.

Buffum, Joseph Curtis. Diary, September 13, 1847, to January, 1855. Bancroft Library, Berkeley, California.

Chamberlain, William E. Diary, April 11 to August 20, 1849. California State Library, Sacramento.

Clark, Anson. Reminiscences. Bancroft Library, Berkeley, California.

Darwin, Charles B. Journal of an Overland Journey Across the Plains in 1849. 3 vols. Henry E. Huntington Library, San Marino, California.

Dewey, Kate Haun. Recollections of Catherine Margaret Haun, A Woman's Trip Across the Plains in 1849. Original typescript. Henry E. Huntington Library, San Marino, California.

DeWolfe, David. Diary, May 2 to October 16, 1849. Typescript. Henry E. Huntington Library, San Marino, California.

Dulaney, Robert. Robert Dulaney to Charles Dulaney, October 16, 1849. Dulaney Papers, Bancroft Library, Berkeley, California.

Evans, George W. B. Notes of the Defiance Gold Hunters' Expedition to California, 1849-1850. 6 vols. Photostatic copy. Henry E. Huntington Library, San Marino, California.

Gorgas, Solomon A. Diary of an Overland Journey from St. Joseph, Mo., to Placerville, California, by way of Fort Laramie, also Return Voyage to New York via the Isthmus of Panama, April 28, 1850, to April 8, 1851. Henry E. Huntington Library, San Marino, California.

Gray, Charles Glass. An Overland Passage from Independence, Missouri, to San Francisco, California, February 26 to November 19, 1849. Henry E. Huntington Library, San Marino, California.

Hart, Robert. Diary of a Journey by way of Santa Fe and the Gila River to the California Gold Fields and Return, 1849-1850. In possession of Mrs. H. L. Donovan, Lexington, Kentucky.

Kent, George F. Journal of a Gold Hunting Expedition to Upper California. Commenced in the Month of February, 1849. Henry E. Huntington Library, San Marino, California.

Markle, John A. Diary, April 18, 1849, to January 9, 1850. Photocopy in Bancroft Library. Original in Auburn Parlor of the Native Sons of the Golden West, Auburn, California.

Means, Harriett Hildreth Perkins. Travel Diary, 1882. University of Kentucky Library, Lexington.

Steuben, William. Diary, 1849-1850. Bancroft Library, Berkeley, California.

Wood, Joseph Warren. Diary of Overland Trip, May 6, 1849, to April 4, 1853. 6 vols. Henry E. Huntington Library, San Marino, California.

Published Journals

Bryant, Edwin. *What I Saw in California, being the Journal of a Tour, by the Emigrant Route and South Pass of the Rocky Mountains, Across the Continent of North America, the Great Desert Basin, and Through California in the Years 1846, 1847.* New York, 1848. (Editions were published in 1848 and 1849.)

————. *What I Saw in California: A Description of Its Soil, Climate, Productions, and Gold Mines;*

with the best Routes and Latest Information for intending Emigrants. 2nd ed., with a map, London, 1849.

Camp, Charles L. (ed.). *John Doble's Journal and Letters from the Mines Mokelumne Hills, Jackson, Volcano, and San Francisco, 1851–1865.* Denver, 1962.

Caughey, John (ed.). *Rushing for Gold.* Berkeley, 1949.

Christman, Florence Morrow (ed.). *One Man's Gold, the Letters & Journal of a Forty-Niner.* New York, 1930.

Emory, William H. *Notes of a Military Reconnoissance from Fort Leavenworth, in Missouri, to San Diego, in California, including Parts of the Arkansas, Del Norte, and Gila Rivers.* U.S., Senate, Exec. Doc. 7, 30th Cong., 1st Sess., 1847. Washington, 1848.

Foreman, Grant (ed.). *Marcy & the Gold Seekers, the Journal of Captain R. B. Marcy with an Account of the Gold Rush over the Southern Route.* Norman, Oklahoma, 1939.

Frémont, John C. *Report of the Exploring Expedition to the Rocky Mountains in the Year 1842, and to Oregon and Northern California in the Years 1843–'44.* U.S., Senate, Exec. Doc. 174, 28th Cong., 2nd Sess., 1845. Washington, 1845.

—————. *The Exploring Expedition to the Rocky Mountains, Oregon and California, to which is added a Description of the Physical Geography of California with recent Notices of the Gold Region from the latest and most Authentic Sources.* New York, 1855.

Gudde, Erwin G. and Elizabeth K. (trans. and eds.). *Charles Preuss, Exploring with Frémont.* Norman, Oklahoma, 1958.

Hammond, George P., and Edward H. Howes (eds.). *Overland to California on the Southeastern Trail 1849. Diary of Robert Eccleston.* Berkeley, 1950.

Howe, Octavius Thorndike. *Argonauts of '49, History and Adventures of the Emigrant Companies from Massachusetts, 1849–1850.* Cambridge, 1923.

Hulbert, Archer Butler. *Forty-Niners, the Chronicle of the California Trail.* Boston, 1931.

Langworthy, Franklin. *Scenery of the Plains, Mountains, and Mines.* Ed. Paul C. Phillips. Narratives

of The Trans-Mississippi Frontier. Princeton, 1932. Originally published in 1855.

Ledyard, Edgar M. (ed.). *A Journal of the Birmingham Emigrating Company by Leander V. Loomis, 1850.* Salt Lake City, 1928.

Leeper, David Rhorer. *The Argonauts of 'Forty Nine, Some Recollections of the Plains and the Diggings.* South Bend, Indiana, 1894.

Longworth, Basil N. *Memorandum of Thoughts, Reflections, and Transactions as Transcribed by Basil Nelson Longworth on His Journey from Washington Township, Guernsey County, Ohio, to Oregon in the Summer, 1853.* Eugene, Oregon, 1959.

Morgan, Dale L. (ed.). *Overland in 1846; Diaries and Letters of the California–Oregon Trail,* 2 vols. Georgetown, California, 1963.

—————, (ed.). *The Overland Diary of James A. Pritchard from Kentucky to California in 1849, with a Biography of Captain James A. Pritchard by Hugh Pritchard Williamson.* Denver, 1959.

Paden, Irene D. (ed.). *The Journal of Madison Berryman Moorman, 1850–1851. Edited with Notes and Introduction together with a Biographical Sketch of the Author by His Granddaughter, Louise Parks Barnes.* San Francisco, 1948.

Pomfret, John E. (ed.). *California Gold Rush Voyages, 1848–1849.* San Marino, California, 1954.

Read, Georgia Willis, and Ruth Gaines (eds.). *Gold Rush: The Journals, Drawings, and other Papers of J. Goldsborough Bruff, Captain, Washington City and California Mining Association, April 2, 1849–July 20, 1851,* 2 vols. New York, 1944.

Richards, Benjamin R. (ed.). *California Gold Rush Merchant, the Journal of Stephen Chapin Davis.* San Marino, California, 1956.

Russ, Carolyn Hale (ed.). *The Log of a Forty-Niner, Journal of Richard L. Hale, Newberry,* Massachussetts. Boston, 1923.

Scamehorn, Howard L. (ed.). *The Buckeye Rovers in the Gold Rush, An Edition of Two Diaries Edited with an Introduction.* Athens, Ohio, 1965.

Sedgely, Joseph. *Overland to California in 1849.* Oakland, California, 1877.

Shaw, R. C. *Across the Plains in Forty-Nine.* Farmland, Indiana, 1896.

Stephens, L. Dow. *Life Sketches of a Jayhawker of '49.* N.p., 1915.

Webster, Kimball. *The Goldseekers of '49. A Personal Narrative of the Overland Trail and Adventures in California and Oregon from 1849–1854.* Manchester, N. H., 1917.

White, Katherine (comp.). *A Yankee Trader in the Gold Rush, the Letters of Franklin A. Buck.* New York, 1930.

Williams, Joseph. *Narrative of a Tour from the State of Indiana to the Oregon Territory in the Years 1841–1842.* Cincinnati, 1843.

Guidebooks

Adams, James Truslow. *Atlas of American History.* New York, 1943.

Baughman, Robert W., *Kansas in Maps.* Topeka, 1961.

Clayton, W. *The Latter-Day Saints' Emigrants' Guide: Being a table of Distances, showing all the Springs, Creeks, Rivers, Hills, Mountains, Camping Places, and all other Notable Places, from Council Bluffs to the Valley of the Great Lake.* St. Louis, 1848. (This guide was republished in Ledyard, cited above).

Hastings, Lansford. *A New Description of Oregon and California: containing Complete Descriptions of those countries, together with the Oregon Treaty, and a Vast Amount of Information Relating to the Soil, Climate, Productions, Rivers and Lakes, and the Various Routes over . . . the Rocky Mountains.* Cincinnati, 1849.

————. *The Eimgrant's Guide to Oregon and California.* Princeton, 1932.

————. *The Emigrants' Guide to Oregon and California, containing . . . Scenes and Incidents of a Party of California Emigrants; and a Description of California. . . .* Cincinnati, 1945.

Marcy, Randolph B. *The Prairie Traveler. A Handbook for Overland Expeditions. With Maps, Illustrations, and Itineraries of the Principal Routes Between the Mississippi and the Pacific.* New York, 1859.

Prucha, Francis Paul. *A Guide to the Military Posts of the United States, 1789–1895.* Madison, 1964.

Stansbury, Howard. *Explorations and Survey of the Valley of the Great Salt Lake of Utah, including a Reconnoissance of a New Route Through the Rocky Mountains.* Philadelphia, 1852.

Ware, Joseph E. *The Emigrant's Guide to California, Containing Every Point of Information for the Emigrant—including Routes, Distances, Water, Grass, Timber, Crossing of Rivers, Passes, Altitudes, with a Large Map of Routes, and Profile of the Country, &c.,—with Full Directions for Testing and Assaying Gold and Other Ores.* St. Louis, 1849. My citations are to the reprint edition, with introduction and notes by John Caughey, Princeton, 1932.

Newspapers

Louisville *Daily Courier,* 1848–1852.

Louisville *Daily Journal,* 1848–1852.

Marietta *Intelligencer,* 1848–1853.

Sacramento *Daily Union,* 1852.

Sacramento *Transcript,* 1850–1851.

San Francisco *Alta California,* 1849–1850.

Periodicals

Barry, Louise. "Kansas before 1854; a Revised Annals, Part Ten, 1838–1839," *Kansas Historical Quarterly,* XXIX (1963), 143-89.

————— (ed.). "Overland to the Gold Fields of California in 1852; The Journal of John Hawkins Clark, Expanded and Revised from Notes Made During the Journey," *Kansas Historical Quarterly,* XI (1942), 227-96.

McCabe, F. S. "The Churches of Kansas (pre-Territorial Period)," *Transactions of the Kansas State Historical Society, Embracing the Third and Fourth Biennial Reports, 1883–1885,* 422-26.

Marshall, Thomas M. (ed.). "The Road to California; Letters of Joseph Price," *Mississippi Valley Historical Review,* XI (1924), 237-57.

Mattes, Merrill, and Esley J. Kirk (eds.). "From Ohio to California in 1849: The Gold Rush Journal of Elijah Bryan Farnham," *Indiana Magazine of History,* XLVI (September and December, 1950), 297-318, 403-20.

Missouri Historical Review, XXVIII (1934), 252-53.

Morgan, Dale L. "The Ferries of the Forty-Niners," *Annals of Wyoming,* XXXI (1959), 145-89, XXXII (1960), 165-203.

—————. "The Mormon Ferry on the North Platte; The Journal of William Empey, May 7 to August 4, 1847," *Annals of Wyoming,* XXI (1949), 111-67.

Niles Weekly Register, LXXXI, September 5, 1846–February 7, 1847.

Paden, Irene D. (ed.). "The Ira J. Willis Guide to the Gold Mines," *California Historical Society Quarterly,* XXXII (September, 1953), 193-207.

Read, Georgia Willis. "Diseases and Doctors on the Oregon–California Trail in the Gold Rush Years," *Missouri Historical Society Review,* XXXVIII (1944), 241-76.

Taggart, Harold (ed.). "The Journal of David Jackson Staples," *California Historical Society Quarterly,* XXII (June, 1943), 119-50.

Western Journal and Civilian (St. Louis), VII (December, 1851), *passim.*

Willman, Lillian M. "The History of Fort Kearney," *Publications of the Nebraska Historical Society,* XXI (1930), 213-326.

Abstracts and Directories

Abstract Book of the Rancho de Youba, and subsequent Transfers. Owned by Mr. Lowell Swetzer, Wheatland, California. This book contains the full land record of this large area in California, including the site of the Johnson Ranch.

Bogardus' Business Directory of San Francisco and Sacramento City, May, 1850.

Secondary Works

Alter, Cecil. *James Bridger, Trapper, Frontiersman, Scout and Guide: A Historical Narrative.* Salt Lake City, 1925.

Bancroft, Hubert Howe. *History of California,* in *the Works of Hubert Howe Bancroft,* XXII, XXIII (1846–1859). San Francisco, 1886.

Biographical Encyclopedia of Ohio of the Nineteenth Century. Cincinnati, 1876.

Birney, Hoffman. *Grim Journey.* New York, 1934.

Branch, E. Douglas. *The Hunting of the Buffalo.* New York, 1929.

Bynington, L. F., and Oscar Lewis. *The History of San Francisco.* 3 vols. San Francisco, 1931.

Chambers, John S. *The Conquest of Cholera, America's Greatest Scourage.* New York, 1938.

Colton, Walter. *Three Years in California.* New York, 1850.

Cowan, Robert Ernest. *The Pioneers of California.* San Francisco, 1929.

Coy, Owen Cochrane. *Gold Days,* San Francisco, 1929.

—————. *The Great Trek.* San Francisco, 1929.

Dale, Harrison Clifford (ed.). *The Ashley-Smith Explorations and the Discovery of a Central Route to the Pacific, 1822–1829.* Glendale, California, 1941.

Dawson, William Leon. *The Birds of California: A Complete, Scientific and Popular Account of the 580 Sub-species of Birds Found in the State.* 4 vols. San Diego, 1923.

Favour, Alpheus. *Old Bill Williams, Mountain Man.* Chapel Hill, N.C., 1936.

Federal Writers' Project. *Missouri, A Guide to the "Show Me State."* New York, 1941.

—————. *The Oregon Trail, the Missouri River to the Pacific.* New York, 1939.

Foreman, Grant (ed.). *Marcy & the Gold Seekers, the Journal of Captain R. B. Marcy, with an Account of the Gold Rush over the Southern Route.* Norman, Oklahoma, 1939.

Hafen, LeRoy R. (ed.). *Central Route to the Pacific by Gwin Harris Heap, with Related Materials on Railway Explorations and Indian Affairs by Edward F. Beale, Thomas Hart Benton, Kit Carson, and Colonel E. A. Hitchcock and other Documents.* Glendale, California, 1957.

Hafen, LeRoy R., and Francis Marion Young. *Fort Laramie and the Pageant of the West, 1834–1890.* Glendale, California, 1938.

Heitman, Francis B. *Historical Register of the United States Army, September 29, 1789 to September 29, 1889.* Washington, 1890.

Hildreth, Samuel P. *Genealogical and Biographical Sketches of the Hildreth Family from the Year 1632 down to the Year 1840.* Marietta, Ohio, 1840.

Hittell, Theodore H. *History of California.* 4 vols. San Francisco, 1885–1897.

Inman, Henry. *The Great Salt Lake Trail.* New York, 1898.

Irving, Washington. *The Adventures of Captain Bonneville.* London, 1837.

Klappholz, Lowell. *Gold! Gold!* New York, 1959.

McGlasham, C. F. *History of the Donner Party, A Tragedy of the Sierra.* Stanford, 1947.

Monoghan, Jay. *The Overland Trail.* Indianapolis, 1947.

Moody, Ralph. *The Old Trails West.* New York, 1963.

Morgan, Dale L. *The Great Salt Lake.* Indianapolis, 1947.

—————. *The Humboldt, Highroad of the West.* New York, 1943.

—————, *Jedediah Smith and the Opening of the West.* Indianapolis, 1953.

Morse, John Frederick. *The First History of Sacramento City, written in 1853, by John Frederick Morse, M.D., with a Historical Note on the Life of Dr. Morse by Caroline Wenzel.* Sacramento, 1945.

Nevins, Allan. *Frémont, the West's Greatest Adventurer: Being a Biography from Certain Hitherto Unpublished Sources of General John C. Frémont, together with His Wife Jessie Benton Frémont.* 2 vols. New York, 1928.

Paden, Irene D. *The Wake of the Prairie Schooner.* New York, 1943.

Parkman, Francis. *The Oregon Trail; Sketches of Prairie and Rocky Mountain Life.* Champlain Ed. Boston, 1898.

Paul, Rodman. *California Gold! the Beginning of Mining in the Far West,* Cambridge, 1947.

Rensch, Hero Eugene. *History of the Mining Districts of California (Columbia Series). Columbia, a Gold Camp of the Old Tolumne; Her Rise and Decline, together with some Mention of Her Social Life and Strivings.* Berkeley, 1936.

Robinson, Alfred. *Life in California: During a Residence of Several Years in that Territory, Comprising a Description of the Country and the Missionary Establishments, with Incidents, Observations, etc., etc.* New York, 1846.

Settle, Raymond W. (ed.). *The March of the Mounted Riflemen, First United States Military Expedition to Travel the Full Length of the Oregon Trail from Fort Leavenworth to Fort Vancouver, May to October, 1849, as Recorded in the Journals of Major Osborne Cross and Geo. Gibbs and the Official Report of Colonel Loring.* Glendale, California, 1940.

Shoemaker, Floyd C. *Missouri and the Missourians; Land of Contrasts and People of Achievements.* Chicago, 1943.

Stewart, George Rippey. *The California Trail, an Epic with Many Heroes.* New York, 1962.

—————. *Ordeal by Hunger, the Story of the Donner Party.* Boston, 1960.

Storer, Tracey L., and Lloyd P. Tevis, Jr. *The California Grizzly.* Berkeley, 1955.

Sullivan, Maurice S. *Jedediah Smith, Trader & Trailbreaker.* New York, 1936.

Summers, Thomas J. *History of Marietta, Ohio.* Marietta, 1903.

Swasey, W. F. *The Early Days and Men of California.* Oakland, California, 1891.

Taylor, Bayard. *Eldorado, or, Adventures in the Path of Empire; Comprising a Voyage to California, via Panama; Life in San Francisco and Monterey; Pictures of the Gold Region, and Experiences of Mexican Travel.* 2 vols. 2nd ed., New York, 1850.

Williams, Walter (ed.). *A History of Northwest Missouri.* 3 vols. New York, 1915.

Unpublished Thesis

Thomas, Melvin Robert. "The Impact of the California Gold Rush on Ohio and Ohioans." Unpublished M.A. thesis, Ohio State University, Columbus, 1949.

Index

OREGON · IDAHO · WYOMING · SOUTH DA

NEBRA

CALIFORNIA · NEVADA · UTAH · COLORADO · KAN

OK

September 15, 1849–February 28, 1850

August 9–September 14, 1849

July 23–August 8, 1849

June 15–July 22, 1849

The Route of
Elisha Douglass Perkins
to the
California Gold Fields